T0271434

# Synthetic
# Spin-Orbit Coupling
# in Cold Atoms

# Synthetic Spin-Orbit Coupling in Cold Atoms

Editors

**Wei Zhang**

Renmin University of China, China

**Wei Yi**

University of Science and Technology of China, China

**Carlos A. R. Sá de Melo**

Georgia Institute of Technology, USA

**World Scientific**

NEW JERSEY · LONDON · SINGAPORE · BEIJING · SHANGHAI · HONG KONG · TAIPEI · CHENNAI · TOKYO

*Published by*

World Scientific Publishing Co. Pte. Ltd.

5 Toh Tuck Link, Singapore 596224

*USA office:* 27 Warren Street, Suite 401-402, Hackensack, NJ 07601

*UK office:* 57 Shelton Street, Covent Garden, London WC2H 9HE

**British Library Cataloguing-in-Publication Data**
A catalogue record for this book is available from the British Library.

SYNTHETIC SPIN-ORBIT COUPLING IN COLD ATOMS

ISBN 978-981-3272-52-1

For any available supplementary material, please visit
https://www.worldscientific.com/worldscibooks/10.1142/11050#t=suppl

Desk Editor: Rhaimie Wahap

Typeset by Stallion Press
Email: enquiries@stallionpress.com

Printed in Singapore

# Preface

Since the first experimental realization of synthetic spin-orbit coupling (SOC) in ultracold atomic gases via Raman processes in 2011, a considerable amount of efforts have been devoted to this research area to reveal exotic effects induced by this novel type of controlling method. Synthetic SOC can greatly modify the system on the single-particle level by mixing the internal state and the external motional degrees of freedom. Together with the already versatile toolbox consisting of, e.g., optical lattice and Feshbach resonance, synthetic SOC significantly broaden the horizon of quantum simulation using cold atoms.

One important direction in this area is to realize and simulate topologically nontrivial phases in ultracold atoms. In the past decade, the study of topological states remains a frontier of research in condensed-matter physics. SOC is believed to play a key role in many interesting phenomena involving nontrivial topology, such as the quantum spin Hall effects, topological insulators, topological superconductors, and Weyl semimetals. Although the forms of the synthetic SOC currently realized in cold atoms differ crucially from those in condensed-matter systems aforementioned, there exist various theoretical proposals on realizing topologically nontrivial phases in different setups.

Furthermore, recent studies suggest that other exotic phases and novel phenomena can be engineered with carefully designed configurations, taking advantage of the extra degrees of freedom and high controllability which are available exclusively in cold atoms. For example, by involving more hyperfine states into Raman processes, one can realize SOC in Bose and Fermi systems with higher spins. Another interesting possibility is considering a hybrid system of cold atoms and optical cavity, such that a dynamic SOC can be induced to study topological effects in non-equilibrium systems. As such, SOC has a great potential of becoming a powerful tool of quantum control in ultracold atomic gases.

This review volume is organized to introduce some recent works in spin-orbit coupled atomic gases. As there already exist some excellent review articles and review volumes covering important progresses from 2011 to 2015, the present work mainly focuses on the new developments emerging after 2015. This review volume is prepared for advanced graduate students, post-doctorate researchers, and colleagues working in the field of cold atoms. We emphasize that the topics included in this review volume are organized in such a way so as to give a reasonably fluent presentation. We did not try to assign credits or priorities in the order of contents.

Editors

*Wei Zhang, Wei Yi, and Carlos A. R. Sá de Melo*

# Contents

# Chapter 1

# Spin-orbit Coupling and Topological Phases for Ultracold Atoms

Long Zhang and Xiong-Jun Liu*

*International Center for Quantum Materials, School of Physics,
Peking University, Beijing 100871, China*

*Collaborative Innovation Center of Quantum Matter,
Beijing 100871, China*

Cold atoms with laser-induced spin-orbit (SO) interactions provide promising platforms to explore novel quantum physics, in particular the exotic topological phases, beyond natural conditions of solids. The past several years have witnessed important progresses in both theory and experiment in the study of SO coupling and novel quantum states for ultracold atoms. Here we review the physics of the SO coupled quantum gases, focusing on the latest theoretical and experimental progresses of realizing SO couplings beyond one-dimension (1D), and the further investigation of novel topological quantum phases in such systems, including the topological insulating phases and topological superfluids. A pedagogical introduction to the SO coupling for ultracold atoms and topological quantum phases is presented. We show that the so-called optical Raman lattice schemes, which combine the creation of the conventional optical lattice and Raman lattice with topological stability, can provide minimal methods with high experimental feasibility to realize 1D to 3D SO couplings. The optical Raman lattices exhibit novel intrinsic symmetries, which enable the natural realization of topological phases belonging to different symmetry classes, with the topology being detectable through minimal measurement strategies. Furthermore, we discuss the realization of novel superfluid phases for SO coupled ultrocold fermions. In particular, we introduce how the non-Abelian Majorana modes emerge in the SO coupled superfluid phases which can be topologically nontrivial or trivial, for which a few fundamental theorems are presented and discussed. The experimental schemes for achieving non-Abelian superfluid

---

*Correspondence addressed to: xiongjunliu@pku.edu.cn

phases are given. Finally, we point out the future important issues in this rapidly growing research field.

## 1.1. Introduction

### 1.A. *Why study spin-orbit coupling?*

Spin-orbit (SO) coupling is a relativistic quantum mechanics effect which characterizes the interaction between the spin and orbital degrees of freedom of electrons when moving in an external electric field. Due to the special relativity, the electron experiences a magnetic field in the rest frame, which is proportional to the electron velocity and couples to its spin by the magnetic dipole interaction, rendering the SO coupling with the following form

$$H_{\rm so} \propto \boldsymbol{\sigma} \cdot \boldsymbol{B}_{\rm eff} \propto \boldsymbol{\sigma} \cdot (\nabla V \times \boldsymbol{p}) = \lambda_{\rm so} \nabla V \cdot (\boldsymbol{p} \times \boldsymbol{\sigma}), \qquad (1.1.1)$$

where $\boldsymbol{\sigma}$ is the spin, $V(\boldsymbol{r})$ is the external electric potential experienced by the electron, $\boldsymbol{p}$ is the electron's momentum, and $\lambda_{\rm so}$ denotes the SO coefficient. In atomic physics the SO coupling is responsible for the fine structure splitting of the optical spectroscopy. In solid state physics the SO interaction of Bloch electrons exhibits several effective forms by taking into account the crystal symmetries and local orbitals around Fermi energy, and can strongly affect the band structure of the system. The typical types of the SO coupling includes the Rashba and Dresselhaus terms, which are due to the structure inversion asymmetry and bulk inversion asymmetry of the materials, respectively, and Luttinger term which describes the SO coupling for the valence hole bands.[1,2] The study of SO coupling for electrons has generated very important research fields in the recent years, including spintronics,[3] topological insulator,[4,5] and topological superconductors (SCs),[6] etc. These studies bring about completely new understanding of the effects of the SO coupling in the condensed matter physics and material science.

*Spintronics.* In a system with SO coupling one can manipulate the electron spins indirectly by controlling the orbital degree of freedom. This lies in the heart of the study of semiconductor spintronics, with the spin degree of freedom of the electron being exploited for improved functionality. From the basic form of the SO interaction, one can see that the components of the spin, momentum, and electric field which couple together are perpendicular to each other: $\boldsymbol{\sigma} \perp \boldsymbol{p} \perp \nabla V$. This implies that applying an external electric field may dynamically drive the electron at opposite spin states to move oppositely in the real space. In particular, consider a Rashba SO coupled system with the presence of an external electric field, described by the potential $V_{ex}(\mathbf{r})$, whose Hamiltonian reads $H = \mathbf{p}^2/2m_e + \lambda_{so}(k_x\sigma_y - k_y\sigma_x) + V_{ex}(\mathbf{r})$, where $m_e$ is the effective mass of Bloch electron. This Rashba SO coupling can be equivalently written in terms of a non-Abelian SU(2) gauge potential and then $H = (\mathbf{p} - e\mathbf{A}_\sigma)^2/2m_e + V_{ex}(\mathbf{r})$, where a trivial constant is neglected and $\mathbf{A}_\sigma = (m_e\lambda_{so}/e\hbar)(-\sigma_y\hat{e}_x + \sigma_x\hat{e}_y)$. Note that the field of a non-Abelian gauge is given by $F_{\mu\nu} = \partial_\mu A_\nu - \partial_\nu A_\mu + (ie/\hbar c)[A_\mu, A_\nu]$, which is associated a spin-dependent magnetic field, given by

$$\mathbf{B} = \frac{m_e^2\lambda_{so}^2}{\hbar^3}\frac{e}{c}\sigma_z\hat{e}_z. \qquad (1.1.2)$$

When electrons are accelerated by the electric field, their spins are tilted to out-of-plane ($\hat{e}_z$) direction. The above formula implies that electron having nonzero spin polarization experiences an effective magnetic field along $+z$ or $-z$ direction depending on its polarization direction. Thus spin-up and spin-down electrons are deflected to opposite sides, leading to a pure spin current and spin accumulation in the edges of the transverse direction. This gives a spin Hall effect (SHE).[7,8] Moreover, in a magnetic semiconductor, the spin of electrons is polarized due to the existence of a Zeeman coupling term $M_z\sigma_z$, which leads to a population imbalance in the spin-up and spin-down states. In this case, while the electrons with opposite spin polarization along $z$ axis are deflected oppositely, a nonzero transverse charge current is also resulted, giving rise to an anomalous Hall effect (AHE).[3] In the anomalous Hall effect both the spin and charge accumulations are obtained. The SHE and AHE may have important applications designing spintronic devices such as SHE transistors and spin injectors.

*Topological insulators.* Taking into account the spin degree of freedom can bring rich nontrivial structure of the Bloch bands for electrons.[4,5] In the presence of SO coupling the spin and momentum of an electron can be entangled, which causes spin texture in the momentum space. For an insulator, all the momentum states of a single band in the $d$-dimension form the first Brillouin zone (FBZ), which is a closed manifold known as torus $T^d$. The spin is attached to each Bloch momentum governed by the SO interaction. A simple illustration of the SO effect on topology is sketched in Fig. 1.1.1. Figure 1.1.1(a) gives a fully spin-polarized configuration, in which case the spin does not wind in the momentum space, thus mapping to a single point of the unit circle. This gives zero winding number ($\mathcal{N}_{1d} = 0$) [Fig. 1.1.1(a)]. In contrast, for case of Fig. 1.1.1(b), the spin winds over all direction of a circle in the real space when the momentum runs over the

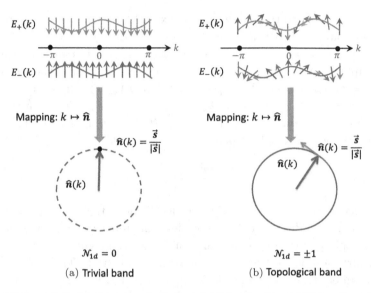

**Fig. 1.1.1.** (a) Sketch for a topologically trivial band without SO coupling. In this case the spin states of the whole Brillouin zone does not wind over a circle, so they map to a single point of the unit ring $S^1$, giving zero winding number. (b) Sketch for a topologically nontrivial band with SO coupling. The spin states of the whole Brillouin zone winds over a circle. Thus they map to the whole unit ring $S^1$, giving a nonzero winding number.

FBZ, giving rise of a nonzero winding number of the Bloch band ($\mathcal{N}_{1d} = 1$). The latter case is referred to as a topological phase, while the former is a trivial one. The transition between a topological phase and a trivial has to experience the band gap closing. As a consequence, in the interface between the topological and trivial regions, e.g the boundary of the material which is in a topological phase, the localized in-gap states emerge, mimicking the gap closing at the interface.

More specifically, the terminology *topological insulator* typically refers to the topological insulating phases in 2D or 3D with time-reversal symmetry,[4,5] which are a new class of insulating topological phases different from the quantum Hall effect discovered in 1980s.[9,10] In such insulators the boundary exhibit helical edge or surface states protected by time-reversal symmetry. Since the concept was introduced in 2005, tremendous proposals and experimental discoveries for the 2D and 3D topological insulators have been performed in the past years.[4,5] Physically, the SO interaction leads to the band inversion of the orbital bands with opposite parity around Fermi energy, accounting for the underlying mechanism for the topological phase transition. The inverted regime corresponds to the topological phase, while the normal un-inverted regime corresponds to the trivial phase.[11,12] The concept of topological insulators with time-reversal symmetry has also been generalized to the phases protected crystal symmetries, dubbed topological crystalline insulators,[13,14,178,179] and also to the topological semimetal or metals.[15–18]

*Topological superconductors.* Topological SC is a topological phase similar to the topological insulators, having a bulk superconducting gap and gapless or midgap excitations in the boundary.[5,6] Unlike the insulating phases, the boundary excitations in a topological SC are known as Majorana modes which are their own antiparticles.[19–22] Mathematically, the annihilation and creation operators of such states are identical $\gamma(x) = \gamma^\dagger(x)$. Especially, the Majorana zero energy modes localized in the point-like topological defects of topological SCs obey non-Abelian statistics,[23–27] and have potential applications to the topological quantum computation,[28–30] which is an essential motivation for the great efforts having been driven in both theory and experiment to search for such exotic modes in the recent years. Note that in a SC the Bogoliubov quasiparticle operators generically takes the form $b_\mu(x) = \alpha_\mu c_\sigma(x) + \beta_\mu c_{\sigma'}^\dagger(x)$, with the spin indices $\sigma, \sigma' = \uparrow, \downarrow$. In an s-wave SC whose pairing order is even one has $\sigma \neq \sigma'$ and the quasiparticle is not a Majorana, while in a spinless (or spin polarized) SC, only a parity-odd (p-wave) pairing phase occurs and then $\sigma = \sigma'$, yielding a Majorana quasiparticle.

An importance consequence of the SO interaction is that the parity of superconducting pairings can be manipulated in the SO coupled materials. In the presence of SO coupling, in general the orbital angular momentum is no longer a good quantum number, nor the spin. Thus in the superconducting pairing channels the parity-even (like s-wave) and parity-odd (like p-wave) pairing states are mixed up. Under some proper conditions, e.g. in the presence of an external Zeeman field which can kill the s-wave pairing, only parity-odd superconducting states shall survive and then the Majorana modes can be realized. This idea has been broadly applied to the recent theoretical and experimental realization of topological SCs based on heterostructures formed by SO coupled materials and conventional s-wave SCs, together with a Zeeman splitting field being applied by magnetic field or ferromagnetic insulators.[22] In the interface the Cooper pairs are forced into p-wave type by the SO coupling and Zeeman field, rendering a topological SC. Nevertheless, while many experimental studies have been performed in observing indirect signatures of Majorana zero modes, based on s-wave

SCs and semiconductor nanowires,[31-34] magnetic chains,[35-37] or topological insulators,[38-40] the rigorous confirmation is yet illusive.

### 1.B. *Brief history of SO coupling for ultracold atoms*

In cold atoms the spin states refer to the internal electronic hyperfine levels, and the spin of an atom is given by the summation of the total spin, orbital angular momentum of all electrons of the atom, and the nuclear spin.[41] Compared with solid state materials, the ultracold atoms are extremely clean systems, with all the parameters being fully controllable in the experiment.[42,43] Such a full controllability enable the ultracold atoms to be ideal platforms to simulate complex quantum physics with exact models which can be beyond or not precisely achievable in solid state materials.

The ultracold atoms do not have intrinsic SO coupling, while an effective SO interaction can be generated by properly coupling the atomic spin states to external fields.[44,45] From the case of a Rashba SO coupling we have seen that the SO interaction is generically equivalent to a non-Abelian gauge potential coupling to the spin degree of freedom of the system. This picture gives birth to the basic ideas in realizing SO coupling for ultracold atoms through the generation of synthetic gauge potentials. Before proposing a SO interaction, the realization of an Abelian gauge potential for a spinless system was first theoretically considered by Jaksch and Zoller[46] in 2003 for optical lattices, and by Juzeliūnas and Öhberg[47] in 2004 for continuum quantum gas. The Abelian gauge potential is associated with an artificial magnetic flux, similar to that in a rotating Bose-Einstein condensate.[48,49] Motivated by the previous study, in 2004 Liu *et al.*[50] proposed to realize a spin-dependent gauge potential of the form $A_3(\mathbf{r})\sigma_z$ by generalizing the previous scheme for magnetic flux to the spin-dependent regime, yielding an early model for realization of SO coupling. Interestingly, this SO term describes a coupling between spin and orbital angular momentum, which attracts particular attention very recently for cold atoms.[51-54] It is noteworthy that the spin-orbital-angular-momentum coupling has been very recently realized in experiment by Lin's group [H.-R. Chen *et al.*, arXiv: 1803.07860], which applied a scheme similar to the one proposed in Ref. 50. Adopting the similar idea of generating spin-dependent gauge potentials, Liu *et al.*[55] and Zhu *et al.*[56] respectively proposed to observe spin Hall effect for ultracold atoms. In both cases, spin-dependent gauge potentials are still Abelian with the form $A = A_3(\mathbf{r})\sigma_z$, associated with a spin-dependent magnetic field $B = B_3\sigma_z\hat{e}_z$, for which the spin-up and spin-down atoms couple to artificial magnetic fields along $+z$ and $-z$ direction, respectively, leading to a (quantum) spin Hall effect. The realization of non-Abelian gauge potentials was first proposed in 2005 by Osterloh *et al.*[57] in optical lattice, and by Ruseckas *et al.*[58] in continuum quantum gas. With the proposed non-Abelian gauge potentials, in principle one can achieve high dimensional SO couplings.

The early proposals for SO coupling are hard to be realized in real experiments. In 2008, Liu *et al.*[59] pointed out that a 1D SO coupling with equal Rashba and Dresselhaus amplitudes can be realized for atoms with a simple Λ-type configuration of internal levels. A similar configuration was considered earlier by Higbie and Stamper-Kern[60] to investigate the properties of a Bose condensate, while not SO coupling. A Lambda type configuration contains two ground hyperfine levels, mimicking the spin-up and spin-down states. A Raman coupling induced by two light beams drives spin-flip transition and transfers momentum simultaneously, giving rise to SO coupling. The basic idea has been broadly applied in experiment to realize the 1D SO coupling for bosons and fermions.[61–75] The first experimental realization was performed with bosons in Spielman's group at NIST.[61] Following this study, Zhang's group at Shanxi University[63] and Zwerlein's group at MIT[64] respectively realized the 1D SO coupling for $^{40}$K and $^{6}$Li Fermi atoms. On the other hand, Chen's group at USTC studied the phonon spectra and observed the roton gap for a SO coupled BEC,[70] and further performed a systematic study of the phase diagram at finite temperature of the SO coupled BEC.[69] Engels' group first realized the 1D SO coupling for bosons in an optical lattice. More recently, the 1D SO coupling has also been realized with lanthanide and alkali earth atoms, including Dy fermions by Lev's group,[73] Yb atoms by Jo's group at HKUST[74] and by Fallani's group,[76] and Sr atoms by Ye's group at JILA.[77]

Realization of a high-dimensional SO coupling (more than 1D) is more significant. The reason is obvious: a high-dimensional SO coupling (e.g. the 2D Rashba term) corresponds to a non-Abelian gauge potential and is associated with nonzero Berry's curvature which could have nontrivial geometric or topological effects, while the 1D SO coupling realized in the aforementioned experiments corresponds to an Abelian potential and does not give a nonzero Berry curvature. The study of broad classes of novel topological states necessitates the realization of high-dimensional SO couplings, including topological superfluid phases.[78,79] Many interesting schemes were proposed for realizing 2D[57,58,80–85] and 3D SO couplings,[86] whereas the experimental realization was not available until very recently. The Rashba type and Dirac type 2D SO couplings are realized by Zhang's group[87] and a collaborative team by Liu's group at PKU and Pan-Chen's group at USTC,[88] respectively. The former realization is based on a tripod scheme in a continuum space, and the latter is based on a scheme called optical Raman lattice with double-$\Lambda$ internal configuration.[85,88] The achieved 2D Dirac Hamiltonian realizes a minimal model for quantum anomalous Hall effect driven by SO coupling, which cannot be achieved with solid state materials but has been firstly obtained for ultracold atoms, and was shown to exhibit novel physics in the bulk and the boundary.[132]

## 1.2. Theory for Synthetic SO Coupling

A natural way to simulate gauge potentials is to change the dynamical behavior of neutral atoms just as moving charges in electromagnetic fields by some external forces, like rotation,[48,49] atom-light interaction[89-92] or laser-assisted-tunneling.[93-95] Juzeliūnas et al.[47] proposed to produce an effective magnetic field by Berry's phase,[96] which arises from the adiabatic motion of a spatial-dependent dark state (a light-dressed eigenstate uncoupled with the excited state). Liu et al.[50] proposed to realize a spin-dependent gauge potential of form $A_3(\mathbf{r})\sigma_z$ by generalizing the previous scheme for magnetic flux to the spin-dependent regime, yielding an early model for realization of SO coupling. This SO term describes a coupling between spin and orbital angular momentum, which attracts particular attention very recently.[51-53] A more generic notion of SO coupling, which corresponds to non-Abelian gauge potentials, can be generated by employing two or more degenerate dressed states.[57,58] The basic idea of generating adiabatic Abelian and non-Abelian gauge fields can, in fact, trace back to Wilczek and Zee's seminal work in 1984.[97]

### 2.A. Gauge potential for continuum gas

We start with the generic theory of producing non-Abelian adiabatic gauge potentials.[98] We consider a $N$-level quantum system which is coupled to external fields, with the Hamiltonian $H = -\frac{\hbar^2}{2m}\nabla^2 + V(\mathbf{r}) + H_{\mathrm{I}}(\mathbf{r})$. Here $V(\mathbf{r})$ is the trapping potential, and $H_{\mathrm{I}}(\mathbf{r})$ denotes a spatially varying interacting term between the system and external fields (e.g. laser addressing). One can diagonalize the coupling Hamiltonian $H_{\mathrm{I}}$ through a unitary transformation $U(\mathbf{r})$ via $H_{\mathrm{I}}^{\mathrm{diag}} = U^\dagger H_{\mathrm{I}} U$, where $H_{\mathrm{I}}^{\mathrm{diag}}$ is diagonal and is written as $H_{\mathrm{I}}^{\mathrm{diag}} = \sum_j^N E_j |\chi_j(\mathbf{r})\rangle\langle\chi_j(\mathbf{r})|$, with $|\chi_j(\mathbf{r})\rangle$ being the eigenstates of $H_{\mathrm{I}}$ and $E_j$ the corresponding energy. With the diagonal bases of $H_I$, the total

Hamiltonian after the transformation can be obtained by

$$H' = U^\dagger H U$$
$$= \frac{1}{2m}[i\hbar\nabla + \boldsymbol{A}(\boldsymbol{r})]^2 + \widetilde{V}(\boldsymbol{r}) + H_I^{\text{diag}}, \qquad (1.2.1)$$

where $\widetilde{V} = U^\dagger V U$ and the gauge $\boldsymbol{A}$ is a $SU(N)$ Berry's connection, taking the $N \times N$ matrix form, and is introduced by

$$\boldsymbol{A}(\boldsymbol{r}) = i\hbar U^\dagger(\boldsymbol{r})\nabla U(\boldsymbol{r}). \qquad (1.2.2)$$

The non-Abelian gauge potential $\boldsymbol{A}(\boldsymbol{r})$ is associated with the $SU(N)$ Berry curvature

$$F_{\mu\nu} = \partial_\mu \boldsymbol{A}_\nu - \partial_\nu \boldsymbol{A}_\mu - \frac{i}{\hbar}[\boldsymbol{A}_\mu, \boldsymbol{A}_\nu], \qquad (1.2.3)$$

and the effective magnetic field $\boldsymbol{B}$ then given by $B_j = \frac{1}{2}\epsilon_{jkl}F_{kl}$. Note that in the above derivative we have not performed adiabatic approximation. In this stage $\boldsymbol{A}$ is a pure gauge given by the $SU(N)$ transformation and the Berry curvature is simply zero $F_{\mu\nu} = 0$. Further applying the adiabatic condition yields adiabatic gauge potential which may lead to nontrivial SO couplings.

Consider that the ground state subspace of $H_I$ has $n$ $(n < N)$ degenerate eigenstates. The off-diagonal couplings within the degenerate ground states are not negligible. However, the adiabatic condition can be satisfied and decouple the ground state subspace from excited levels when the ground state energy, $E_g$, is well separated from those of the remaining $N-n$ states, namely, when $|\boldsymbol{v} \cdot \boldsymbol{A}_{ij}| \ll |E_g - E_j|$ holds for $i = 1, 2, \ldots, n$ and $j = n+1, n+2, \ldots, N$. In this case, we neglect the couplings between the ground state subspace and excited states and reach the reduced column vector of wave functions only for the degenerate subspace as $\bar{\Psi} = (\Psi_1, \Psi_2, \ldots, \Psi_n)^\mathsf{T}$, which satisfies

$$i\hbar\frac{\partial\bar{\Psi}}{\partial t} = \left[\frac{1}{2m}\left(-i\hbar\nabla - \boldsymbol{A}^{(n)}\right)^2 + V_{\text{eff}}\right]\bar{\Psi}, \qquad (1.2.4)$$

where $\boldsymbol{A}^{(n)}$ is a $n \times n$ matrix reduced from the pure gauge $\boldsymbol{A}$, rendering a $U(n)$ non-Abelian gauge potential if $n > 1$. The effective trapping potential matrix $V_{\text{eff}}$ is given by $V_{\text{eff},jl} = E_j\delta_{jl} + \langle\chi_j(\boldsymbol{r})|V|\chi_l(\boldsymbol{r})\rangle + \Phi_{jl}$, with the scalar potential $\Phi$ being given by

$$\Phi_{jk} = \frac{1}{2m}\sum_{l>n}\boldsymbol{A}_{jl} \cdot \boldsymbol{A}_{lk}. \qquad (1.2.5)$$

In the case of $n = 1$, i.e. the ground state is non-degenerate, the reduced adiabatic gauge potential becomes Abelian.

*The Lambda-type configuration.* A minimal scheme to realize SO coupling by generating spin-dependent gauge potential was proposed based on a $\Lambda$-type configuration.[59] As shown in Fig. 1.2.1(a), the there-level atoms are coupled to two laser beams, with the interaction Hamilonian

$$H_{\mathrm{I}} = \hbar\Delta - (\hbar\Omega_1|g_\uparrow\rangle\langle e| + \hbar\Omega_2|g_\downarrow\rangle\langle e| + \text{h.c.}), \qquad (1.2.6)$$

where $\Omega_1 = \Omega\sin\theta e^{iS_1}$ and $\Omega_2 = \Omega\cos\theta e^{iS_2}$ with $\Omega = \sqrt{|\Omega_1|^2 + |\Omega_2|^2}$, with $S_{1,2}$ being determined by the phases of laser beams. When the detuning $\Delta$ is much larger than the coupling strength $\Omega$, the above Hamiltonian has two nearly degenerate ground states

$$|\chi_1(\boldsymbol{r})\rangle = \cos\theta e^{iS_2(\boldsymbol{r})}|g_\uparrow\rangle - \sin\theta e^{iS_1(\boldsymbol{r})}|g_\downarrow\rangle, \qquad (1.2.7)$$

$$|\chi_2(\boldsymbol{r})\rangle \simeq \sin\theta e^{iS_2(\boldsymbol{r})}|g_\uparrow\rangle + \cos\theta e^{iS_1(\boldsymbol{r})}|g_\downarrow\rangle, \qquad (1.2.8)$$

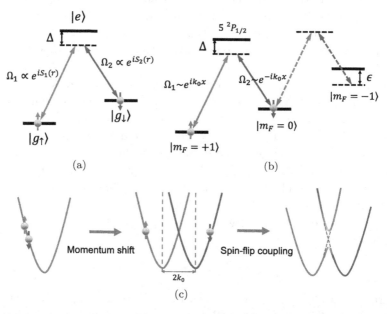

**Fig. 1.2.1.** A minimal scheme for 1D SO coupling. (a) The $\Lambda$-type configuration, with the atom levels (dubbed spin up and spin-down) coupled to two light beams. (b) A practical layout for $^{87}$Rb atoms, coupling to two beams propagating oppositely in $x$ direction. The $F = 1$ manifold is employed with one hyperfine state well separated by a quadratic Zeeman shift $\epsilon$. (c) An intuitive picture for the 1D SO coupling. The two-photon Raman coupling imprints a momentum difference to the two spin states due to momentum transfer in the two-photon process, and induce a spin-flip transition which opens a Zeeman gap at zero momentum. The whole effects generate a 1D SO coupling together with a Zeeman term.

which make up a spin-1/2 system providing the realization of SO coupling with proper spatial-dependent phases $S_{1,2}(\mathbf{r})$. The following is an alternative equivalent picture. Due to the condition that $|\Delta| \gg \Omega$, the single-photon transitions between the ground states $|g_{\uparrow,\downarrow}\rangle$ and the excited state $|e\rangle$ are greatly suppressed, while the two-photon transition between the two ground states becomes dominant. In this regime, by adiabatically removing the excited state the system is effectively regarded as a two-state ($|g_{\uparrow}\rangle$ and $|g_{\downarrow}\rangle$) configuration coupled by a two-photon Raman process. Consider a simple case where the laser fields are counter-propagating along $x$-direction so that the Rabi frequencies $\Omega_1 = \Omega_0 e^{ik_0 x}$ and $\Omega_2 = \Omega_0 e^{-ik_0 x}$ [see Fig. 1.2.1(b)], where $k_0$ denotes the wave number of the two lasers. As mentioned above, when $|\Delta|$ is large enough, one can adiabatically eliminate the excited state and obtain the effective spin-1/2 Hamiltonian in $x$-direction

$$H_{\text{eff}} = \begin{pmatrix} \frac{\hbar^2 k_x^2}{2m} + \frac{\delta}{2} & \frac{\Omega_R}{2} e^{i2k_0 x} \\ \frac{\Omega_R}{2} e^{-i2k_0 x} & \frac{\hbar^2 k_x^2}{2m} - \frac{\delta}{2} \end{pmatrix}, \quad (1.2.9)$$

where $\delta$ is a small two-photon detuning for the Raman coupling and the Raman Rabi frequency $\Omega_R = \Omega_0^2/\Delta$. After the transformation $U = \exp(-ik_0 x \sigma_z)$, the Hamiltonian renders a 1D SO coupling (with equal amplitudes of Rashba and Dresselhaus terms)

$$H_{1D} = U H_{\text{eff}} U^\dagger = \frac{\hbar^2 (k_x + k_0 \sigma_z)^2}{2m} + \frac{\delta}{2}\sigma_z + \frac{\Omega_R}{2}\sigma_x. \quad (1.2.10)$$

From the above SO Hamiltonian one can see that the generated gauge potential $\mathbf{A} = \hbar k_0 \sigma_z \hat{e}_x$ is spin-dependent, but still Abelian, rather than non-Abelian.

The SO coupling can naturally emerge from the view of momentum transfer, as illustrated in Fig. 1.2.1(c). Due to the laser polarization, the transition from $|g_{\uparrow}\rangle$ to $|g_{\downarrow}\rangle$ or the other way round amounts to absorbing a photon from one laser and then emitting a photon to the other, accompanied with momentum transfer by $2k_0$. To be specific, the atomic momentum increases by $k_0$, the so-called recoil momentum, after absorption of one photon; it gains another $k_0$ in the subsequent emission since the two coupling beams are counter-propagating. As a result, by Raman process, the spectra of atoms in the two states $|g_{\uparrow}\rangle$ and $|g_{\downarrow}\rangle$ exhibit a relative $2k_0$-momentum shift along $k_x$ direction [the former two pictures in Fig. 1.2.1(c)]. Furthermore, a finite Raman coupling strength leads to a spin-flip transition, which corresponds to the $\sigma_x$ term in Eq. (1.2.10) and opens a gap at $k_x = 0$, eventually leading to the spectra shown in the last picture of Fig. 1.2.1(c). It is the spin-flip transition associated with momentum transfer that accounts for the basic mechanism of generating SO couplings.

It is noteworthy that simply applying another pair of laser beams along say $\pm y$ directions to induce additional Raman coupling and momentum transfer along $y$ direction cannot yield 2D SO coupling. Instead, in this case, the both Raman couplings along $x$ and $y$ axes will combine and induce the 1D SO coupling along $\hat{e}_x + \hat{e}_y$ direction. As we shall see later, to realize high-dimensional SO couplings based on $\Lambda$-type configurations, one needs to consider optical lattices.

*The tripod-configuration.* An early scheme to generate non-Abelian gauge potentials is based on a tripod-type configuration with four-level atoms coupled to spatially varying laser fields,[58] as depicted in Fig. 1.2.2. The atom-light coupling Hamiltonian reads

$$H_{\mathrm{I}} = \hbar\Delta|0\rangle\langle 0| - \hbar\sum_{j=1}^{3}(\Omega_1|0\rangle\langle j| + \mathrm{h.c.}), \qquad (1.2.11)$$

where the Rabi frequencies $\Omega_j$ characterize the corresponding atomic transition amplitudes and are defined by $\Omega_1 = \Omega\sin\theta\cos\phi e^{iS_1}$, $\Omega_2 = \Omega\sin\theta\sin\phi e^{iS_2}$, and $\Omega_3 = \Omega\cos\theta e^{iS_3}$, with $\Omega = \sqrt{|\Omega_1|^2 + |\Omega_2|^2 + |\Omega_3|^2}$. The above Hamiltonian has two degenerate dark states

$$|\chi_1\rangle = \sin\phi e^{iS_{31}}|1\rangle - \cos\phi e^{iS_{32}}|2\rangle, \qquad (1.2.12)$$

$$\begin{aligned}|\chi_2\rangle = {}& \cos\theta\cos\phi e^{iS_{31}}|1\rangle + \\ & + \cos\theta\sin\phi e^{iS_{32}}|2\rangle - \sin\theta|3\rangle,\end{aligned} \qquad (1.2.13)$$

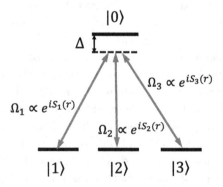

**Fig. 1.2.2.** The tripod configuration for non-Abelian gauge fields. The three ground states couple to three beams which drive transitions to the excited level.

with $S_{ij} = S_i - S_j$. One can embed spatial dependence into dark states by using standing-wave fields (associated with the angles $\phi$ and $\theta$) or non-parallel propagating running waves (associated with the relative phases $S_{ij}$) or vortex beams with orbital angular momentum (phases $S_{ij}$). The two dark states form a pseudospin-1/2 system, for which a U(2) non-Abelian gauge potential $\mathbf{A}^{(2)}$ can arise in the adiabatic motion, with the elements $\mathbf{A}_{jk}^{(2)} = i\hbar\langle\chi_j|\nabla|\chi_k\rangle$ given by

$$\mathbf{A}_{11}^{(2)} = \hbar(\cos^2\phi\nabla S_{23} + \sin^2\phi\nabla S_{13}),$$
$$\mathbf{A}_{22}^{(2)} = \hbar\cos^2\theta(\cos^2\phi\nabla S_{13} + \sin^2\phi\nabla S_{23}), \qquad (1.2.14)$$
$$\mathbf{A}_{12}^{(2)} = \hbar\cos\theta[1/2\sin(2\phi)\nabla S_{12} - i\nabla\phi].$$

Properly choosing the parameters $\theta, \phi$ and/or $S_{ij}$ can readily yield Rashba or Dresselhaus type SO couplings. For example, let $\theta = \phi = \pi/4, S_{12} = k_0 x, S_{23} = k_0 x + k_0 y$, and $S_{13} = -k_0 x + k_0 y$. The gauge potential becomes

$$\mathbf{A}^{(2)} = (\sqrt{2}\hbar/4)k_0\sigma_x\hat{e}_x + (\hbar/4)k_0\sigma_z\hat{e}_y, \qquad (1.2.15)$$

where a trivial constant has been neglected. The above gauge potential is of $U(2)$ form and leads to a 2D SO coupling. The tripod scheme provides a natural realization of non-Abelian gauge potentials for an effective pseudospin-1/2 system, while it also suffers challenges in experimental study. For example, the tripod configuration includes three ground states $|j\rangle$ ($j = 1, 2, 3$), with one of the three states having to be a metastable state which may have relatively short lifetime, particularly in the presence of interactions.[99] On the other hand, the realization of tripod scheme necessitates the resonant Raman couplings between each two of the three ground states, which is vulnerable to the fluctuation of external magnetic field which is applied for experiments of generation SO couplings. Finally, unlike the $\Lambda$ scheme, in the tripod system the SO coupling is defined for pseudospin states (i.e. dark states), rather than real atom spins (hyperfine eigenstates), which could be hard to be directly measured or engineered in experiments. As a result, while the first proposal of tripod scheme was introduced over a decade ago,[58] it was successfully demonstrated in experiment only recently by Shanxi University in collaboration with CUHK.[87]

## 2.B. *Gauge potential in optical lattices*

In optical lattices the motion of atoms can be described by hopping between lattice sites. Accordingly, the SO coupling in optical lattices corresponds to the spin-flip hopping between neighboring sites, similar to the spin-flip transition associated with momentum transfer in the continuum gas. For example, for a 2D square optical lattice, the tight-binding model for the case with a Rashba SO coupling can be written as

$$H = \sum_{\langle j,l \rangle, s, s'} \left[ -t_0 c_{j,s}^{\dagger} (e^{i\frac{\theta}{2}\mathbf{A}\cdot\mathbf{d}_{jl}})_{ss'} c_{l,s'} + h.c. \right], \qquad (1.2.16)$$

where $\mathbf{A} = (\sigma_y, -\sigma_x)$, $c_{l,s}$ and $c_{l,s}^{\dagger}$ are annihilation and creation operators, respectively, $\langle j, l \rangle$ represents the hopping between nearest-neighbor sites, and $\mathbf{d}_{jl}$ denotes the unit vector from $l$-site to $j$-site. The above tight-binding Hamiltonian describes a spin-conserved hopping with coefficient $t_s = t_0 \cos \theta/2$ along $x$ and $y$ directions, and spin-conserved hoppings with coefficient $t_{so}^x = t_0 \sin \theta/2$ ($t_{so}^y = -t_0 \sin \theta/2$) coupling to $\sigma_y$ ($\sigma_x$) along $x$ ($y$) direction. Transforming the Hamiltonian into $\mathbf{k}$ space yields $H = \sum_{\mathbf{k}} c_{\mathbf{k},s}^{\dagger} \mathcal{H}_{\mathbf{k}}^{ss'} c_{\mathbf{k},s}$, with the Bloch Hamiltonian

$$\mathcal{H}(\mathbf{k}) = -2t_s(\cos k_x a + \cos k_y a)\hat{I}$$
$$- 2t_{so}\left[\sin k_x a \sigma_y - \sin k_y \sigma_x\right], \qquad (1.2.17)$$

where $\hat{I}$ is a $2 \times 2$ identity matrix and $t_{so} = t_0 \sin \theta/2$. While a purely Rashba SO coupling has not been realized in experiment in an optical lattice, Engels' group first demonstrated the 1D SO coupling for bosons in optical lattices.[71]

The recent studies show that instead of realizing a purely Rashba SO coupling, it is more natural to realize the Dirac-type 2D SO coupling based on the so called optical Raman lattice scheme,[85,100] where besides a 2D Rashba-type SO term induced by Raman coupling lattice, the spin-conserved hopping couples to the $\sigma_z$-component rather than identity matrix $\hat{I}$. In the lowest $s$-band regime, the Bloch Hamiltonian becomes

$$\mathcal{H}(\mathbf{k}) = \left[m_z - 2t_s(\cos k_x a + \cos k_y a)\right]\sigma_z$$
$$- 2t_{so}\left[\sin k_x a \sigma_y - \sin k_y \sigma_x\right], \qquad (1.2.18)$$

which describes a minimal quantum anomalous Hall (QAH) model driven by SO coupling and has been first realized by the collaborating team at PKU and USTC.[88] The optical Raman lattice schemes exhibit high feasibility in experimental realization with multiple advantages, and are becoming a dominating technique in investigating novel high-dimensional SO effects and topological physics in experiment. The detailed discussions will be presented in the next sections.

## 1.3. Experimental Issues of Realizing the 1D SO Coupling

### 3.A. *Realization of Λ-type configuration*

The Λ-type coupling scheme can be readily realized in real cold atom systems. Fig. 1.2.1(b) illustrates a typical hyperfine-level configuration in $^{87}$Rb bosons, with ground manifold $F = 1$. The excited state $|e\rangle$ indeed includes all relevant states corresponding to $D_1$ and $D_2$ transitions. To separate the Λ-type configuration from other state, one can apply an external bias magnetic field $\mathbf{B} = B_z \hat{e}_z$ to split up the ground states. The energy shift is nonlinear due to the quadratic Zeeman splitting

$$\Delta E_{|F,m_F\rangle} = \mu_B g_F m_F B_z + \frac{(g_J - g_I)^2 \mu_B^2}{2\hbar \Delta E_{\text{hfs}}} B^2, \tag{1.3.1}$$

where $g_{I,J,F}$ are Landé factors and $\Delta E_{\text{hfs}}$ denote hyperfine splitting. As a result, when two of the three ground states are coupled in two-photon resonance, they are detuned from the third state, and an isolated Λ-system is resulted. For example, with a bias field of strength 14G, the Zeeman splitting between two neighboring hyperfine states $|m_F = 0\rangle$ and $m_F = -1$ is about 10.2MHz, which has a discrepancy of $8E_r$ with respect to the Zeeman splitting between $|m_F = 0\rangle$ and $m_F = 1$ if applying $\lambda = 786$nm Raman beams [Fig. 1.2.1(b)]. With the simple scheme the 1D SO coupling can now be routinely realized in both Bose-Einstein condensates (BECs)[61,65] and Fermi gases.[63,64]

### 3.B. *Cancellation of Raman couplings through D1 and D2 transitions*

Note that the net Raman coupling is obtained by taking into account the contributions through both the $D_1$ and $D_2$ lines. The total Raman coupling between two ground states, e.g. $|m_F = +1\rangle$ and $|m_F = 0\rangle$ for $^{87}$Rb, is given by

$$\Omega_R = \sum_F \frac{\Omega_{2F,D_1}^* \Omega_{1F,D_1}}{\Delta} + \sum_F \frac{\Omega_{2F,D_2}^* \Omega_{1F,D_2}}{\Delta + E_s}, \tag{1.3.2}$$

where $\Delta$ is one-photon detuning for $D_1$ line, $E_s$ is the fine-structure splitting, $\Omega_{1F,D_j}$ and $\Omega_{2F,D_j}$ represent the one-photon Rabi-frequencies induced by the two laser beams corresponding to the transitions from $|m_F = +1\rangle$ and $|m_F = 0\rangle$ to an excited state of quantum number $F$ in the $D_j$ ($j = 1, 2$) lines, respectively.

Let $|e_{F,D_j}\rangle$ denote an excited state corresponding to the $D_j$ $(j = 1, 2)$ line. We can verify the following identity

$$\sum_F \Omega^*_{2F,D_1}\Omega_{1F,D_1} + \sum_F \Omega^*_{2F,D_2}\Omega_{1F,D_2}$$
$$= \sum_{F,j=1,2} \langle g_\uparrow |d_{2F} \cdot E_2 | e_{F,D_j}\rangle\langle e_{F,D_j}|d_{1F} \cdot E_1|g_\downarrow\rangle,$$
$$= \langle g_\uparrow |(d_{2F} \cdot E_2)(d_{1F} \cdot E_1)|g_\downarrow\rangle,$$
$$= 0, \tag{1.3.3}$$

where $d_{jF}$ is the corresponding dipole vector. In the above derivative we have applied the identity that $\sum_{F,j} |e_{F,D_j}\rangle\langle e_{F,D_j}| = I$. Thus the couplings through the $D_1$ and $D_2$ lines in alkali atoms contribute oppositely to the Raman transition. We obtain that

$$\Omega_R = \sum_F \frac{E_s}{\Delta(\Delta + E_s)}\Omega^*_{2F,D_1}\Omega_{1F,D_1}. \tag{1.3.4}$$

Assuming that the deunings for both $D_1$ and $D_2$ lines are in the same sign, we consider the following two situations. First, consider the red-detuned regime (similar for blue detunings) and when $\Delta < E_s$, we have from the above result that

$$\Omega_R \propto |\Omega_1\Omega_2/\Delta| \propto (1/\tau_{\text{life}})(\Delta/\Gamma), \tag{1.3.5}$$

where the lifetime satisfies $1/\tau_{\text{life}} \propto |\Omega_j|^2\Gamma/\Delta^2$ and $\Gamma$ denotes the natural linewidth of the relevant transition.[101] Secondly, when $|\Delta| \gg E_s$, we have

$$\Omega_R \propto |\Omega_1\Omega_2|/\Delta^2 \propto (1/\tau_{\text{life}})(E_s/\Gamma). \tag{1.3.6}$$

In this case, the Raman coupling strength $\Omega_R$ and life time $\tau_{\text{life}}$ cannot be enhanced at the same time by increasing $\Delta$ and Rabi-frequencies $\Omega_j$. Thus we have the following conclusions: (1) To induce an appreciable Raman coupling strength $\Omega_R$, the detuning $\Delta$ cannot be much larger than fine structure splitting $E_s$ of the excited states. (2) A large enough life time $\tau_{\text{life}}$, however, requires that $\Delta$ should be much larger than $\Omega_j$. (3) For alkali atoms, the proper parameter regime is that $|\Delta| \sim E_s$. (4) The atomic candidates with large fine structure splitting $E_s$ are preferred for the generation of spin-orbit coupling.

It is noteworthy that, if the optical transitions are applied between $D_1$ and $D_2$ lines, namely $0 < -\Delta < E_s$, the both $D_{1,2}$ transitions contribute in the same sign to the Raman coupling. This implies that the Raman coupling can be largely enhanced in this regime. Nevertheless, the magnitude of $\Delta$ is restricted by the fine-structure splitting $E_s$. The typical magnitudes of $E_s$ for alkali atoms are that $E_s \sim 7.1$THz for $^{87}$Rb,[102] $E_s \sim 1.8$THz for $^{40}$K,[103] and $E_s \sim 10$GHz for $^6$Li,[103] implying that the realization is favorable for $^{87}$Rb, marginally feasible for $^{40}$K, while suffers strong heating for $^6$Li atoms.

### 3.C.  *Realization of 1D SO coupling for bosons*

Lin *et al.*[61] first reported the experimental realization of 1D SO coupling in a Bose-Einstein condensate of $^{87}$Rb atoms, and in particular, demonstrated two quantum phase transitions driven by Raman coupling strength [Fig. 1.3.1(a,d)]. One is two minima of the dressed energy dispersion merging into a single minimum [Fig. 1.3.1(b,c)]; the other is a transition from a

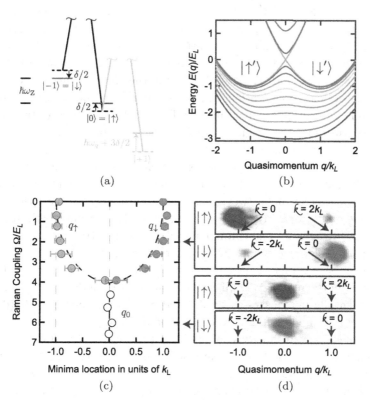

(a)     (b)

(c)     (d)

**Fig. 1.3.1.** The NIST experiment of the realization of 1D SO coupling with $^{87}$Rb bosons.[61] (a) The the Raman coupling configuration for the ground manifold. (b) The spectra of the SO coupled dressed states with different magnitudes of the Raman coupling strengths. (c) The band minima of the SO coupled spectra with different Raman coupling strengths. (d) The TOF imaging of atom clouds due to SO coupling.

spatially spin-mixed state to a separated state. The latter one is induced by the modified interactions between the two dressed states (the two minima of dispersion).[104] The spin-mixed phase is also called "stripe" phase,[105] where atoms condense into a superposition of the two minima [Fig. 1.3.2(a)], thus exhibiting density fringes,[106–108] while the separated phase is known as "plane-wave" or "magnetized" phase since the condensation occurs at only one minimum [Fig. 1.3.2(a)].

Ji et al.[69] extended the phase diagram of $^{87}$Rb BECs to the finite-temperature case by observing the evolution of condensate magnetization with the temperature. Unlike the "magnetized" phase, the condensate in the stripe state should exhibit no magnetization at zero temperature; it is not always true in experiments due to undesired circumstances (e.g. the forming of domain wall). However, the histogram of magnetization distribution should have a Gaussian-like peak around zero magnetization after a large number of measurements, as shown in Fig. 1.3.2(c). By contrast, the histogram in the magnetized phase shows two sharp peaks around $\pm 1$ [Fig. 1.3.2(c)], indicating the spontaneous $Z_2$-symmetry breaking in Bose condensation. Experimental measurements determines the finite-temperature phase diagram as in Fig. 1.3.2(b), where the stripe phase will first turn into the magnetized phase before becoming the normal state as the temperature increases.

Another related issue is about excitation spectrum. Because the stripe phase breaks both U(1) symmetry (superfluid phase) and translational symmetry, there will be two linear Goldstone modes in the spectrum[109] [schematically shown in the insets of Fig. 1.3.2(b)]. The magnetized phase, however, breaks only one continuous (phase) symmetry, and there is only one linear Goldstone mode. Furthermore, the tendency towards periodic order (the stripe phase) near the transition indicates that the excitation spectrum in the magnetized phase has a roton-type mode,[104,110] which was experimentally demonstrated by Ji et al.[70] [see inset of Fig. 1.3.2(d)]. In Ref. 70, roton-mode softening was observed by Bragg spectroscopy, showing the roton gap is vanishingly small in the critical regime [Fig. 1.3.2(d)]. Thus the magnetized phase has a greater low-energy density-of-states than the stripe phase, meaning the magnetized phase can gain more entropy from thermal fluctuations and become more favorable. This gives an explanation for why the phase boundary bends towards the stripe phase side in the finite-temperature phase diagram [Fig. 1.3.2(b)].

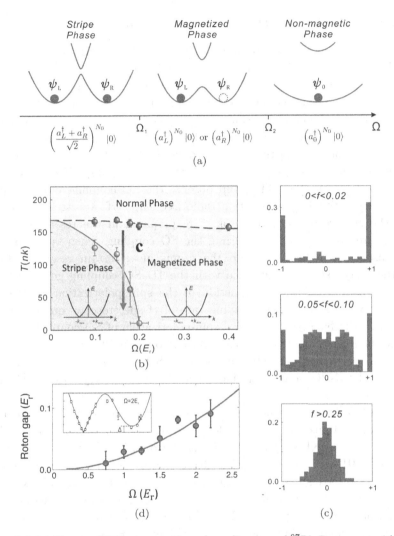

(a)

(b)

(c)

(d)

**Fig. 1.3.2.** Zero- and finite-temperature phase diagrams of $^{87}$Rb Bose gases with 1D SO coupling. (a) Single-particle dispersion and zero-temperature phase diagram as a function of the Raman coupling $\Omega$, which shows a stripe phase, a magnetized phase and a non-magnetic phase as $\Omega$ increases. (b) Finite-temperature phase diagram of spin-orbit coupled bosons. Insets: Schematic low-energy spectrum for the stripe and magnetized phases. (c) Evolution of the magnetization histogram as the temperature is lowered (equivalent to increasing the condensate fraction $f$). (d) Softening of roton mode. The measured roton gap becomes smaller as $\Omega$ decreases and the vanishing indicates the phase transition to the stripe phase. Inset: The low-energy excitation spectrum, which clearly shows a roton-maxon structure.

### 3.D. *Realization of SO coupling for fermions*

Wang *et al.*[63] at Shanxi group and Cheuk *et al.*[64] at MIT group respectively realized 1D SO coupling in atomic Fermi gases using $^{40}$K and $^6$Li atoms. The 1D SO coupling induced momentum transfer along $x$ direction leads to a spin-momentum lock along the this direction. This spin imbalance was observed by time-of-flight (TOF) expansion in the experiment with $^{40}$K atoms. On the other hand, note that $^6$Li fermions have a tiny fine structure splitting in excited levels. The strong heating brings about serious challenging in realizing the SO coupling with degenerate $^6$Li gas. Instead, the MIT group observe the SO coupling induced spin imbalance through spin injection radio-frequency (rf) spectroscopy. During the rf spectroscopy, the atoms are prepared in the reservoir states which are not involved in realizing the SO coupling. Then, a low-power rf pulse was applied to pump the atoms at a certain reservoir state to the target spin-1/2 subspace with the 1D SO coupling generated by Raman couplings. Being a function of the spin polarization of dressed states, from the resonant pumping rate of rf spectroscopy one can read out both the band structure and spin-polarization of SO coupled system. Since the realization, SO coupling induced novel superfluids has been a hot issue. In particular, the interplay between SO coupling and strong interactions (BEC-BCS crossover) has been analyzed theoretically[111–114] and experimentally.[115,116]

The 1D SO coupling has also been realized in fermionic lanthanide and alkali earth atoms, including Dy fermions by Lev's group,[73] Yb atoms by Jo's group at HKUST[74] and by Fallani's group,[76] and Sr atoms by Ye's group at JILA.[77] Compared with alkali atoms, the lanthanide and alkali earth atoms have an effective large fine structure splitting $E_s$ while a small natural linewidth of transition $\Gamma$. As a result, the SO coupling generated for such atom candidates in principle can have a long lifetime. In particular, for the $^{161}$Dy fermions, a lifetime up to 400ms was observed in the experimental realization.[73] Such life is limited by dipolar decay of the spin, rather than by light-induced heating.

### 1.4. SO Couplings Beyond One Dimension

Until 2016, only the 1D SO couplings had been reported in experiments. Realization of SO couplings beyond 1D regime is much more important for quantum simulation.[78,79,82,114,117–121] It is because a high-dimensional

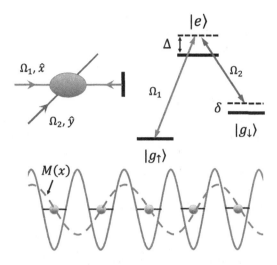

**Fig. 1.4.1.** The Raman lattice scheme in 1D case. Atoms are trapped in a 1D optical lattice with an internal three-level Λ-type configuration coupled to radiation by a standing wave (red line) and a running wave (blue line). The induced Raman potential $M(x)$ is anti-symmetric with respect to each lattice site.

SO coupling (e.g. the 2D Rashba term) corresponds to a non-Abelian gauge potential, which is associated with nonzero Berry's curvature and nontrivial geometric or topological effects, while 1D SO coupling is an Abelian potential and does not give a nonzero Berry curvature. The study of broad classes of novel topological states necessitates the realization of high-dimensional SO couplings, such as the 2D and 3D topological insulating states, Weyl semimetals, and topological superfluid phases. The previous schemes for realizing 2D and 3D SO couplings include the tripod scheme[58] and its multipod extension,[80] ring-structure coupling scheme,[83] and realization with magnetic pulses.[84,86] Nevertheless, except for the tripod scheme successfully demonstrated in the recent experiment,[87] most of these schemes are hard to achieve in real cold atom experiments.

*Optical Raman lattice.* Recently, the so called optical Raman lattice schemes are proposed to realize high-dimensional SO couplings and novel topological quantum phases. The essential idea of the optical Raman lattice is that one combines the realization of conventional optical lattice, which governs the normal hopping of atoms, and the periodic Raman lattice, which determines spin-flip hopping couplings, in the system. Moreover, such realization of optical and Raman lattices are achieved through transitions induced by *the same* optical fields. A consequence of this combination is that the generated conventional optical lattice and Raman lattice exhibit an automatically fixed relative spatial configuration, satisfying certain intrinsic relative space symmetries without any fine tunings. It turns out that, with such space symmetries, not only the realization of high-dimensional SO couplings can be greatly simplified, but also the resulted SO coupled systems naturally host various types of topological phases. The experimental realizations of optical Raman lattices have been successfully achieved in 1D and 2D regimes. In the following we shall review the details of optical Raman lattice schemes, and the recent important experimental progresses.

### 4.A. *Optical Raman lattice: 1D case*

#### 4.A.1. *Model*

To illustrate how the optical Raman lattice scheme works, we first consider 1D regime, as proposed in Ref. 100,122. As shown in Fig. 1.4.1, the 1D blue-detuned optical Raman lattice is formed through a simple Λ-type configuration, with a standing-wave beam $\Omega_1$ (red line, along $x$ direction) and a running wave beam $\Omega_2$ (blue line, along $y$ direction) being applied.[100] The standing-wave beam applies to both the spin-up $|g_\uparrow\rangle$ and spin-down $|g_\downarrow\rangle$ states. For linearly polarized leaser beams, the standing wave field generates a 1D optical lattice which is spin-independent. Namely, the lattice potential reads $V_{\text{latt}}(x) = V_0 \cos^2 k_0 x$, with $V_0 \propto |\Omega_1|^2/\Delta$. Further applying the running wave beam generates the Raman coupling potential through the two photon process, given by $M(x) = M_0 \cos k_0 x$, with the amplitude $M_0 \propto |\Omega_1\Omega_2|/\Delta$. All the irrelevant phase factors, including the initial phases of beams and $e^{ik_0 y}$ in $\Omega_2$, are ignored. The realized 1D Hamiltonian reads

$$H_{1D} = \frac{\hbar^2 k_x^2}{2m} + V_0 \cos^2 k_0 x + M_0 \cos k_0 x \sigma_x + \frac{\delta}{2}\sigma_z, \qquad (1.4.1)$$

where the tunable two-photon detuning $\delta$ plays the role of an effective Zeeman splitting. Before proceeding, we give a few remarks on the above realization. First, the whole setting applies only two laser beams which can be generated from a single laser source, and no fine tuning is required. Secondly, both the optical lattice and Raman lattice are generated through the same standing wave beam $\Omega_1$. This leads to a fixed relative configuration between $V_{\text{latt}}$ and $M(x)$, namely, (a) the periodicity of $M(x)$ is one-half of that of $V_{\text{latt}}$; (b) the Raman potential is antisymmetric with respect to each lattice site of $V_{\text{latt}}$. This relative configuration is topologically stable, since any phase fluctuations, if existing in the laser beams, only lead to global shift of the optical Raman lattice, but cannot affect the relative configuration. We shall see that with this relative space symmetry the optical Raman lattice can naturally realizes nontrivial topological phases.

The topological physics can be best seen with the tight-binding model, while we emphasize that the following results are not restricted in the tight-binding regime. Consider the lowest $s$-band model, the tight-binding Hamiltonian generally takes the form

$$
H_{\mathrm{TB}} = \sum_{\langle i,j \rangle, \sigma} t_0^{ij} c_{i\sigma}^{\dagger} c_{j\sigma} + \sum_i \frac{\delta}{2}(n_{i\uparrow} - n_{i\downarrow})
$$
$$
+ \sum_{\langle i,j \rangle} (t_{\mathrm{so}}^{ij} c_{j\uparrow}^{\dagger} c_{j+1\downarrow} + \mathrm{h.c.}), \tag{1.4.2}
$$

where $c_{j\sigma}$ $(c_{j\sigma}^{\dagger})$ are the annihilation (creation) operators of $s$-orbit for spin $\sigma = \uparrow, \downarrow$ at lattice site $j$ and the atom number operator $n_{j\sigma} \equiv c_{j\sigma}^{\dagger} c_{j\sigma}$. Here the spin-conserved hopping couplings $t_0^{ij}$ are induced by the lattice potential, given by

$$
t_0^{ij} = \int dx \phi_{s\sigma}^{(i)}(x) \left[ \frac{\hbar^2 k_x^2}{2m} + V_{\mathrm{latt}}(x) \right] \phi_{s\sigma}^{(j)}(x), \tag{1.4.3}
$$

with $\phi_{s\sigma}^{(j)}(x)$ being the Wannier functions for s-bands, and the Raman-coupling driven spin-flip hopping coefficients $t_{\mathrm{so}}^{ij}$ are

$$
t_{\mathrm{so}}^{ij} = \int dx \phi_{s\uparrow}^{(i)}(x) M(x) \phi_{s\downarrow}^{(j)}(x). \tag{1.4.4}
$$

Since the wavefunctions $\phi_{s\sigma}^{(i)}(x)$ are spin-independent (due to the spin-independent lattice) and satisfy $\phi_{s\sigma}^{(j)}(x) = \phi_{s\sigma}^{(0)}(x - x_j)$, it follows that $t_0^{ij} = -t_0$ with

$$
t_0 = - \int dx \phi_s^{(0)}(x) \left[ \frac{\hbar^2 k_x^2}{2m} + V_{\mathrm{latt}}(x) \right] \phi_s^{(0)}(x - a). \tag{1.4.5}
$$

On the other hand, from the relative space antisymmetry of Raman coupling potential, one can find that

$$
t_{\mathrm{so}}^{j,j\pm1} = \pm(-1)^j t_{\mathrm{so}}, \tag{1.4.6}
$$

with

$$
t_{\mathrm{so}} = M_0 \int dx \phi_s^{(0)}(x) \cos(k_0 x) \phi_s^{(0)}(x - a). \tag{1.4.7}
$$

The staggered property of the spin-flip hopping can be absorbed by a gauge transformation that $c_{j\downarrow} \to -e^{i\pi x_j/a} c_{j\downarrow}$, where $a$ denotes the lattice constant. The tight-binding model is finally written as

$$H_{\text{TB}} = -t_0 \sum_{\langle i,j \rangle} (c_{i\uparrow}^\dagger c_{j\uparrow} - c_{i\downarrow}^\dagger c_{j\downarrow}) + \sum_i \frac{\delta}{2}(n_{i\uparrow} - n_{i\downarrow})$$

$$+ \left[ \sum_j t_{\text{so}} (c_{j\uparrow}^\dagger c_{j+1\downarrow} - c_{j\uparrow}^\dagger c_{j-1\downarrow}) + \text{h.c.} \right]. \tag{1.4.8}$$

It can be seen that the gauge transformation effectively reverses the sign of the hopping coefficient of spin-down states. Transforming the Hamiltonian to momentum space yields (with $m_z = \frac{\delta}{2}$)

$$H_{\text{TB}} = \sum_k c_{k,s}^\dagger \mathcal{H}_{ss'}(k_x) c_{k,s'}, \tag{1.4.9}$$

$$\mathcal{H}(k_x) = (m_z - 2t_0 \cos k_x a)\sigma_z + 2t_{\text{so}} \sin k_x a \sigma_y.$$

From the above results one can see that the Raman coupling has two novel effects: (a) it induces $\pi/a$-momentum transfer between spin-up and spin-down states; (b) it further induces the SO coupling. This is a direct consequence of the relative configuration between Raman and optical lattices. The two effects can also be pictorially described in Fig. 1.4.2. In particular, the $\pi/a$-momenta transfer effectively reverses the sign of hopping coefficients of one spin states (say spin-up), and the remaining SO coupling further opens a gap, giving an insulating phase with nontrivial topology.

### 4.A.2. *1D AIII class topological insulator*

By a direct check one can see that the time-reversal symmetry $\mathcal{T} = i\sigma_y K$ is broken for the Hamiltonian, where $K$ is the complex conjugate. On the other hand, the charge-conjugation (particle-hole) symmetry, defined by

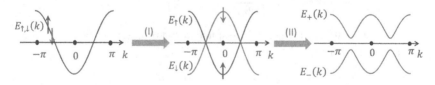

**Fig. 1.4.2.** Illustration of band structure of 1D optical Raman lattice in the $s$-band regime. (I) The $\pi/a$-momentum transfer effectively reverses the sign of hopping coefficient of one spin state (spin-up); (II) The SO coupling opens a topological gap, leading to a 1D AIII class topological insulator.

$\mathcal{C} : (c_s, c_s^\dagger) \to (\sigma_z)_{ss'}(c_{s'}^\dagger, c_{s'})$, is also broken. The chiral symmetry, defined as the product of the charge-conjugation and time-reversal symmetry is conserved, with $(\mathcal{CT})H(\mathcal{CT})^{-1} = H$, and $(\mathcal{CT})^2 = 1$. Note that the charge-conjugation is a unitary symmetry, and the chiral symmetry is anti-unitary. The complete symmetry group of the Hamiltonian is $U(1) \times Z_2^T$, with $Z_2^T$ being an anti-unitary group formed by $\{I, \mathcal{CT}\}$. In the first quantization picture the chiral symmetry is imply $S = \sigma_x$, and is satisfied even when a $m_y \sigma_y$ term exists in the Hamiltonian. From the symmetry analysis we know that the Hamiltonian $H$ belongs to chiral unitary (AIII) class according to the Altland-Zirnbauer ten-fold classification,[123] and its topology is characterized by an integer $Z$, namely the 1D winding number, given by

$$N_{1D} = \frac{1}{4\pi i} \int dk_x \mathrm{Tr}\left[\sigma_x \mathcal{H}^{-1}(k_x)\partial_{k_x}\mathcal{H}(k_x)\right]$$
$$= \frac{1}{2\pi} \int dk_x \hat{h}_z \partial_{k_x} \hat{h}_y, \tag{1.4.10}$$

where $(\hat{h}_y, \hat{h}_z) = (h_y, h_z)/h$, with $h_y = 2t_{\mathrm{so}} \sin k_x a, h_z = m_z - 2t_0 \cos k_x a$, and $h = (h_y^2 + h_z^2)^{1/2}$. By a straightforward calculation one can find that

$$N_{1D} = \begin{cases} 1, & \text{for } |m_z| < 2t_0, \\ 0, & \text{for } |m_z| > 2t_0. \end{cases} \tag{1.4.11}$$

The nonzero winding number can simply interpreted that the spin axis of the lower subband winds one complete circle when the momentum $k$ runs over the FBZ, as sketched in Fig. 1.4.3.

The nontrivial topology with $|m_z| < 2t_0$ can support degenerate boundary modes. Considering open boundaries located at $x = 0, L$, respectively

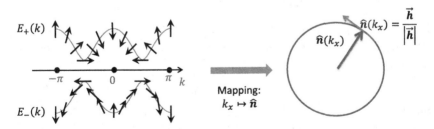

**Fig. 1.4.3.** The nonzero winding number implies that the spin polarization axis $h$ of the Bloch states winds over one circle when $k$ runs over the FBZ. This lead to a mapping between $k$ space and the $S^1$ circle and covers the $S^1$ completely.

and diagonalizing $H$ in position space $H = \sum_{x_i} \mathcal{H}(x_i)$ with $\mathcal{H}(x_i) = -(t_s \sigma_z + i t_{so}^{(0)} \sigma_y) \hat{c}_{x_i}^\dagger \hat{c}_{x_i + a} + m_z \sigma_z \hat{c}_{x_i}^\dagger \hat{c}_{x_i} + h.c.$, we obtain the edge state for the boundary $x = 0$ as

$$\psi_L(x_i) = \frac{1}{\sqrt{\mathcal{N}}} [(\lambda_+)^{x_i/a} - (\lambda_-)^{x_i/a}] |\chi_+\rangle, \qquad (1.4.12)$$

and accordingly the one on $x = L$ by $\psi_R(x_i) = \frac{1}{\sqrt{\mathcal{N}}} [(\lambda_+)^{(L-x_i)/a} - (\lambda_-)^{(L-x_i)/a}] |\chi_-\rangle$, with $\mathcal{N}$ being the normalization factor, the spin eigensates $\sigma_x |\chi_\pm\rangle = \pm |\chi_\pm\rangle$, and $\lambda_\pm = (m_z \pm \sqrt{m_z^2 - 4t_s^2 + 4|t_{so}^{(0)}|^2})/(2t_s + 2|t_{so}^{(0)}|)$. Thus the two edge modes are polarized to the opposite $\pm x$ directions. Note $\psi_L$ and $\psi_R$ span the complete Hilbert space of *one single* 1/2-spin or spin-qubit. Each edge state equals *one-half* of a single spin, namely, a 1/4-spin. Furthermore, the edge state to 1/2-fractionalization. A convenient way is to consider the semi-infinite geometry which has the open boundary at $x = 0$. We then calculate the particle number of the zero mode localized on this boundary. Note the total number of quantum states in the system is given by

$$N = \sum_{E<0} \langle \psi_E | \hat{n}_E | \psi_E \rangle + \sum_{E>0} \langle \psi_E | \hat{n}_E | \psi_E \rangle + \langle \psi_0 | \hat{n}_E | \psi_0 \rangle, \qquad (1.4.13)$$

where we denote by $\hat{n}_E = \mathbb{I}$ the state number operator and $|\psi_E\rangle$ is the eigenstate with energy $E$. Since the Hamiltonian satisfies $\{\mathcal{H}, \sigma_x\} = 0$, the energy spectrum is symmetric. We have then $\sum_{E<0} \langle \psi_E | \hat{n}_E | \psi_E \rangle = \sum_{E>0} \langle \psi_E | \hat{n}_E | \psi_E \rangle$. It follows that

$$\sum_{E<0} \langle \psi_E | \hat{n}_E | \psi_E \rangle = \frac{1}{2} [N - \langle \psi_0 | \hat{n}_E | \psi_0 \rangle]. \qquad (1.4.14)$$

The particle number of the zero mode depends on its occupation. If the zero mode is unoccupied, the particle number of it is given by

$$n_0 = \sum_{E<0} [\langle \psi_E | \hat{n}_E | \psi_E \rangle_1 - \langle \psi_E | \hat{n}_E | \psi_E \rangle_0]. \qquad (1.4.15)$$

Here $\langle \rangle_1$ and $\langle \rangle_0$ represents the cases with one (topological phase) and zero (trivial phase) bound modes, respectively. Using the Eq. (1.4.14) one finds directly

$$n_0 = -\frac{1}{2} \langle \psi_0 | \hat{n}_E | \psi_0 \rangle = -\frac{1}{2}. \qquad (1.4.16)$$

Similarly, if the zero mode is occupied, the particle number is $n_0 = 1/2$. It is trivial to know that this result can be applied to the case with two boundaries located far away from each other, say respectively at $x = 0$ and $x = L$. Since the two zero modes are obtained independently, each of them carries $+1/2$ ($-1/2$) particle if it is occupied (unoccupied).

### 4.A.3. *Interacting effects on topology*

The $Z$ classification implies that single-particle couplings respecting $U(1)$ and $\mathcal{CT}$ cannot gap out the edge modes in arbitrary $N$-chain system of 1D lattices. The topological classification can be reduced with interactions. In other words, in the presence of interaction, the edge states may split out even the system respects the $U(1)$ and $\mathcal{CT}$ symmetry. It was shown the Ref. 100 that with the following generic interactions between edge modes at one side (left or right) of $N$ chains

$$H_{\text{int}} = \sum_{j,j'}^{N} \sum_{\{12...j;1'2'...j'\}} V_{12...j}^{1'2'...j'} c_1^\dagger c_2^\dagger \ldots c_j^\dagger c_{j'} \ldots c_{2'} c_{1'}$$
$$+ \text{H.c.}, \tag{1.4.17}$$

where $V_{12...j}^{1'2'...j'} = (V_{12...j}^{1'2'...j'})^*$ to ensure the chiral symmetry, the edge-state degeneracy of this side is still protected until the number of chains reaches $N = 4$. This implies that under (arbitrarily small) interactions the topological classification of 1D AIII insulator reduces from $Z$ to $Z_4$. This result can also be understood from the projective representations of the $U(1) \times Z_2^T$ symmetry group. It can be shown that the $U(1) \times Z_2^T$ group has 4 inequivalent projective representations, with three being topological and one trivial.[124] This also confirms that the topological classification reduces to $Z_4$ by interactions.

On the other hand, for a single chain of 1D AIII class insulator, increasing interaction can lead to topological phase transition. The relevant interaction includes the onsite Hubbard interaction

$$H_U = U \sum_j n_{j\uparrow} n_{j\downarrow}, \tag{1.4.18}$$

which is shown to renormalize the mass $u = 2t_0$ and magnetization $w = m_z$, and thus can induce phase transition when the interaction exceeds critical value.[100] From the previous discussion we know that in the single-particle regime the critical point of phase transition is described by the scaling relation $u = w$. Under repulsive interaction, one can expect that $u$ increases compared with the linear scaling due to the detrimental effects of interaction. The new phase transition scaling under interaction can be studied with standard bosonization method, together with renormalization group (RG) flow.[125] For convenience we rotate that $\sigma_y, \sigma_z \to (\sigma_z, -\sigma_y)$. The Hamiltonian can then be written as $H \to H' = -t_{so}^{(0)} \sum_{<i,j>,\sigma} \hat{c}_{i\sigma}^\dagger \hat{c}_{j\sigma} + H_{\text{SO}} + H_Z$, with $H_{\text{SO}} = t_s \sum_j (-1)^j (c_{j\uparrow}^\dagger c_{j+1,\downarrow} - c_{j\uparrow}^\dagger c_{j-1,\downarrow} + \text{h.c.})$ and

$H_Z = \Gamma_y \sum_j (-1)^j (i c_{j\downarrow}^\dagger c_{j\uparrow} - \text{h.c.})$. Note that the SO term and spin-conserved hopping term are exchanged. The low-energy physics can be well captured by the continuum approximation ($x = ja$):

$$c_{j\sigma} \approx \sqrt{a} [\psi_{R\sigma}(x) e^{ik_F x} + \psi_{L\sigma}(x) e^{-ik_F x}], \qquad (1.4.19)$$

with $k_F = \pi/2$. Neglecting the fast oscillating terms we obtain the continuum representation of the two mass terms by:

$$H_{\text{SO}} \approx iu \int dx \, (\psi_L^\dagger \sigma^x \psi_R - \psi_R^\dagger \sigma^x \psi_L), \qquad (1.4.20)$$

$$H_Z = w \int dx \, (\psi_R^\dagger \sigma^y \psi_L + \psi_L^\dagger \sigma^y \psi_R). \qquad (1.4.21)$$

Here $u = 2t_{\text{SO}}, w = \Gamma_y$. Using the standard bosonization formula $\psi_{rs} = \frac{1}{\sqrt{2\pi a}} e^{-\frac{i}{\sqrt{2}} [r\phi_\rho - \theta_\rho + s(r\phi_\sigma - \theta_\sigma)]}$ with $r = R, L$ and $s = \uparrow, \downarrow$, we reach the bosonized Hamiltonian densities

$$\mathcal{H}_{\text{SO}} = \frac{u}{\pi a} \sin \sqrt{2} \phi_\rho \cos \sqrt{2} \theta_\sigma, \qquad (1.4.22)$$

$$\mathcal{H}_Z = \frac{w}{\pi a} \cos \sqrt{2} \phi_\rho \sin \sqrt{2} \theta_\sigma. \qquad (1.4.23)$$

The topology of the system depends on which of $u$ and $w$ flows to the strong-coupling regime first under RG. A direct power counting shows the same RG flow for the masses $u$ and $w$ in the first-order perturbation. Therefore the next-order perturbation expansion is necessary to capture correctly the fate of the topological phase transition. By deriving the RG flow equations up to one-loop order,[100] the renormalization to $u, w$, the umklapp scattering $g_\rho$ and spin backscattering $g_\sigma$ by:

$$\frac{du}{dl} = \frac{3 - K_\rho}{2} u - \frac{g_\rho u}{4\pi v_F} + \frac{g_\sigma u}{4\pi v_F},$$

$$\frac{dw}{dl} = \frac{3 - K_\rho}{2} w + \frac{g_\rho w}{4\pi v_F} + \frac{g_\sigma w}{4\pi v_F}, \qquad (1.4.24)$$

$$\frac{dg_\rho}{dl} = \frac{g_\rho^2}{\pi v_F}, \quad \frac{dg_\sigma}{dl} = \frac{g_\sigma^2}{\pi v_F},$$

where the bare values of the coupling constants $g_\rho = -g_\sigma = U, u = 2t_{\text{SO}}^{(0)}, w = \Gamma_y$, and $l$ is the logarithm of the length scale. Being a higher order correction, the renormalization of Luttinger parameter $K_\rho$ has been neglected. For $U > 0$, $g_\sigma$ marginally flows to zero and can be dropped off. This is consistent with the result that repulsive interaction cannot gap out the spin sector in the 1D Hubbard model. $g_\rho$ is marginally relevant and can be solved by $g_\rho(l) = \frac{\pi v_F g_\rho(0)}{\pi v_F - g_\rho(0)l}$. Substituting this result into RG equations of $u$ and $w$ yields after integration $u(l) = u(0)[1 - \frac{g_\rho(0)l}{\pi v_F}]^{\frac{1}{4}} e^{(3-K_\rho)l/2}, w(l) = w(0)[1 + \frac{g_\rho(0)l}{\pi v_F}]^{\frac{1}{4}} e^{(3-K_\rho)l/2}$. As having been analyzed previously, the repulsive interaction ($g_\rho > 0$) suppresses SO induced mass term $u$ while enhances the trivial mass term $w$. With these results one can find that that the scaling of topological phase transition is given by[100]

$$u(0) = [w(0)]^\gamma, \quad \gamma \approx 1 - \frac{g_\rho(0)}{\pi v_F(3 - K_\rho)}, \tag{1.4.25}$$

where $u(0)$ and $w(0)$ denote their bare values (before having interactions), $g_\rho(0)$ is the umklapp scattering coefficient, and $K_\rho$ is the Luttinger coefficient of the charge sector. Note that in the original paper of Ref. 100 a "1/4"-factor was wrongly included in the second term of $\gamma$ by carelessness. For repulsive interaction $U > 0$, one has $\gamma < 1$. The above scaling relation implies that a repulsive interaction suppresses the topological phase (note that the bare value $u(0), w(0) < 1$ for the weak coupling regime). Accordingly, if initially the noninteracting system is topologically nontrivial with $u(0) > w(0) > 0$, increasing $U$ to the regime $u(0) < [w(0)]^\gamma$ can drive the system into a trivial phase.

### 4.A.4. *Further studies*

With optical Raman lattice many interesting physics can be explored besides the studies introduced above. In particular, it was shown that if adding an $s$-wave pairing potential, the AIII class topological phase enters into a 1D BDI class topological SC/superfluid. A resonant cross Andreev reflection was predicted in such topological SC,[126] as a consequence of oppositely spin-polarized edge modes in the left and right hand ends of the 1D system. On the other hand, by putting the current system in a cavity a novel topological phase, called topological superradiant phase,[127] was proposed, and was further generalized to the superfluid regimes.[128] Finally, with the relative configuration between Raman and optical lattices a Hidden nonsymmorphic symmetry was pointed out for the optical Raman lattice system, and was shown to be responsible for the degeneracy at the first Brillouin zone.[129,130]

### 4.A.5. *Experimental realization*

The optical Raman lattice scheme for 1D topological state was realized very recently by HKUST group with the alkali earth $^{173}$Yb fermions.[131]

The realization with alkali earth atoms is of explicit advantage in having a long lifetime due to the absence of relatively small fine-structure splitting between $D_1$ and $D_2$ lines which limits the realization of SO coupling in alkali atoms to apply near-resonant optical transitions. Another peculiar property of the realization in $^{173}$Yb atoms is that the optical lattice is spin-dependent, rather than spin-independent, namely $V_{\text{latt}}^\uparrow(x) \neq V_{\text{latt}}^\downarrow(x)$. In this case, the chiral symmetry discussed in the above subsection is generically not satisfied. Interestingly, it was shown in the work that the 1D topological phase and degenerate end states can still be obtained, and are protected by two hidden symmetries called magnetic group and non-local chiral symmetries.[131] The quenching dynamics are also investigated in the experiment, with a nontrivial topology-dependent spin relaxation dynamics being observed.

### 4.B.   *Optical Raman lattice: 2D case*

4.B.1.   *Model*

The optical Raman lattice scheme can be generalized to the 2D case. As was proposed in,[85] a basic model of 2D optical Raman lattice for spin-1/2 system reads

$$H_{2D} = \frac{\hbar^2 \mathbf{k}^2}{2m} + V_{\text{latt}}(x,y) + m_z\sigma_z + M_x(x,y)\sigma_x$$
$$+ M_y(x,y)\sigma_y, \qquad (1.4.26)$$

where the spin-independent lattice potential $V_{\text{latt}}(x,y) = V_0(\cos^2 k_0 x + \cos^2 k_0 y)$, the Raman lattice potentials $M_x = M_0 \cos k_0 x \sin k_0 y$ and $M_y = M_0 \cos k_0 y \sin k_0 x$, with $(V_0, M_0)$ being the amplitudes, and the constant Zeeman term $m_z$ relates to the two-photon detuning by $m_z = \delta/2$. Similar to the 1D model, in the present 2D optical Raman lattice each Raman potential ($M_x$ or $M_y$) is antisymmetric with respect to lattice potential $V_{\text{latt}}(x,y)$ along one direction (e.g. the $x$ direction for $M_x$), while it is symmetric along another direction (the $y$ direction for $M_x$). We shall see that this fundamental model Hamiltonian naturally realizes 2D SO coupling and can bring about rich topological phases.[85,88]

We introduce now how to realize the Hamiltonian (1.4.26). The following realization is based on $^{87}$Rb bosons, as proposed in Ref. 88. However, the results are generically valid for both bosons and fermions. The 2D plane is set in the $x - z$ plane, as considered in Ref. 88. The 2D Raman lattice scheme, sketched in Fig. 1.4.4(a), consists of a blue-detuned

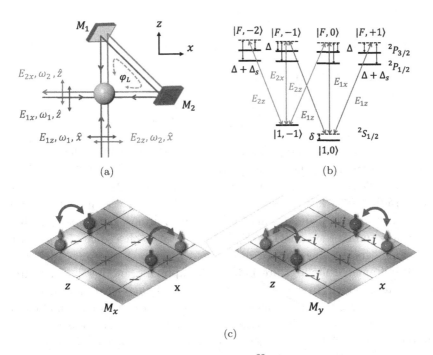

**Fig. 1.4.4.** The Raman lattice scheme in 2D case.[88] (a) Sketch of the setup for realization. The light components $E_{1x,1z}$ (blue lines) form a spin-independent square optical lattice in the intersecting area and generate two periodic Raman potentials, together with the light components $E_{2x,2z}$ (red lines). (b) All the relevant optical transitions for $^{87}$Rb atoms, including the $D_2$ ($5^2S_{1/2} \rightarrow 5^2P_{3/2}$) and $D_1$ ($5^2S_{1/2} \rightarrow 5^2P_{1/2}$) lines of transitions. The spin is defined by $|\uparrow\rangle = |1,-1\rangle$ and $|\downarrow\rangle = |1,0\rangle$, $\Delta_s$ denotes the fine-structure splitting, and $\delta$ is a two-photon detuning. (c) Spin-flip hopping driven by Raman potential $M_x$ (left), and $M_y$ (right). The Raman potential $M_x$ ($M_y$) drives spin-flip hopping along $x$ ($z$) direction, with the forwarding and backward hopping from a site having opposite signs.

square lattice created by two standing-wave light components (blue lines), and two periodic Raman potentials generated with additional plane-wave lights (red lines). The standing waves in the intersecting area are $\boldsymbol{E}_{1x} = \hat{z}\bar{E}_{1x}e^{i\varphi_L/2}\cos(k_0x - \varphi_L/2)$ and $\boldsymbol{E}_{1z} = \hat{x}\bar{E}_{1z}e^{i\varphi_L/2}\cos(k_0z - \varphi_L/2)$, where $\bar{E}_{1x/1z}$ are amplitudes and the phase $\varphi_L = k_0L$ is acquired through the optical path $L$ from intersecting point to mirror $M_1$, then to $M_2$, and back to the intersecting point [Fig. 1.4.4(a)]. Here the initial phases have been ignored. For alkali atoms, the optical potential generated by linearly polarized lights is spin-independent when the detuning $\Delta$ is much larger than the hyperfine structure splittings. The square lattice potential then takes the form

$$V_{\text{latt}}(x, z) = V_{0x} \cos^2(k_0 x - \varphi_L/2) + V_{0z} \cos^2(k_0 z - \varphi_L/2), \qquad (1.4.27)$$

where the amplitudes are given by

$$V_{0x/0z} = \frac{3\Delta + 2\Delta_s}{3\Delta(\Delta + \Delta_s)} \alpha_{D_1}^2 \bar{E}_{1x/1z}^2. \qquad (1.4.28)$$

Here $\Delta_s$ denotes the energy splitting between the $D_1$ and $D_2$ lines, as discussed in section III. The dipole matrix elements $\alpha_{D_2} = |\langle J = 1/2 || e\mathbf{r} || J' = 3/2 \rangle| = \sqrt{2}\alpha_{D_1} = \sqrt{2}|\langle J = 1/2 || e\mathbf{r} || J' = 1/2 \rangle| = 4.227 e a_0$, with $a_0$ the Bohr radius. The lattice potentials have been taken into account all relevant transitions from $D_1$ and $D_2$ lines for the alkali atoms. It can be seen that by simply tuning the strengths $|\bar{E}_{1x}| = |\bar{E}_{1z}|$, we reach a square lattice with the same depths along $x$ and $z$ directions.

The Raman couplings are induced by applying another beam of frequency $\omega_2$, incident from the $z$ direction. This light along with lattice beams can be generated from a single laser source via an acoustic-optic modulator which controls their frequency difference $\delta\omega = \omega_2 - \omega_1$ and amplitude ratio. Two plane-wave fields are formed at the intersecting point $\mathbf{E}_{2z} = \hat{x}\bar{E}_{2z}e^{ik_0 z}$ and $\mathbf{E}_{2x} = \hat{z}\bar{E}_{2x}e^{i(-k_0 x + \varphi_L + \delta\varphi_L)}$. The relative phase $\delta\varphi_L = L\delta\omega/c$, acquired by $\mathbf{E}_{2x}$, can be precisely manipulated by changing the optical path $L$ or $\delta\omega$, and it controls the dimensionality of the realized SO coupling. The standing-wave and plane-wave beams form a *double-$\Lambda$* type configuration as shown in Fig. 1.4.4(b), with $\mathbf{E}_{1x}$ and $\mathbf{E}_{2z}$ generating one Raman potential via $|e_1\rangle$ in the form $M_{0x} \cos(k_0 x - \varphi_L/2)e^{i(k_0 z - \varphi_L/2)}$, and $E_{1z}$ and $E_{2x}$ producing another one via $|e_2\rangle$ as $M_{0y} \cos(k_0 x - \varphi_L/2)e^{-i(k_0 z - \varphi_L/2) + i\delta\varphi_L}$, with $M_{0x(0y)} \propto \bar{E}_{1x(1z)}\bar{E}_{2z(2x)}$. Note that for the present blue-detuned lattice, atoms are located in the region of minimum intensity of lattice fields. It follows that terms like $\cos(k_0 x - \varphi_L/2)\cos(k_0 z - \varphi_L/2)$, which are antisymmetric with respect to each lattice site in both $x$ and $z$ directions, have negligible contribution to the low-band physics. Neglecting such terms yields the Raman coupling potentials

$$\begin{aligned}
\mathcal{M}_x(x, z)\sigma_x &= (M_x - M_y \cos\delta\varphi_L)\sigma_x, \\
\mathcal{M}_y(x, z)\sigma_y &= M_y \sin\delta\varphi_L \sigma_y.
\end{aligned} \qquad (1.4.29)$$

Here $M_x = M_{0x} \cos(k_0 x - \varphi_L/2)\sin(k_0 z - \varphi_L/2)$ and $M_y = M_{0y} \cos(k_0 z - \varphi_L/2)\sin(k_0 x - \varphi_L/2)$. Similar to the calculation of the lattice potentials given above, the amplitudes of the Raman potentials are obtained straight-forwardly that

$$M_{0x/y} = \frac{\bar{E}_{1x/z}\bar{E}_{2z/x}\alpha_{D_1}^2}{6\sqrt{2}}\left(\frac{1}{\Delta} - \frac{1}{\Delta + \Delta_s}\right). \tag{1.4.30}$$

With the above results for square lattice and Raman potentials, the total Hamiltonian is followed by

$$\begin{aligned}H_{2D} &= \frac{\hbar^2 \mathbf{k}^2}{2m} + V_{\text{latt}}(x, z) + \mathcal{M}_x(x, z)\sigma_x \\ &\quad + \mathcal{M}_y(x, z)\sigma_y + m_z\sigma_z,\end{aligned} \tag{1.4.31}$$

where the effective Zeeman term $m_z = \delta/2$ is considered [Fig. 1.4.4(b)]. The above Hamiltonian returns to that in Eq. (1.4.26) when the phase difference takes the optimal value $\delta\varphi_L = \pi/2$, and replace $z \to y$. The present scheme is of topological stability, namely, the realization is intrinsically immune to any phase fluctuations in the setting. Moreover, the relative phase $\delta\varphi_L$ determines the relative strength of $\sigma_x$ and $\sigma_y$ terms in the Raman potential, which generates SO coupling along $x$ and $y$ directions, respectively. Tuning $\delta\varphi$ between $\pi/2$ and $\pi$ can reach a crossover between the 2D and 1D SO couplings. If tuning $\delta\omega = 50\text{MHz}$, we have $\delta\varphi_L = \pi/2$ for $L = 1.5\text{m}$, which gives a 2D SO coupling, while $\delta\varphi_L = \pi$ if increasing to $L = 3.0$ m, and the SO coupling becomes 1D form. Note that the fluctuations, e.g. due to the mirror oscillations, have very tiny effect on $L$ and thus the relative phase. These advantages ensure that this proposal can be realized in experiment.

### 4.B.2. Chern insulator for s-bands

Similar to the 1D regime, the present 2D optical Raman lattice can bring about nontrivial topological quantum states. We consider the lowest $s$-band model for the current consideration, while we emphasize that for the lattice system, the generic high-band model can be naturally obtained and may give rise to novel new topological physics. For convenience, we focus our study on the isotropic case with $V_{0x} = V_{0z} = V_0$, $M_{0x} = M_{0y} = M_0$ and $\delta\phi_L = \pi/2$. The topological physics can be best understood with tight-binding model, while all the results derived below in this section is not restricted by tight-binding approximation, namely, they are valid for non-tight-binding regime. Taking into account the spin-conserved and spin-flip hopping terms between the nearest-neighbor sites, we obtain the tight-binding Hamiltonian for the $s$-bands by

$$H_{\text{TI}} = -\sum_{<i,j>,\sigma} t_0^{ij} \hat{c}_{i\sigma}^\dagger \hat{c}_{j\sigma} + \sum_{<i,j>} (t_{\text{so}}^{ij} \hat{c}_{i\uparrow}^\dagger \hat{c}_{j\downarrow} + \text{H.c.})$$
$$+ \sum_i m_z (\hat{n}_{i\uparrow} - \hat{n}_{i\downarrow}), \qquad (1.4.32)$$

where $i = (i_x, i_z)$ is the 2D lattice-site index, the particle number operators $\hat{n}_{i\sigma} = \hat{c}_{i\sigma}^\dagger \hat{c}_{i\sigma}$. The spin-conserved hopping couplings $t_0^{ij}$ are induced by the lattice potential, calculated by

$$t_0^{ij} = \int d^2\mathbf{r} \phi_{s\sigma}^{(i)}(\mathbf{r}) \Big[ \frac{p_x^2 + p_z^2}{2m} + V_{\text{latt}}(\mathbf{r}) \Big] \phi_{s\sigma}^{(j)}(\mathbf{r}), \qquad (1.4.33)$$

where $\phi_{s\sigma}^{(j)}(\mathbf{r})$ denotes the $s$-orbital Wannier function at the $j$-th site at the spin state $\sigma$. On the other hand, the spin-flip hopping couplings $t_{\text{so}}^{ij}$ are driven by the Raman potentials $M_x(x, z)$ and $M_y(x, z)$, which are anti-symmetric with respect to each square lattice site along $x$ and $z$ directions, respectively, and are obtained by

$$t_{\text{so}}^{ij} = \int d^2\mathbf{r} \phi_{s\uparrow}^{(i)}(\mathbf{r})[M_x(x, z)\sigma_x + M_y(x, z)\sigma_y]\phi_{s\downarrow}^{(j)}(\mathbf{r}). \qquad (1.4.34)$$

It is trivial to know that $t_0^{ij}$ for the nearest-neighbor hopping is a constant in all directions, namely $t_0^{ii\pm1} = t_0$. On the other hand, from the Raman potential configurations we have that

$$\int d^2\mathbf{r} \phi_{s\uparrow}^{(i)}(\mathbf{r}) M_x(x, z)\phi_{s\downarrow}^{(j)}(\mathbf{r})$$
$$= \int d^2\mathbf{r} \phi_{s\uparrow}^{(i)}(\mathbf{r}) M_0 \cos(k_0 x) \sin(k_0 z)\phi_{s\downarrow}^{(j)}(\mathbf{r})$$
$$= (-1)^{i_x + i_z} M_0 \int d^2\mathbf{r} \phi_{s\uparrow}^{0,0}(\mathbf{r}) \cos(k_0 x)$$
$$\times \sin(k_0 z)\phi_{s\downarrow}^{j_x - i_x, j_z - i_z}(\mathbf{r})\delta_{i_z j_z}(1 - \delta_{i_x, j_x}), \qquad (1.4.35)$$

$$\int d^2\mathbf{r} \phi_{s\uparrow}^{(i)}(\mathbf{r}) M_y(x, z)\phi_{s\downarrow}^{(j)}(\mathbf{r})$$
$$= \int d^2\mathbf{r} \phi_{s\uparrow}^{(i)}(\mathbf{r}) M_0 \cos(k_0 z) \sin(k_0 x)\phi_{s\downarrow}^{(j)}(\mathbf{r})$$
$$= (-1)^{i_x + i_z} M_0 \int d^2\mathbf{r} \phi_{s\uparrow}^{0,0}(\mathbf{r}) \cos(k_0 z) \sin(k_0 x)$$
$$\phi_{s\downarrow}^{j_x - i_x, j_z - i_z}(\mathbf{r})\delta_{i_x j_x}(1 - \delta_{i_z, j_z}). \qquad (1.4.36)$$

The above formulas show that the Raman potential $M_x(x, z)$ only induces the spin-flip hopping along $x$ direction, while $M_y(x, z)$ only induces the spin-flip hopping along $z$ direction, both of them cannot drive onsite spin-flip transitions. These properties are resulted from the antisymmetry of $M_x(x, z)$ ($M_y(x, z)$) along $x$ ($z$) direction [Fig. 1.4.4(c)]. Note that we have neglected the term $M_0 \cos(k_0 x) \cos(k_0 z)$ in the originally realized Hamiltonian, which is antisymmetric with respect to lattice site along both $x$ and $z$ directions, cannot induce hopping along either $x$ or $z$ direction, nor induce the onsite spin-flip coupling within the $s$-band. The leading-order nonzero contribution of this term is the onsite transition between $s$-band and $p_x p_z$-band (not $p_x$ or $p_z$ band), which is negligible when the Raman lattice is weak compared with the square lattice potential $V_{\text{latt}}$.

From the Eqs. (1.4.35) and (1.4.36), and together with the antisymmetry of Raman potentials, it can be read that the spin-flip hopping terms between two neighboring sites satisfy

$$t_{\text{so}}^{j_x, j_x \pm 1} = \pm(-1)^{j_x + j_z} t_{\text{so}},$$
$$t_{\text{so}}^{j_z, j_z \pm 1} = \pm i(-1)^{j_x + j_z} t_{\text{so}}, \tag{1.4.37}$$

where $t_{\text{so}} = M_0 \int d^2\mathbf{r} \phi_s^{0,0}(x, z) \cos(k_0 x) \cos(k_0 z) \phi_s^{0,0}(x - a, z)$ is proportional to Raman coupling strength $M_0$. Note that the staggered property of the SO terms are a consequence of the relative spatial configuration of the lattice and the Raman potentials, namely, the periodicity (unit cell) of the Raman lattice is one-half (double) of the square lattice periodicity (unit cell), and the Raman potential $M_x(x, z)$ ($M_y(x, z)$) is antisymmetric along $x(z)$ direction, while symmetric along $z(x)$ direction [Fig. 1.4.4(c)]. Since the relative spatial profiles of the Raman potentials and the square lattice are determined by the same standing wave lights and are automatically fixed, these properties are topologically stable against any kind of fluctuations in the system.

The staggered spin-flip hopping terms described in Eq. (1.4.37) bring about two important effects. First, the staggered property implies that the coupling between spin-up and spin-down states transfers $\pi/a$ momentum between them along both $x$ and $z$ direction, which effectively shifts the Brillouin zone by half for the spin-down relative to spin-up states. This is exactly similar to the situation in the 1D optical Raman lattice, as discussed previously. Moreover, in additional to the relative half Brillouin zone shift, the remaining effect of the spin-flip hopping leads to the normal Rashba type SO coupling in the $x - z$ plane. Note that the former effect can be absorbed by redefining the spin-down operator $\hat{c}_{j\downarrow} \to e^{i\pi r_j/a} \hat{c}_{j\downarrow}$. We recast the tight-binding Hamiltonian into

$$H_{\mathrm{TB}} = -t_0 \sum_{\langle i j \rangle} \left( c_{i\uparrow}^{\dagger} c_{j\uparrow} - c_{i\downarrow}^{\dagger} c_{j\downarrow} \right) + \sum_{i} m_z (n_{i\uparrow} - n_{i\downarrow})$$

$$+ \left[ \sum_{j_x} t_{\mathrm{so}} (c_{j_x\uparrow}^{\dagger} c_{j_x+1\downarrow} - c_{j_x\uparrow}^{\dagger} c_{j_x-1\downarrow}) + \mathrm{h.c.} \right]$$

$$+ \left[ \sum_{j_z} i t_{\mathrm{so}} (c_{j_z\uparrow}^{\dagger} c_{j_z+1\downarrow} - c_{j_z\uparrow}^{\dagger} c_{j_z-1\downarrow}) + \mathrm{h.c.} \right]. \tag{1.4.38}$$

Note that the Raman potential $M_x(x, z)$ only induces the spin-flip hopping along $x$ direction, while $M_y(x, z)$ only induces the spin-flip hopping along $z$ direction. These properties are resulted from the antisymmetry of $M_x(x, z)$ $[M_y(x, z)]$ along $x(z)$ direction. Just like the 1D Hamiltonian [Eqs. (1.4.8)], the opposite signs in $t_0$ terms are due to the relative $(\pi, \pi)$-momentum transfer between spin-up and spin-down Bloch states. The staggered property of the SO terms are a consequence of the relative spatial configuration of the lattice and the Raman potentials, namely, the Raman potential $M_x(x, z)[M_y(x, z)]$ is antisymmetric along $x(z)$ direction, while symmetric along $z(x)$ direction [Fig. 1.4.4(c)].

Transforming the tight-binding Hamiltonian into momentum space yields $H_{\mathrm{TB}} = \sum_{\mathbf{q}, s\sigma, \sigma'} c_{\mathbf{q}\sigma}^{\dagger} \mathcal{H}_{\sigma\sigma'}(\mathbf{q}) c_{\mathbf{q}\sigma'}$, with the Bloch Hamiltonian

$$\mathcal{H}(\mathbf{q}) = [m_z - 2t_0(\cos q_x a + \cos q_z a)]\sigma_z$$
$$+ 2t_{\mathrm{so}} \sin q_x a \sigma_y + 2t_{\mathrm{so}} \sin q_z a \sigma_x, \tag{1.4.39}$$

which is a typically two-band model of the form $\mathcal{H}(\mathbf{q}) = \sum_{\alpha=x,y,z} h_\alpha(\mathbf{q}) \sigma_\alpha$, with $h_x(\mathbf{q}) = 2t_{\mathrm{so}} \sin q_z a$, $h_y(\mathbf{q}) = 2t_{\mathrm{so}} \sin q_x a$ and $h_z(\mathbf{q}) = m_z - 2t_0(\cos q_x a + \cos q_z a)$, giving the spectra $E_\pm = \pm\sqrt{\sum_\alpha h_\alpha^2(\mathbf{q})}$. When $|m_z| \neq 4t_0$ or $0$, the system is fully gapped in the bulk; otherwise, the system has one or two Dirac points. This Bloch Hamiltonian around $\Gamma$ point takes the form

$$\mathcal{H}(\mathbf{q}) \approx [m_z - 4t_0 + t_0(q_x^2 a + q_z^2 a)]\hat{\sigma}_z$$
$$+ \lambda_{\mathrm{so}} q_x \hat{\sigma}_y + \lambda_{\mathrm{so}} q_z \hat{\sigma}_x, \tag{1.4.40}$$

with $\lambda_{\mathrm{so}} = 2a t_{\mathrm{so}}$. Thus the lowest $s$-band model is described by a normal Rashba type SO coupling in a square lattice, plus the kinetic energy term coupling to spin component $\sigma_z$ term. This is in sharp contrast to a purely Rashba SO coupled system which is topologically trivial in the single-particle regime. The present realization gives a minimal two-band

QAH model driven by 2D SO coupling, which was shown to exhibit novel topological features in both the bulk and the edge.[132] We note that this model cannot be precisely realized with solid state materials, where the minimal case is a four-band model,[133] but has been firstly achieved in the ultracold atoms.

The topology of the present QAH model is characterized by Chern number, which can be calculated by

$$\mathrm{Ch}_1 = -\frac{1}{24\pi^2} \int d\omega dk_x dk_y \,\mathrm{Tr}\big[GdG^{-1}\big]^3, \tag{1.4.41}$$

where the Green's function $G^{-1}(\omega, \mathbf{q}) = \omega + i\delta^+ - \mathcal{H}(\mathbf{q})$, and the trace is operated on the spin space. Integrating over the frequency space yields that

$$\mathrm{Ch}_1 = \frac{1}{4\pi} \int dq_x dq_z \hat{\mathbf{h}} \cdot \partial_{q_x} \hat{\mathbf{h}} \times \partial_{q_z} \hat{\mathbf{h}}, \tag{1.4.42}$$

where $\hat{\mathbf{h}} = (h_x, h_y, h_z)/|\mathbf{h}(\mathbf{q})|$. By a straightforward calculation one can find that

$$\mathrm{Ch}_1 = \begin{cases} \mathrm{sgn}(m_z), & \text{for } 0 < |m_z| < 4t_0, \\ 0, & \text{for } |m_z| > 4t_0, \text{or } m_z = 0. \end{cases} \tag{1.4.43}$$

While the Chern number can in principle be determined by measuring the Berry's curvature at the whole $\mathbf{q}$ space, the precise measurement could be quite challenging in the real experiment. Interestingly, the Chern number of the current system can be measured by a much simpler way as introduced below, due the symmetry of the present optical Raman lattice.

### 4.B.3. *Detecting band topology by minimal measurements*

As the topology of a Chern band is classified by Chern numbers, and also characterized by chiral edge states in the boundary, the detection of Chern bands can in principle be performed by measuring the bulk Chern number or chiral edge states. For the edge states, the measurement strategy includes the light-Bragg scattering which can detect the spectra of the edge modes.[134-137] On the other hand, for a Chern band in a synthetic dimension, the edge states can be easily imagined directly.[138,139] The bulk Chern number can be in principle be measured by imaging the Berry curvature over the Brillouin zone,[140,141] e.g. through the Hall transport measurement. The Chern number is given by the integral of the Berry curvature over Brillouin zone. The measurements of Bulk Chern number were performed in the recent experiments.[142-145]

In the present Chern insulator realized with 2D optical Raman lattice, a minimal measurement of the Chern number can be performed due to the novel inversion symmetry existing in the model, as proposed in Ref. 146. The inversion symmetry of the Hamiltonian $H_{2D}$ is defined by $\mathcal{P} = \sigma_z \otimes R_{2D}$, giving $\mathcal{P}H_{2D}\mathcal{P}^{-1} = H_{2D}$, where the 2D spatial operator $R_{2D}$ transforms the Bravais lattice vector $\mathbf{R} \to -\mathbf{R}$. Thus, the Bloch Hamiltonian satisfies $\mathcal{P}\mathcal{H}(\mathbf{q})\mathcal{P}^{-1} = \mathcal{H}(-\mathbf{q})$, which gives $[\sigma_z, \mathcal{H}(\mathbf{\Lambda}_j)] = 0$ at four highly symmetric momenta $\{\mathbf{\Lambda}_j\} = \{\Gamma(0,0), X_1(0,\pi), X_2(\pi,0), M(\pi,\pi)\}$. It is then indicated that in the tight-binding regime the Bloch states $|u_{\pm}(\mathbf{\Lambda}_j)\rangle$ are also eigenstates of the parity operator $\sigma_z$ with eigenvalues $\xi^{(\pm)} = +1$ or $-1$. Therefore, similar to topological insulators,[147] one can define the following invariant

$$\Theta = \prod_j \text{sgn}\left[\xi^{(-)}(\mathbf{\Lambda}_j)\right].\tag{1.4.44}$$

It has been proven rigorously that $\Theta = +1$ when the band is in trivial, and $\Theta = -1$ when the band is topological.[146] Such generic proof needs to construct a nontrivial connection between the inversion symmetric QAH system and the time-reversal invariant topological insulating system. Moreover, the Chern number of the lower band is given by[146]

$$\text{Ch}_1 = -\frac{1-\Theta}{4}\sum_i \text{sgn}\left[\xi^{(-)}(\mathbf{\Lambda}_j)\right].\tag{1.4.45}$$

For the present specific system, it is straightforward to check that when the Zeeman term varies from $m_z \gtrsim 0$ to $m_z \lesssim 0$, two parity eigenvalues $\xi^{(-)}(X_{1,2})$ change sign and then $\text{Ch}_1$ changes from 1 to $-1$. Also, the Chern number vanishes for $|m_z| > 4t_0$.

The fact that the topology of the inversion symmetric bands can be determined by only the Bloch states at four symmetric momenta can greatly simplify the experimental detection of the topological bands. In the experiment at low but finite temperature $T$, the parity eigenvalues can be measured through the spin polarizations at the corresponding Bloch momenta. The measured spin polarization at momentum $\mathbf{q}$ is given by

$$P(\mathbf{q}) = \frac{n_\uparrow(\mathbf{q}, T) - n_\downarrow(\mathbf{q}, T)}{n_\uparrow(\mathbf{q}, T) + n_\downarrow(\mathbf{q}, T)},\tag{1.4.46}$$

with $n_{\uparrow,\downarrow}(\mathbf{q}, T)$ being the density of atoms of the corresponding spin state at temperature $T$ in the first Brillouin zone. At the four symmetric momenta, the spin-polarization measured in the experiment is

$$P(\mathbf{\Lambda}_j) \approx \xi^{(-)}(\mathbf{\Lambda}_j) f(E^{(-)}(\mathbf{\Lambda}_j), T) + \xi^{(+)}(\mathbf{\Lambda}_j) f(E^{(+)}(\mathbf{\Lambda}_j), T),$$
$$\approx \xi^{(-)}(\mathbf{\Lambda}_j) [f(E^{(-)}(\mathbf{\Lambda}_j), T) - f(E^{(+)}(\mathbf{\Lambda}_j), T)], \qquad (1.4.47)$$

where the distribution function $f(E) = 1/[e^{(E-\mu)/k_B T} - 1]$ if the topological band is simulated with bosons, and $f(E) = 1/[e^{(E-\mu)/k_B T} + 1]$ if it is simulated with fermions, with $\mu$ being the chemical potential, and $E^{(-)}(\mathbf{q})$ and $E^{(+)}(\mathbf{q})$ are the energy of the lower and upper $s$-bands, respectively. Note that $f(E^{(-)}(\mathbf{\Lambda}_j), T) \geqslant f(E^{(+)}(\mathbf{\Lambda}_j), T)$, which ensures

$$\text{sgn}[\xi^{(-)}(\mathbf{\Lambda}_j)] = \text{sgn}[P(\mathbf{\Lambda}_j)]. \qquad (1.4.48)$$

The above result implies that the measurement of the spin-polarization at low but finite temperature can be applied to determine the Chern number of the Bloch bands. From the expectation values of $\sigma_z$ one can find that the Bloch bands are topological in Fig. 1.4.5(a,b) and trivial in Fig. 1.4.5(c,d). Further, together with the Eq. (1.4.45), in the former topological regime we have the Chern number $\text{Ch}_1 = +1$ for the lower band and $\text{Ch}_1 = -1$ for the upper band [Fig. 1.4.5(b)].

## 4.C. *Experimental realization of 2D SO coupling and topological bands*

Wu *et al.*[88] performed the first experimental realization of the 2D Raman lattice scheme in a Bose gas of $^{87}$Rb atoms. In the setup, the frequency difference is set at $\delta\omega = 35$MHz, and the relative phase $\delta\varphi_L$ is controlled by the propagating length between the two mirrors [Fig. 1.4.4(a)]. Two observations are mainly preformed: (i) The crossover between 1D and 2D SO couplings, which is reflected by the atom distributions in spin-resolved TOF images. (ii) The measurements of spin texture and band topology. The Chern number $\text{Ch}_1$ [Eq. (1.4.45)] can be readily read from the measured spin texture at a well-chosen temperature.

### 4.C.1. *1D-2D crossover*

The atoms are first prepared in the spin-up state and then adiabatically loaded into the $\Gamma$ point. The spin-resolved TOF expansion is performed to projects Bloch states onto free momentum states with fixed spin polarizations. Figure 1.4.6(a) shows the TOF images for various values of $\delta\varphi_L$. For spin-up ($|\uparrow\rangle$) state, five atom clouds are observed: besides the

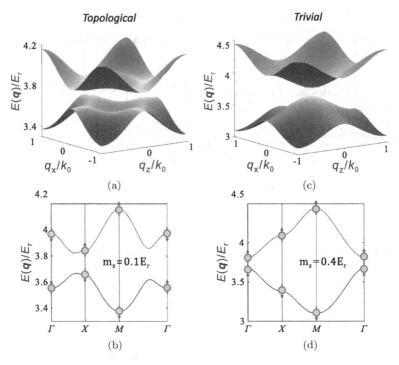

**Fig. 1.4.5.** Band structure and spin texture of the $s$-bands.[88] (a-c) Two examples of gapped band structure with nontrivial (a) or trivial (c) topology. (b,d) The band structure and spin polarization distributions $\langle\sigma_z\rangle$ for the topological (b) and trivial (d) cases. Here $V_0 = 5E_{\rm r}$, $M_0 = 1.2E_{\rm r}$ and $m_z = 0.1E_{\rm r}$ (a,b), or $0.4E_{\rm r}$ (c,d).

major BEC cloud retained at momentum $(k_x, k_z) = (0,0)$, four small fractions of BEC clouds are transferred to momenta $(\pm 2k_0, 0)$ and $(0, \pm 2k_0)$ by the first-order transition due to the lattice potential $V_{\rm latt}$. The SO coupling is reflected by the two or four small BEC clouds in the state $|\downarrow\rangle$, depending on $\delta\varphi_L$, at the four diagonal corners with momenta $(\pm k_0, \pm k_0)$. These atom clouds are generated by the Raman transitions, which flip spin and transfer momenta of magnitude $\sqrt{2}k_0$ along the diagonal directions. As given in Eq. (1.4.29), the Raman terms $\mathcal{M}_x$ and $\mathcal{M}_y$ depend on $\delta\varphi_L$. For $\delta\varphi_L = \pi/2$, four small clouds in state $|\downarrow\rangle$ with TOF momentum $\boldsymbol{k} = (\pm k_0, \pm k_0)$ are observed, reflecting the 2D SO coupling. On the other hand, by tuning the relative phase to $\delta\varphi_L = 3\pi/4$, the population of atom clouds in the two diagonal directions becomes imbalanced. Furthermore, the system reduces to 1D SO couplings when $\delta\varphi_L = \pi$ and $2\pi$, with $\mathcal{M}_x = M_x \pm M_y$ and

$\mathcal{M}_y = 0$. In this case, the Raman pumping only generates a single diagonal pair of BEC clouds, as shown in Fig. 1.4.6a for $\delta\varphi_L = \pi, 2\pi$. This is similar to the 1D SO coupling in the free space, where the Raman coupling flips the atom spin and generates a pair of atom clouds with opposite momenta. Fig. 1.4.6(a) also shows a difference of distribution between lower left and upper right BEC clouds at $\mid\downarrow\rangle$, which is due to non-tight-binding correction. A simple analysis reveals that while the fully antisymmetric Raman terms $\cos(k_0 x + \alpha)\cos(k_0 z + \beta)$ have negligible effects in the tight-binding limit of the lattice, they give finite contributions in the moderate lattice regime and are responsible for such difference of distribution. To quantify crossover effect, we define $W = (\mathcal{N}_{\hat{x}-\hat{z}} - \mathcal{N}_{\hat{x}+\hat{z}})/(\mathcal{N}_{\hat{x}-\hat{z}} + \mathcal{N}_{\hat{x}+\hat{z}})$ to characterize the imbalance of the Raman coupling induced atom clouds, with $\mathcal{N}_{\hat{x}\pm\hat{z}}$ denoting the atom number of the two BEC clouds along the diagonal $\hat{x} \pm \hat{z}$ direction. The result of $W$ is shown in Fig. 1.4.6(b) and is characterized by a simple cosine curve $\cos\delta\varphi_L$, signifying the crossover between 2D and 1D SO couplings realized in the present BEC regime.

### 4.C.2. Band topology

To detect the topology of the bands, the spin distribution in the first Brillouin zone is measured for the 2D isotropic SO coupling with $\delta\varphi_L = \pi/2$. For this purpose, a cloud of atoms needs to prepared at a temperature such that the lowest band is occupied by a sufficient number of atoms, whereas the population of atoms in the higher bands is small. Measurements of spin texture at different temperatures[88] suggest that a temperature around $T = 100$ nK is preferred to extract the spin texture information of the lowest band. In comparison, if the temperature is too high, atoms are distributed over several bands and the visibility of the spin polarization will be greatly reduced, while a too low temperature can also reduce the experimental resolution since the atoms will be mostly condensed at the band bottom.

The spin polarization is then measured as a function of detuning $m_z$ to reveal the topology of the lowest energy band, with $V_0 = 4.16 E_r$ and $M_0 = 1.32 E_r$. The numerical calculations and TOF measured images of $P(\mathbf{q})$ are given in Fig. 1.4.6(c), which also show agreement between the theoretical and experimental results. In Fig. 1.4.6(d), the values of polarization $P(\mathbf{\Lambda}_j)$ are plotted for the four highly symmetric momenta $\Gamma$, $X_1$, $M$ and $X_2$. It can be seen that $P(X_1)$ and $P(X_2)$ always have the same sign, while the signs of $P(\Gamma)$ and $P(M)$ are opposite for small $|m_z|$ and the same for large $|m_z|$, with a transition occurring at the critical value of $|m_z^c|$ which is a

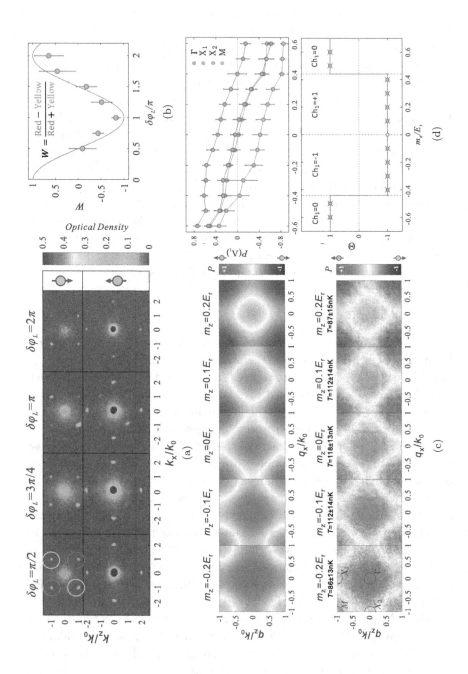

←——————————————————————————————

**Fig. 1.4.6.** (*figure on facing page*) Experimental demonstration.[88] (a-b) The crossover between 1D and 2D SO couplings. Spin-resolved TOF images of BEC atoms are shown for different relative phase $\delta\varphi_L$ (a) and the imbalance $W$ between the Raman coupling induced atoms in the two diagonal directions is measured as a function of $\delta\varphi_L$ (b). (c-d) Spin texture and band topology. Spin texture at different $m_z$ is shown by tuning the two-photon detuning (c). Experimental measurements (lower row) are compared to numerical calculations at $T = 100$nK (upper row). Spin polarization $P(\Lambda_j)$ at the four symmetric momenta $\{\Lambda_j\} = \{\Gamma, X_1, X_2, M\}$ can be readily read as a function of $m_z$ (c), which determines the product $\Theta = \Pi_{j=1}^{4}\mathrm{sgn}[P(\Lambda_j)]$ and the Chern number $\mathrm{Ch}_1$. In all the cases, $V_0 = 4.16E_r$ and $M_0 = 1.32E_r$.

bit larger than $0.4E_r$. At transition points the spin-polarization $P(X_1)$ or $P(X_2)$ vanishes due to the gap closing and thermal equilibrium. From the measured spin polarizations, the product $\Theta$ and the corresponding Chern number are readily read off and also plotted [Fig. 1.4.6(d)]. The results agree well with numerical calculations which predict two transitions between the topologically trivial and nontrivial bands near $m_z^c \approx \pm 0.44E_r$. Note that around $m_z = 0$, the spin-polarizations at $X_{1,2}$ change sign through zero, implying the gap closing at $X_{1,2}$ and a change of Chern number by 2. This confirms that for the 2D SO coupled system realized in the present experiment, the energy band is topologically nontrivial when $0 < |m_z| < |m_z^c|$, while it is trivial for $|m_z| > |m_z^c|$.

## 4.D. *Recent improvement of the realization and generalization to 3D SO coupling*

While the blue-detuned optical Raman lattice exhibits high feasibility in realizing the 2D SO coupling and topological physics, it still suffers a couple of limitations for the study. First, the realization is restricted in the blue-deunted regime, which has a limited tunability in the relative strength of the lattice and Raman potentials, since the optical transitions from $D_1$ and $D_2$ lines cancel out the Raman potentials. Moreover, the present scheme lacks the precise $C_4$ and inversion symmetry $\mathcal{P}$ due to the presence of the term like $\cos(k_0 x - \varphi_L/2)\cos(k_0 z - \varphi_L/2)$ in the Raman potentials [[Fig. 1.4.7(b)]]. Such symmetry-breaking term, negligible only when the Raman potential is weak compared with the square lattice depth, can generically induce the coupling between the $s$-band and higher band states. This effect is shown to be detrimental and can reduce the topological region of the $s$-band [[Fig. 1.4.7(d,f)]].[148] Finally, the realization necessitates a relatively large bias magnetic field to split the hyperfine levels, so that the phase difference

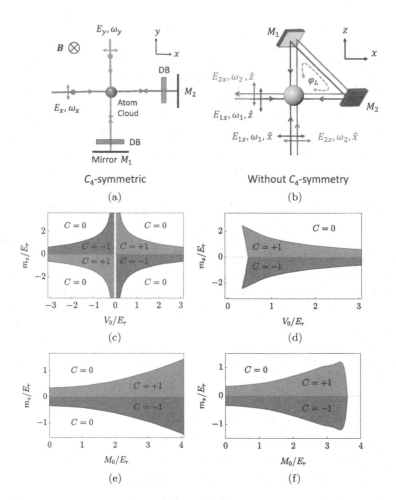

**Fig. 1.4.7.** The new optical Raman lattice scheme for 2D Dirac type SO coupling,[149] and comparison to the previous realization.[88] (a) The new scheme applies two independent laser beams incident along $x$ and $y$ directions, through which the lattice and Raman potentials are generated simultaneously. The complete setup is as simple as the generation of the conventional square lattice for ultracold atoms, and the system is of inversion and $C_4$ symmetries. (b) The previous realization adopts the laser beams long $x$ and $z$ directions which are correlated by a long triangle loop formed though lattice area, mirrors $M_1$ and $M_2$. This system does not have precise inversion symmetry or $C_4$ symmetry, which are approximately valid only in the weak Raman coupling regime.[88] (c,e) With the new scheme the broad topological regime is obtained in the parameter space. (d,f) In the previous realization only a finite relatively narrow area of the parameter space supports topological phase.[148]

$\delta\varphi_L$ with required magnitudes can be achieved. Such a strong bias magnetic field can bring up heating due to the uncontrollable fluctuations.

To solve the challenges pointed above, a new optical Raman lattice scheme was proposed very recently for the realization of high-dimensional SO couplings with high controllability.[149] In the new realization, the system has a rigorous relative reflection symmetry/antisymmetry between lattice and Raman potentials, rendering a precise inversion symmetry of the QAH model [Fig. 1.4.7(a)]. This symmetry is irrespective of the type of detuning of optical transitions, namely, the realization is valid for both blue and red optical transitions, and also for transitions between $D_1$ and $D_2$ lines which render the optimal regime to realize SO couplings. In particular, in the case with $\delta\varphi = \pi/2$ and $M_{01} = M_{02} = M_0$, the system exhibits a precise $C_4$ symmetry defined in the 2D lattice plane: $(x, y; \sigma_x, \sigma_y) \to (y, -x; \sigma_y, -\sigma_x)$, giving $C_4 H C_4^{-1} = H$. The precisely controllable symmetry, the phase diagram with topological region much broader than that in the previous case without exact inversion or $C_4$ symmetry has been predicted [Fig. 1.4.7(c-f)]. The new scheme was realized in a latest experiment with $^{87}$Rb atoms, where all the key advantages of the new scheme have been confirmed.[150] Especially, a long lifetime up to several seconds is observed in the realized 2D SO coupled gas.

Moreover, the new optical Raman lattice scheme can be extended to the realization of 2D Rashba type and 3D Weyl type SO couplings. Compared with the scheme for the 2D Dirac type SO coupling which has been focused on in the above discussions, the The generation of 2D Rashba and 3D Weyl types has an exquisite request that two sources of laser beams have distinct frequencies of factor-two difference. Interestingly, it was found that the $^{133}$Cs atoms provide an ideal candidate for such realization.[149] The new nontrivial topological physics were also predicted. The new optical Raman lattice schemes solve the essential challenges in exploring high-dimensional SO coupled quantum gases and shall advance the research in this direction, particularly in the quantum many-body physics and quantum far-from-equilibrium dynamics with novel topology for ultracold atoms.

## 1.5. Topological Superfluid and Majorana Zero Modes

The realization of SO couplings beyond 1D regime for ultracold atoms advances an important step to explore in experiment the topological superfluid, which is a highly-sought-after phase hosting the exotic Majorana modes, as introduced in this section.

The discussion in this section is organized as follow. First we introduce the basics of Majorana modes, clarifying what kind of systems can host such modes. Then we introduce the spinless 1D $p$-wave and 2D chiral $p + ip$ SCs, and discuss the emergence of the Majorana modes in such intrinsic topological SCs. The realization of topological SC/superfluid from a hybrid system formed by SO coupled system with $s$-wave pairing order is briefly discussed. After the introduction to the background, we turn to showing a generic theory for the 2D chiral SCs/superfluids, with which the topology of a 2D SC/superfluid, characterized by Chern numbers, can be simply determined by Fermi surface (FS) properties and pairing symmetry. This theorem provides a simple but generic criteria to identify the topology of a 2D SC/superfluid phase, even the real system might be complicated. The application of this generic theory to various SO coupled systems is considered. Moreover, we introduce a generic theory for the existence of non-Abelian Majorana zero modes (MZMs), from which we show that the MZMs can exist in 2D trivial SCs. The results can be further generalized to 3D Weyl semimetal with superconductivity/superfluidity.

### 5.A. *Basics of Majorana zero modes*

Eighty years ago, Ettore Majorana proposed a new fermion which is a real solution to the Dirac equation, and identical to its antiparticle and now called Majorana fermion (MF) ,[151] and speculated that it might interpret neutrinos. The self-hermitian property of the Majorana particle indicates that the operator of a MF in the real space satisfies

$$\gamma(\boldsymbol{x}) = \gamma^\dagger(\boldsymbol{x}). \tag{1.5.1}$$

A direct consequence is that the Majorana field breaks $U(1)$ gauge symmetry and conserves no electric charge. Alternatively, the expectation value of a MF is zero, namely, it is charge neutral.

While the evidence of MFs as elementary particles in high energy physics is yet elusive, the search for MFs, or MZMs, has energetically revived in condensed matter physics and become an exciting pursuit in recent years.[19-21] The quest for Majorana modes in solid state physics is mostly driven by both the exploration of the fundamental physics and the promising applications of such modes, obeying non-Abelian statistics, to a building block for fault-tolerant topological quantum computer .[23-30] Note that in condensed matter materials the only elementary particle is the electron which has an effective 'antiparticle' called hole in solid state physics. A MF can then emerge as a quasiparticle in a solid state material, e.g. such quasiparticle can be formed as a superposition of electron and hole

$$\gamma = uc + vc^\dagger, \quad u = v^*. \tag{1.5.2}$$

A natural hosting material for the Majorana-like quasiparticles could be SCs (or superfluids), where the superconducting pairing couples the electron and hole, leading to the qusiparticles in SCs being superpositions of electrons and holes. Nevertheless, for an $s$-wave SC, the pairing occurs between spin-up and spin-down electrons

$$H^s_{\text{pair}} = \sum_{\mathbf{k}} \Delta_s c_{\mathbf{k},\uparrow} c_{-\mathbf{k},\downarrow} + h.c., \tag{1.5.3}$$

or equivalently, in the Nambu space it couples the spin-up electron and spin-down hole in the form $H_{\text{pair}} = \sum_{\mathbf{k}} \Delta_s d^\dagger_{\mathbf{k},\uparrow} c_{\mathbf{k},\downarrow} + h.c.$, where the hole operator is defined as $d_{\mathbf{k},\uparrow} = c^\dagger_{-\mathbf{k},\uparrow}$. Thus the quasiparticle in the $s$-wave SC in general renders the superposition form of the electron and hole $b_{\mathbf{k}} = u c_{\mathbf{k},\uparrow} + v d_{\mathbf{k},\downarrow} = u c_{\mathbf{k},\uparrow} + v c^\dagger_{-\mathbf{k},\downarrow}$, which is not a MF due to the distinction of the spin states of the electron and hole. One can soon find that the Majorana quasiparticle can be realized once the spin degree of freedom can be effectively removed in the SC. For a (an effective) spinless (or spin-polarized) fermion system, the basic superconducting pairing order is $p$-wave and

$$H^p_{\text{pair}} = \sum_{\mathbf{k}} \Delta_p(\mathbf{k}) c_{\mathbf{k}} c_{-\mathbf{k}} + h.c., \tag{1.5.4}$$

with the parity-odd pairing $\Delta_p(\mathbf{k}) = -\Delta_p(-\mathbf{k})$. Similar to the analysis for the $s$-wave SC, the quasiparticle in this case takes the form $\gamma_{\mathbf{k}} = u c_{\mathbf{k}} + v c^\dagger_{-\mathbf{k}}$, which in real space gives $\gamma(\boldsymbol{x}) = u(\boldsymbol{x}) c(\boldsymbol{x}) + v(\boldsymbol{x}) c^\dagger(\boldsymbol{x})$. The MF is then resulted for $u = v^*$.

Due to the self-hermitian property, a single MZM has no well defined Hilbert space spanned by usual complex fermion quantum states. Instead, a complex fermion mode, whose Hilbert space defines a single qubit and is spanned by two fermionic quantum states $|0\rangle$ and $|1\rangle$, can be formed by two independent Majorana quasiparticles. This follows that the Hilbert space of two MFs equals to that of a single complex fermion mode, namely, the quantum dimension of a MF $d^2_\gamma = 2$, which leads to a highly usual consequence that a single Majorana mode has an irrational quantum dimension[29]

$$d_\gamma = \sqrt{2}. \tag{1.5.5}$$

The non-integer quantum dimension leads to an exotic property, namely, the non-Abelian statistics for the MZMs, which is the essential motivation in the recent years of extensive studies of topological SCs and in condensed matter physics and related topics in ultracold atoms.

## 5.B. *Intrinsic p-wave SCs*

### 5.B.1. *1D spinless p-wave SC*

The simplest toy model hosting Majorana modes is the 1D spinless $p$-wave SC, as proposed by Kitaev.[152] In the topologically nontrivial phase, at each end of the 1D system is located a Majorana zero bound mode. The Hamiltonian of the model is given below

$$H = -\mu \sum_j c_j^\dagger c_j - \frac{1}{2} \sum_j (t c_j^\dagger c_{j+1} + \Delta e^\phi c_j c_{j+1} + h.c.), \tag{1.5.6}$$

where $\mu$ is chemical potential, $t$ is hopping coefficient, and $\Delta$ is the $p$-wave pairing with a phase $\phi$. Transforming the above Hamiltonian into momentum space yields the Bogoliubov de Gennes (BdG) form

$$H = \frac{1}{2} \sum_k \mathcal{C}_k^\dagger \mathcal{H}_k \mathcal{C}_k, \tag{1.5.7}$$

$$\mathcal{H}_k = (-t\cos k - \mu)\tau_z - \sin k(\cos\phi\,\tau_y - \sin\phi\,\tau_x),$$

where the operator $\mathcal{C}_k = [c_k, c_{-k}^\dagger]^T$ in the Nambu space, and the Pauli matrices $\tau_{x,y,z}$ act on the Nambu space. It is convenient to rotate $\cos\phi\,\tau_y - \sin\phi\,\tau_x \to \tau_y$, so that $\mathcal{H}_k = -(t\cos k + \mu)\tau_z - \sin k\,\tau_y$. The present Hamiltonian is very similar to the case for the 1D AIII class topological insulator, as obtained in the 1D optical Raman lattice. The topology of the present system can then be studied in the similar way. First, the bulk of the present 1D superconductor is gapped when $\mu \neq \pm t$, and is gapless at $\mu = \pm t$, which corresponds to transition between topological and trivial phases. Note the Hamiltonian $\mathcal{H}_k$ has time-reversal symmetry ($T$) and charge-conjugation (particle-hole $\mathcal{C}$) symmetry, defined by

$$T = K, \quad \mathcal{C} = \tau_x K, \tag{1.5.8}$$

with $K$ the complex conjugate. It follows that symmetries transform $T\mathcal{H}(k)T^{-1} = \mathcal{H}(-k)$ and $C\mathcal{H}(k)C^{-1} = -\mathcal{H}^*(-k)$. Thus the present system belongs to the so-called BDI class, with the topology being classified by 1D winding number. Similar to the 1D AIII class topological insulator, the 1D winding number is obtained straightforwardly by

$$N_{1D} = \frac{1}{4\pi i} \int dk \operatorname{Tr}\left[\tau_x \mathcal{H}^{-1}(k) \partial_k \mathcal{H}(k)\right]$$
$$= \frac{1}{2\pi} \int dk \hat{h}_z \partial_k \hat{h}_y, \tag{1.5.9}$$

where $(\hat{h}_y, \hat{h}_z) = (h_y, h_z)/h$, with $h_y = -\sin k, h_z = -t\cos k - \mu$, and $h = (h_y^2 + h_z^2)^{1/2}$. It further gives that

$$N_{1D} = \begin{cases} 1, & \text{for } |\mu| < t, \\ 0, & \text{for } |\mu| > t. \end{cases} \tag{1.5.10}$$

The topological number has a simple intuitive picture that the vector $\boldsymbol{h} = (\hat{h}_y, \hat{h}_z)$ winds $2\pi$ over a circle when $k$ runs over the FBZ.

Similar to the insulating phase, the nontrivial topology with $|\mu| < t$ can a boundary mode at each end of the 1D system if considering open boundary condition. The only difference is that here the boundary mode is a MF, rather than a Dirac fermion mode. Let the open boundaries locate at $x = 0, N$, respectively and diagonalizing $H$ in position space, we obtain the Majorana edge modes for the boundaries $x = 0, N$ as

$$\gamma_L(x_j) = \frac{1}{\sqrt{\mathcal{N}}}[(\lambda_+)^{x_j/a} - (\lambda_-)^{x_j/a}]\big[c(x_j)e^{i\phi/2}$$
$$+ c^\dagger(x_j)e^{-i\phi/2}\big], \tag{1.5.11}$$

$$\gamma_R(x_j) = \frac{i}{\sqrt{\mathcal{N}}}[(\lambda_+)^{(L-x_j)/a} - (\lambda_-)^{(L-x_j)/a}]\big[c(x_j)e^{i\phi/2}$$
$$- c^\dagger(x_j)e^{-i\phi/2}\big], \tag{1.5.12}$$

with $\mathcal{N}$ being the normalization factor and $\lambda_\pm = (\mu \pm \sqrt{\mu^2 - t^2 + |\Delta|^2})/(t + |\Delta|)$. It is easy to verify that the both MZMs satisfy the self-hermitian property: $\gamma_{L,R} = \gamma_{L,R}^\dagger$. Another important property of the MZMs is that when the phase factor $\phi$ varies by $2\pi$, one gets $\gamma_{L,R} \to -\gamma_{L,R}$. say each MZM acquires only $\pi$ phase. This property closely related to the fractional Josephson effect for $p$-wave SCs and the non-Abelian statistics of MZMs.[20]

5.B.2.  *2D chiral $p + ip$ SC*

Similar to the connection between the 1D $p$-wave SC and the 1D AIII class insulator, the 2D $p + ip$ SC is a *superconducting version* of the quantum Hall effect.[153] The Hamiltonian of the 2D $p + ip$ SC can be described by

$$H = \int d^2r \left\{ \psi^\dagger(\mathbf{r}) \left( -\frac{\hbar^2}{2m} \nabla^2 - \mu \right) \psi(\mathbf{r}) \right.$$

$$\left. + \frac{\Delta}{2} \left[ e^{i\phi} \psi (\partial_x + i\partial_y)\psi + h.c. \right], \right. \tag{1.5.13}$$

where $\psi(\mathbf{r})$ denotes the spinless fermion field operator at position $\mathbf{r}$ in the 2D space, $m$ is the mass of the fermion, and $\phi$ is the phase of the SC order $\Delta$. Again, transforming the Hamiltonian to $\mathbf{k}$ space we obtain

$$H = \frac{1}{2} \int d^2k \Phi_{\mathbf{k}}^\dagger \mathcal{H}_{\mathbf{k}} \Phi_{\mathbf{k}}, \tag{1.5.14}$$

$$\mathcal{H}_{\mathbf{k}} = \xi(\mathbf{k})\tau_z + k_x \Delta(\cos\phi\tau_y - \sin\phi\tau_x)$$

$$- k_y \Delta(\cos\phi\tau_x + \sin\phi\tau_y),$$

where $\Phi_{\mathbf{k}} = [\psi_{\mathbf{k}}, \psi_{-\mathbf{k}}^\dagger]^T$ and $\xi(\mathbf{k}) = \frac{\hbar^2 k^2}{2m} - \mu$. It is also convenient to rotate $\cos\phi\tau_y - \sin\phi\tau_x \to \tau_y$ and $\cos\phi\tau_x + \sin\phi\tau_y \to \tau_x$, so that $\mathcal{H}_{\mathbf{k}} = \xi(\mathbf{k})\tau_z + k_x \Delta\tau_y - k_y \Delta\tau_x$. It is straightforward to know that for nonzero $\Delta$, the bulk of the SC is gapped when $\mu \neq 0$. Accordingly, the gap is closed at $\mathbf{k} = 0$ when $\mu = 0$, which implies that the phase transition occurs, with the topology of the two regions with $\mu > 0$ and $\mu < 0$ being different. The regime of $\mu > 0$ is called the 'BCS' type weakly paired phase, while $\mu < 0$ corresponds to the 'BEC' type strongly paired phase.[153]

It can be verified that the time-reversal symmetry $T = K$ is broken for the Hamiltonian, while the charge conjugation symmetry $\mathcal{C} = \tau_x K$ keeps. Thus the $p + ip$ SC belongs to the D class in the AZ ten-fold classification. The topology of the present $p + ip$ SC is then characterized by Chern number, which can be calculated in the same way as done for QAH effect

$$\text{Ch}_1 = -\frac{1}{24\pi^2} \int d\omega dk_x dk_y \text{Tr} \left[ G dG^{-1} \right]^3, \tag{1.5.15}$$

where the Green's function $G^{-1}(\omega, \mathbf{q}) = \omega + i\delta^+ - \mathcal{H}(\mathbf{k})$, and the trace is operated on the Nambu space. Integrating over the frequency space yields that

$$\text{Ch}_1 = \frac{1}{4\pi} \int dk_x dk_y \hat{\mathbf{h}} \cdot \partial_{k_x} \hat{\mathbf{h}} \times \partial_{k_y} \hat{\mathbf{h}}, \tag{1.5.16}$$

where $\hat{\mathbf{h}} = (h_x, h_y, h_z)/|\mathbf{h}(\mathbf{k})|$, with $h_x = -k_y \Delta$, $h_y = k_x \Delta$ and $h_z = \xi(\mathbf{k})$.
By a straightforward calculation one can find that

$$\text{Ch}_1 = \begin{cases} +1, & \text{for } \mu > 0, \\ 0, & \text{for } \mu < 0. \end{cases} \tag{1.5.17}$$

In the topologically nontrivial phase $\mu > 0$, the 2D SC supports chiral edge states in the boundary, which are analogy to the chiral edge states obtained in the boundary of the 2D quantum Hall effect. Nevertheless, in the present $p+ip$ SC, the edge states are Majorana modes, while in quantum Hall effect they are chiral Dirac fermions. Furthermore, when the SC order is attached with a vortex with $\Delta \to \Delta e^{i\theta(\mathbf{r})}$, where $\theta(\mathbf{r})$ is the azimuthal angle, a Majorana zero mode can be obtained in the vortex core. It is also noteworthy that the Pfaffian state of the $\nu = 5/2$ fractional quantum Hall state can be mapped to a $p + ip$ SC ground state, hence hosting the MZMs.[154] For more discussions about the MZMs localized in SC vortices the readers can refer to the nice review article by Alicea.[20]

### 5.C.  Topological superconductor/superfluid from a conventional s-wave pairing order

While the intrinsic $p$-wave SCs naturally host the Majorana modes in the boundary and vortex cores, the materials with such intrinsic superconducting pairings are delicate and hard to synthesize. More recently, it has been proposed that hybrid systems of $s$-wave SC and SO coupled matters with odd number of FSs can favor effective $p$-wave pairing states, bringing the realization of MZMs in realistic solid state experiments.[19–22,78,79,155–160] In such hybrid systems, the the superconductivity is induced on the SO coupled material by proximity effect. Due to the presence of SO coupling, the parity-even ($s$-wave) and parity-odd ($p$-wave) pairing orders generically mix in the helical (eigenstate) bases at the interface. Under proper condition, e.g. by applying an external Zeeman field which kills the $s$-wave pairing while keeps the $p$-wave, the purely effective $p$-wave SC and Majorana modes can be obtained (More details for the topological superconductivity from proximity effect will be introduced in the later section after we present a generic theory for chiral topological superconductors). Motivated by these proposals, experimental studies have been performed to observe Majorana induced zero bias conductance anomalies with different heterostructures formed by $s$-wave SCs and semiconductor nanowires,[31–34] magnetic chains,[35–37] or topological insulators.[38–40] Nevertheless, the current experimental observations are not fully unambiguous, and the rigorous proof of Majorana modes in experiment is yet to be available.

Along with the exciting progresses made in the solid state physics, the exploration of MFs with ultracold atom systems has been also proposed and extensively studied, see e.g.[78,79,161-164,180] The motivation is quite straightforward. The $s$-wave superfluid phase can be achieved in ultracold fermions by tuning the $s$-wave Feshbach resonance, which is a mature technology in ultracold atoms.[165] Together with the SO coupling synthesised for the cold atom systems, the effective $p$-wave superfluid from an $s$-wave Feshbach resonance can be obtained. Nevertheless, for ultracold atoms the superfluid should be realized intrinsically, rather than by proximity effect. As such an intrinsic $s$-wave superfluid cannot be achieved in 1D system, but at least for 2D or 3D systems. As a result, to observe MZMs in a topological superfluid from $s$-wave Feshbach resonance, to realize a 2D SO coupling for Fermi gas is necessary.[85]

### 5.D. Chiral topological superfluids/superconductors: a generic theory

Instead of studying the topological superconductivity realized with various different platforms through proximity effect, we introduce in this subsection a generic theorem to determine the topology of a generic 2D system, as characterized by Chern numbers, after opening a gap through having superfluid/superconductivity.[166] Then we shall investigate the application of this theorem to various experimental systems. To simplify the description, we first classify the normal bands of the system without pairing into three groups: (1) the upper bands which are above the Fermi energy; (2) the lower bands which are below the Fermi energy; (3) the middle bands which are crossed by Fermi energy. In the most generic case, each middle band may have multiple FSs (Fermi loops), and we denote by $(i_M, j)$ the $j$-th FS loop of the $i_M$-th middle band. Let the total Chern number of the upper (lower) normal bands be $n_U$ ($n_L$). It can be shown that the Chern number the superfluid pairing phase induced in the system is given by[166]

$$
\begin{aligned}
\text{Ch}_1 = n_L - n_U \\
+ \sum_{i_M} \left[ (-1)^{q_{i_M}} n_F^{(i_M)} - \frac{1}{2\pi} \sum_j (-1)^{q_{i_M,j}} \oint_{\partial S_{i_M,j}} \nabla_{\mathbf{k}} \theta_{\mathbf{k}}^{(i_M,j)} \cdot d\mathbf{k} \right].
\end{aligned}
$$

$$(1.5.18)$$

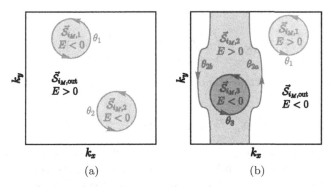

**Fig. 1.5.1.** (a)-(b) Configurations of FSs of normal bands.[165] The solid lines with arrows denote FSs, which are boundaries of the vector areas $\mathbf{S}_{i_M,j}$ in $\mathbf{k}$ space. The blue and green colors denote the closed FSs, and the red color denotes open FSs. From the definition for the formula (1.5.18), we have for (a) that $q_{i_M,1} = q_{i_M,2} = q_{i_M} = 1$, and for (b) that $q_{i_M} = q_{i_M,1} = q_{i_M,2} = 0$, while $q_{i_M,3} = 1$.

Here $n_F^{(i_M)}$ is the Chern number of the $i_M$-th middle band and $\theta_{\mathbf{k}}^{(i_M,j)} = \arg[\Delta_{\mathbf{k}}^{(i_M,j)}]$ is the phase of the pairing order projected onto the $(i_M,j)$-th FS loop. Note that the pairing can occur between two different Fermi surfaces, say between $(i_M,j)$ and $(i'_M,j')$. In this case the Eq. (1.5.18) is still valid, but the integral will be performed on both Fermi surfaces at the same time. More details can be found in Ref. 166. The integral direction is specified by arrows along FS lines in Fig. 1.5.1(a,b), which defines the boundary $\partial \mathcal{S}_{i_M,j}$ of the vector area $\mathbf{S}_{i_M,j}$ in $\mathbf{k}$ space. The quantities $\{q_{i_m,j}, q_{i_M}\} = \{0,1\}$ are then determined by "right-hand rule" specified below. The quantity $q_{i_M,j} = 0$ (or 1) if the energy of normal states within the area $\mathcal{S}_{i_M,j}$ is positive (or negative), while $q_{i_M} = 1$ (or 0) if the region $\mathcal{S}_{i_M,\text{out}}$, which is complementary to the sum of $\mathcal{S}_{i_M,j}$, have positive (or negative) energy. Some typical examples are shown in Fig. 1.5.1. With this theorem the Chern number of the superfluid phase can be simply determined once we know the properties of lower and upper bands, and the normal states at the Fermi energy, which govern the phases of $\Delta_{\mathbf{k}}^{(i_M,j)}$.

We introduce the proof of the above theorem for a multiband system. However, to facilitate the discussion, here we focus on the case with a single FS. The generalization to the multiple FSs will be assessed, with the detailed generic proof can be found in Ref. 166 (while some typos are corrected here). We write down the BdG Hamiltonian for a generic multi-band system by

$$\mathcal{H}_{\mathrm{BdG}}(\mathbf{k}) = \begin{pmatrix} \mathcal{H}(\mathbf{k}) - \mu & \Delta(\mathbf{k}) \\ \Delta^\dagger(\mathbf{k}) & -\mathcal{H}^T(-\mathbf{k}) + \mu \end{pmatrix}, \qquad (1.5.19)$$

where $\mathcal{H}(\mathbf{k})$ is the normal Hamiltonian, $\Delta(\mathbf{k})$ is the pairing order matrix, with

$$\Delta_\mathbf{k}^{(j_1, j_2)} = \langle u_\mathbf{k}^{(j_1)} | \Delta(\mathbf{k}) | u_{-\mathbf{k}}^{(j_2)*} \rangle$$

for the two normal bands $j_1$ and $j_2$, and $\mathbf{k}$ is the local momentum measured from FS center. Note that if FS is not symmetric with respect to its center, one can continuously deform it to be symmetric without closing the gap. In this process the topology of the system is not changed. Finally we can always write down $\mathcal{H}_{\mathrm{BdG}}$ in the above form to study the topology. Denote by $u_\mathbf{k}^{(i_M)}$ the eigenvector of the normal band crossing Fermi energy and

$$\mathcal{H}(\mathbf{k})u_\mathbf{k}^{(i_M)} = \epsilon_\mathbf{k}^{(i_M)}u_\mathbf{k}^{(i_M)}, \qquad (1.5.20)$$

where $\epsilon_k^{(i_M)}$ denotes the normal dispersion relation. Focusing on the pairing on FS, the eiegenstates of $\mathcal{H}_{\mathrm{BdG}}$ takes the generic form

$$|u(\mathbf{k})\rangle = [..., \alpha_k^{(i_M)}u_k^{(i_M)}; \beta_k^{(i_M)}u_{-k}^{(i_M)*}, ...]^T, \qquad (1.5.21)$$

where '...' denotes the components from the subbands other than $i_M$-th one. The eigenstate is associated with a Berry's connection calculated by

$$\mathcal{A}_\mathbf{k} = i\hbar\langle u(\mathbf{k})|\nabla_\mathbf{k}|u(\mathbf{k})\rangle. \qquad (1.5.22)$$

The key process is that we consider the weak pairing potential limit with $\Delta(k) \to \gamma\Delta(k)$ with $\gamma \to 0^+$ (without closing bulk gap), in which case we can expect that the contribution from superconducting pairing to the Berry connection is fully dominated by the states right at FS. We shall then extract the $\mathcal{A}_\mathbf{k}^{(i_M)}$ component our of $\mathcal{A}_\mathbf{k}$, namely, the Berry connection corresponding to the middle $i_M$-th band, given by

$$\mathcal{A}_\mathbf{k}^{(i_M)} = i[\alpha^{(i_M)}(\mathbf{k})u_\mathbf{k}^{(i_M)}]^\dagger \nabla_\mathbf{k}[\alpha^{(i_M)}(\mathbf{k})u_\mathbf{k}^{(i_M)}]$$
$$+ i[\beta^{(i_M)}(\mathbf{k})u_{-\mathbf{k}}^{(i_M)*}]^\dagger \nabla_\mathbf{k}[\beta^{(i_M)}(\mathbf{k})u_{-\mathbf{k}}^{(i_M)*}]. \qquad (1.5.23)$$

From the generic BdG equation we can obtain the coefficients in the weak pairing order regime by

$$\alpha^{(i_M)} = \frac{\bar{\epsilon}_\mathbf{k}^{(i_M)} - \mu - \left[|\gamma\Delta_k^{(i_M)}|^2 + \left(\frac{\epsilon_k^{(i_M)} + \epsilon_{-k}^{(i_M)}}{2} - \mu\right)^2\right]^{1/2}}{N^{(i_M)}(k)},$$

$$\beta^{(i_M)} = \frac{\gamma \Delta_k^{(i_M)*}}{N^{(i_M)}(k)},$$

$$N^{(i_M)} = \sqrt{\left|\alpha^{(i_M)}(k)\right|^2 + \left|\beta_\pm^{(i_M)}(k)\right|^2},$$

where $\bar{\epsilon}_{\mathbf{k}}^{(i_M)} = (\epsilon_{\mathbf{k}}^{(i_M)} + \epsilon_{-\mathbf{k}}^{(i_M)})/2$ and $\Delta_k^{(i_M)} = \Delta_k^{(i_M,i_M)}$. The Chern number of the superfluid is then obtained by

$$\mathrm{Ch}_1 = n_L - n_U + (2\pi)^{-1} \oint \nabla_{\mathbf{k}} \times \mathcal{A}_k^{(i_M)} d^2k. \tag{1.5.24}$$

By a straightforward calculation we obtain the Berry curvature associated with $\mathcal{A}_k^{(i_M)}$ in the weak pairing order limit by

$$\mathcal{B}_k^{(i_M)} = \left[(2\Theta_{\mathcal{S}} - 1)\widetilde{\mathcal{B}}_k^{(i_M)} + 2\nabla_k \Theta_{\mathcal{S}} \times \widetilde{\mathcal{A}}_k^{(i_M)}\right]$$
$$- \nabla_{\mathbf{k}} \times \frac{\Theta_{\mathcal{S}} \Im[\Delta_k^{(i_M)} * \nabla_k \Delta_k^{(i_M)}]}{|\Delta_k^{(i_M)}|^2}. \tag{1.5.25}$$

Here $\Theta_{\mathcal{S}}$ denotes the step function that is 1 for the area where the normal states have positive energies, and 0 for the area where the normal states have negative energies, representing a step change when crossing the FS. Let $\theta_k^{(i_M)} = \arg[\langle u_k^{(i_M)}|\Delta(k)|u_{-k}^{(i_M)*}\rangle]$ be the phase of order parameter on FS. With the above results we get further the Chern number of the superfluid phase by

$$\mathrm{Ch}_1 = n_L - n_U + (-1)^{q_{i_M}}\left[n_F^{(i_M)} + \frac{1}{\pi}\left(\oint_{\partial \mathcal{S}_{i_M}} \widetilde{\mathcal{A}}_k^{(i_M)} \cdot dk\right.\right.$$
$$\left.\left. - \int_{\mathcal{S}_{i_M}} \nabla_{\mathbf{k}} \times \widetilde{\mathcal{A}}_k^{(i_M)} d^2k - \frac{1}{2}\oint_{\partial \mathcal{S}_{i_M}} \nabla_k \theta_k^{(i_M)} \cdot dk\right)\right] \tag{1.5.26}$$

with $\widetilde{\mathcal{A}}_k^{(i_M)} = i[u_{\mathbf{k}}^{(i_M)}]^\dagger \nabla_{\mathbf{k}} u_{\mathbf{k}}^{(i_M)}$ being the Berry connection for the normal band states. If we choose a gauge so that $\widetilde{\mathcal{A}}_k^{(i_M)}$ is smooth on $\mathcal{S}_{i_M}$, the two terms regarding $\widetilde{\mathcal{A}}_k^{(i_M)}$ in the right hand side of Eq. (1.5.26) cancels. We then reach the formula (1.5.18) for the case with a single FS.

The proof can be generalized to the case with generic multiple FSs which may be closed or open [Fig. 1.5.1(b)], with multiple bands crossed by Fermi energy, and with the pairing within each FS or between two different FSs, given that the pairing fully gaps out the bulk.[166] Actually, for the case with multiple FSs, if the pairing occurs within each FS, which renders the Fulde-Ferrell-Larkin-Ovchinnikov (FFLO) or pair density wave (PDW) type orders,[167,168,181] the total Chern number is simply a summation of the contribution from all FSs. On the other hand, if the pairing occurs between two different FSs, the integration in the formula (1.5.18) is performed on both FSs simultaneously. The generic theorem introduced here is not restricted by pairing types, is powerful to quantitatively determine the topology of the superfluid phases, and can be particularly useful for condensed matter materials when the system is complicated. We also note that this theorem is best applied to judge the topology of phases with relatively weak pairing orders. This is because a strong pairing order may fully deform FSs of the original system and then change Chern number governed by Eq. (1.5.18), driving a topological phase transition.[166] However, monitoring such phase transitions with increasing pairing orders can determine the whole topological phase diagram.

### 5.E. *Applications to topological superfluids (superconductors) for SO coupled systems*

In this subsection we introduce the applications of the generic theorem given in formula (1.5.18) to various types of SO coupled systems with $s$-wave superfluid/supercontuctor pairing.

### 5.E.1. *TI & s-wave SC*

We first consider the 2D hybrid system formed by TI surface states and an $s$-wave SC (or superfluid).[155] The TI surface states are described by $(2+1)$d Dirac Hamiltonian. Together with the $s$-wave pairing order induced by the substrate SC through proximity effect, the effective Hamiltonian can be written down as

$$H = H_{\text{TI-surf}} + H_{\text{pair}}^s,$$

$$H_{\text{TI-surface}} = \int d^2k c_{\mathbf{k},s}^\dagger \left[ v_F(k_x \sigma_y - k_y \sigma_x) - \mu \right]_{ss'} c_{\mathbf{k},s'},$$

$$H_{\text{pair}}^s = \int d^2k \Delta_s c_{\mathbf{k},\uparrow} c_{-\mathbf{k},\downarrow} + h.c., \tag{1.5.27}$$

where $v_F$ is the Fermi velocity of the surface states. The normal states of the surface Hamiltonian $\mathcal{H}_0(\mathbf{k}) = v_F(k_x \sigma_y - k_y \sigma_x)$ read $|u_+(\mathbf{k})\rangle = [1, e^{i\varphi(\mathbf{k})}]/\sqrt{2}$ and $|u_-(\mathbf{k})\rangle = [-e^{-i\varphi(\mathbf{k})}, 1]/\sqrt{2}$, with $\varphi(\mathbf{k}) = \tan^{-1}(-k_x/k_y)$, and the energy $\epsilon_\pm = \pm v_F(k_x^2 + k_y^2)^{1/2}$. If the Fermi energy is located at the upper

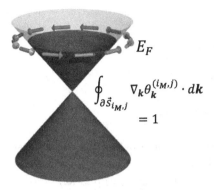

**Fig. 1.5.2.** Surface states of the 3D topological insulator, with spin-momentum locking being shown at the Fermi surface. The phase winding of the pairing order projected onto the Fermi surface gives +1.

band (Fig. 1.5.2), the projected pairing on the FS is given by

$$\Delta_+(\mathbf{k}) = \Delta_s \langle u_+(\mathbf{k})|T|u_+(-\mathbf{k})\rangle$$
$$= \frac{\Delta_s}{2}\left[-e^{-i\varphi(-\mathbf{k})} + e^{-i\varphi(\mathbf{k})}\right]$$
$$= \Delta_s e^{-i\varphi(\mathbf{k})}. \tag{1.5.28}$$

We can apply the formula (1.5.18) to obtain the Chern number of the gapped superconducting phase. Note that here we only have a single band crossing the Fermi energy (i.e. one middle band). We have that $n_U = n_L = 0$, and $n_F = 0$. For the Fermi energy located above the Dirac point, we have all the factors $q_{iM} = q_{iM,j} = 1$. Thus the Chern number

$$\mathrm{Ch}_1 = -\frac{1}{2\pi}\oint_{\partial \mathcal{S}_{i_M,j}} \nabla_{\mathbf{k}}\varphi_{\mathbf{k}} \cdot d\mathbf{k}$$
$$= -1. \tag{1.5.29}$$

It is easy to confirm that when the Fermi energy is located below the Dirac point, the Chern number is again $\mathrm{Ch}_1 = -1$, so the phase is the same.

### 5.E.2. *Rashba SO system & s-wave SC & Zeeman splitting*

Now we consider the 2D hybrid system formed by 2D Rashba SO coupled system with an external Zeeman field and the *s*-wave SC.[79,82] The effective

Hamiltonian can be written down as

$$H = H_0 + H_{\text{pair}}^s,$$

$$H_0 = \int d^2k c_{\mathbf{k},s}^{\dagger} \left[ \frac{\hbar^2 \mathbf{k}^2}{2m} - \mu + \lambda_R(k_x \sigma_y - k_y \sigma_x) + V_z \sigma_z \right]_{ss'} c_{\mathbf{k},s'},$$

$$H_{\text{pair}}^s = \int d^2k \Delta_s c_{\mathbf{k},\uparrow} c_{-\mathbf{k},\downarrow} + h.c., \tag{1.5.30}$$

where $V_z$ is the external Zeeman field along the $z$ direction, and $\lambda_R$ is the Rashba SO coefficient. The normal states of the surface Hamiltonian $\mathcal{H}_0 = \frac{\hbar^2 \mathbf{k}^2}{2m} - \mu + \lambda_R(k_x \sigma_y - k_y \sigma_x) + V_z \sigma_z$ read $|u_+(\mathbf{k})\rangle = [\cos(\theta/2), \sin(\theta/2)e^{i\varphi(\mathbf{k})}]$ and $|u_-(\mathbf{k})\rangle = [-\sin(\theta/2)e^{-i\varphi(\mathbf{k})}, \cos(\theta/2)]$, with $\varphi(\mathbf{k}) = \tan^{-1}(-k_x/k_y)$ and $\theta = \tan^{-1}(\lambda_R k/V_z)$, and the energy $\epsilon_\pm = \frac{\hbar^2 \mathbf{k}^2}{2m} - \mu \pm \lambda_R(k_x^2 + k_y^2)^{1/2}$. Assume that the Fermi energy is located at the lower band. The projected pairing on the FS is given by

$$\begin{aligned}
\Delta_-(\mathbf{k}) &= \Delta_s \langle u_-(\mathbf{k})|T|u_-(-\mathbf{k})\rangle \\
&= \Delta_s \sin(\theta/2)\cos(\theta/2)\left[ -e^{i\varphi(\mathbf{k})} + e^{i\varphi(-\mathbf{k})} \right] \\
&= -\Delta_s \sin(\theta/2)\cos(\theta/2)e^{i\varphi(\mathbf{k})}. \tag{1.5.31}
\end{aligned}$$

We can also apply the formula (1.5.18) to obtain the Chern number of the gapped superconducting phase with different parameter conditions by considering the weak pairing order regime. First we take that the Fermi energy is located inside the Zeeman gap opened at $\mathbf{k} = 0$, so that there is only one middle band [Fig. 1.5.3(a)]. For the Rashba system we have that $n_U = n_L = 0$, and $n_F = 0$. For the present Fermi energy configuration, we

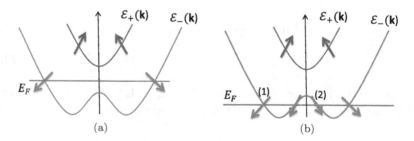

**Fig. 1.5.3.** Sketch of the normal bands for a Rashba SO coupled system. (a) The Fermi energy locates within the Zeeman gap. (b) The Fermi energy locates below the Zeeman gap and crosses the lower subband at four Fermi points, leading to two Fermi surfaces: (1) the outer one and (2) the inner one.

have all the factors $q_{iM} = q_{iM,j} = 1$. Thus the Chern number

$$\text{Ch}_1 = -\frac{1}{2\pi} \oint_{\partial S_{iM,j}} \nabla_{\mathbf{k}} \varphi_{\mathbf{k}} \cdot d\mathbf{k}$$
$$= +1. \tag{1.5.32}$$

On the other hand, if the Fermi energy crosses two FSs, as in the configuration shown in Fig. 1.5.3(b), we have for the two FSs that have that $n_U = n_L = 0$, and $n_F = 0$, while $q_{iM} = q_{iM,1} = 1$ and $q_{iM,2} = 0$. Then the Chern number reads

$$\text{Ch}_1 = \frac{1}{2\pi} \oint_{\partial S_{iM,2}} \nabla_{\mathbf{k}} \varphi_{\mathbf{k}} \cdot d\mathbf{k} - \frac{1}{2\pi} \oint_{\partial S_{iM,1}} \nabla_{\mathbf{k}} \varphi_{\mathbf{k}} \cdot d\mathbf{k}$$
$$= 1 - 1 = 0. \tag{1.5.33}$$

Thus the cases in Fig. 1.5.3(a) and Fig. 1.5.3(b) correspond to two different basic phases, with one being topologically nontrivial and one trivial (similarly, if the Fermi energy is located above the Zeeman gap, the phase is also trivial). The quantitative transition between the two different phase occurs when the gap is closed. This can be obtained by examining the bulk gap, giving the critical point equation by $V_z^2 = \mu^2 + \Delta_s^2$. The topologically nontrivial regime corresponds to $V_z^2 > \mu^2 + \Delta_s^2$.

### 5.E.3. *PDW state for a Dirac metal*

We consider the effective tight-binding Hamiltonian $H_{\text{TB}}$ on a square lattice proposed in Ref. 166

$$H_{\text{TB}} = \sum_{\mathbf{k}} \left( c_{\mathbf{k}\uparrow}^\dagger, c_{\mathbf{k}\downarrow}^\dagger \right) \mathcal{H}_{\text{TB}} \begin{pmatrix} c_{\mathbf{k}\uparrow} \\ c_{\mathbf{k}\downarrow} \end{pmatrix};$$
$$\mathcal{H}_{\text{TB}} = \left( m_z - 2t_x \cos k_x - 2t_z \cos k_z \right) \sigma_z$$
$$+ 2t_{so} \sin k_z \sigma_y + t_{xI} \sin k_x \sigma_0 + m_x \sigma_x, \tag{1.5.34}$$

where $s = \uparrow, \downarrow$, $t_{x/z}$ is the hoping constant along $x/z$ direction, $t_{so}$ is the strength of spin-flip hopping, and $m_{z,x}$ denote the effective Zeeman couplings. As described in Fig. 1.5.4(a), the above Hamiltonian describes a Dirac semimetal if $m_x = 0$ and $|m_z| < 2(t_x + t_z)$, with two Dirac points at $\Lambda_\pm = (\pm \cos^{-1}[(m_z - 2t_x)/2t_z], 0)$. The term $t_{xI} \sin k_x \sigma_0$ breaks the inversion symmetry and leads to an energy difference between the two Dirac points. Finally, a nonzero $m_x$ opens a local gap at the two Dirac points. For simplicity we take that $t_z = t_0, t_x = t_0 \cos \theta_0$ and $t_{xI} = 2t_0 \sin \theta_0$ to facilitate the further discussion. The realization with a new optical Raman

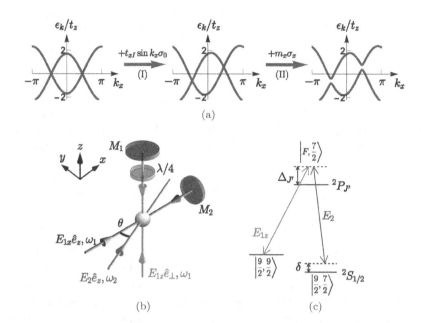

**Fig. 1.5.4.** Sketch for band structure and realization of a 2D SO coupled Dirac metal.[166] (a) Schematic diagram of the band structure, with the inversion symmetry broken (process I) and gap opening at Dirac points (II) for a 2D SO coupled Dirac semimetal. (b) Proposed experimental setting for realization. The standing wave lights formed by $\mathbf{E}_{1x,1z}$ generate a blue-detuned square lattice. The incident polarization of $\mathbf{E}_{1z}$ is $\hat{e}_\perp = \alpha\hat{e}_x + i\beta\hat{e}_y$, and the $\lambda/4$-wave plate changes the polarization of the reflected field to $\hat{e}'_\perp = \alpha\hat{e}_x - i\beta\hat{e}_y$. The Raman coupling, illustrated in (c) for $^{40}$K fermions, is generated by $\mathbf{E}_{1z}$ and an additional running light $\mathbf{E}_2$ which has tile angle $\theta$ with respect to $x$-$z$ plane.

lattice is sketched in Fig. 1.5.4(b,c), where only a single Raman coupling is applied in the scheme. The details of the realization are neglected here, but the readers can refer to Ref. 166. Note that here the 2D Dirac metal is driven by SO interaction, and is distinct from graphene, of which the Dirac points are protected by symmetry only if SO coupling is absent.[169]

The superfluid phase can be induced by considering an attractive Hubbard interaction. The total Hamiltonian

$$H = H_{\mathrm{TB}} - U\sum_i n_{i\uparrow}n_{i\downarrow}. \tag{1.5.35}$$

Due to the existence of multiple FSs corresponding to different Dirac cones, in general we can have intra-cone pairing (FFLO) and inter-cone (BCS) pairing orders, defined by $\Delta_{2q} = (U/N)\sum_k \langle c^\dagger_{q+k\uparrow} c_{q-k\downarrow}\rangle$, with $q = \pm Q$

or 0. Note that for the present Dirac system, a BCS pairing cannot fully gap out the bulk but leads to nodal phases, while the FFLO or FF order can.[166] On the other hand, since the inversion symmetry is broken, the BCS pairing would be typically suppressed. From the mean field results of a uniform system shown in Fig. 1.5.5(a,b), we find that the BCS pairing nearly vanishes, and $\Delta_{-2Q}$ dominates over $\Delta_{2Q}$, with $\pm Q \approx \Lambda_{\pm}$ for positive $\mu$. The topology of the superfluid phase can be characterized by the Chern number.

The topology of the superfluid PDW phase can be determined with the theorem presented in formula (1.5.18). When there is only one FS, the system with an FF order is topological since $n_L = n_U = n_F^{(i_M)} = 0$ and $\int_{\partial S_{(i_M)}} \nabla \theta_k^{(i_M)} \cdot dk = \pm 2\pi$, giving the Chern number

$$\text{Ch}_1 = \frac{1}{2\pi} \int_{\partial S_{(i_M)}} \nabla \theta_k^{(i_M)} \cdot dk \mp 1 \qquad (1.5.36)$$

for the Fermi energy crossing the left and right Dirac cones, respectively. In contrast, when there are two FSs (at both Dirac cones), from the same or different bands, one can readily find that the contributions from both FSs cancels out and the Chern number $\text{Ch}_1 = 0$, rendering a trivial phase. This result implies that the system can be topological when it is in an FF phase.

A rich phase diagram is given in Fig. 1.5.5(c,d), where the topological and trivial FF phases with one of $\Delta_{\pm 2Q}$ being nonzero, and FFLO phase with both $\Delta_{\pm 2Q}$ being nonzero, are obtained. The gapless phase with nonzero $\Delta_{\pm 2Q}$ may be obtained due to imperfectly nesting Fermi surfaces. It is particularly interesting that in Fig. 1.5.5(d) a broad topological region is predicted when $m_z$ is away from $m_z = 2t_z$. The broad topological region implies that the upper critical value $\Delta_{2q}^{(c)}$, characterizing the transition from topological FF state to other phases, is largely enhanced compared with the case for $m_z = 2t_z$. This is a novel effect due to the following mechanism. Note that the pairing $\Delta_{-2Q}$ also couples the particle-hole states at $\tilde{Q} = (\pi - Q, 0)$. Increasing $\Delta_{-2Q}$ to $\Delta_{-2Q}^{(c)}$ closes the bulk gap at $\tilde{Q}$ momentum, with the critical value being solved from BdG Hamiltonian as

$$\Delta_{-2Q}^{(c)} = \sqrt{m_x^2 + m_p^2 - \left( \frac{t_{xI}}{t_x} \sqrt{t_x^2 - \frac{m_p^2}{16}} - \mu \right)^2}, \qquad (1.5.37)$$

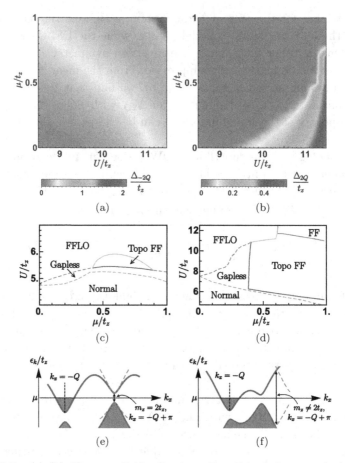

**Fig. 1.5.5.** (a)–(b) The superfluid order by self-consistent theory and phase diagram.[166] The superfluid order $\Delta_{-2Q}$ dominates over others for $\mu > 0$. (c) Phase diagram with normal-size topological region when $t_x = t_{so} = t_z, t_{xI} = 0.7t_z, m_z = 2t_z, m_x = 0.3t_z$. (d) Phase diagram with broad topological region. The values of the parameters in diagrams (a,b,d) are $m_z = 2.92t_z, t_x = 0.92t_z, t_{so} = t_z, t_{xI} = 0.8t_z, m_x = 0.3tz$. Schematic diagrams showing the underlying mechanism for the normal-size (c) and broad (d) topological regions.

where $m_p = 2(m_z - 2t_z)$. For $m_z = 2t_z$, we have $Q = \pi/2$ and a small critical value $\Delta_{-2Q}^{(c)} \lesssim m_x$. This is because the gap closes at the right hand Dirac point $\tilde{\mathbf{Q}} = (\pi/2, 0)$, where the original bulk gap less than $2m_x$ before having superfluid pairing [Fig. 1.5.5(e)]. Importantly, for $m_z = 3t_z$, we find that $\Delta_{-2Q}^{(c)} \sim 2t_z$, which is of the order of band width. In this regime $\tilde{\mathbf{Q}}$ is away from the right hand Dirac point, and corresponds to a large bulk gap before adding $\Delta_{-2Q}$ [Fig. 1.5.5(f)]. As a result, a large $\Delta_{-2Q}$ is necessary to drive the phase transition, giving a broad topological region, as shown in Fig. 1.5.5(d). Note that an equivalent picture for the phase transition is that increasing the order $\Delta_{-2Q}$ deforms the band structure, which pushes one band (another band) at $\tilde{\mathbf{Q}} = (\pi - Q, 0)$ toward (away from) the Fermi energy. When $\Delta_{-2Q}$ reaches the critical value, the band is pushed across the Fermi energy, leading to an additional contribution from Fermi surface to the Chern number given in Eq. (1.5.18), and the phase transition occurs.

*Further discussions.* The generic theorem given Eq. (1.5.18) is powerful to determine the topology of the 2D SC/superfluid characterized by Chern numbers. The application can be go much beyond the examples introduced above, which are relatively simple. For example, the topological superfluid phase based on SO coupled QAH model as discussed in Section IV.B exhibit rich phase diagram with different Chern numbers.[85] To determine the full topological phase diagram through the conventional computation of the Chern numbers is some tedious, while using our present generic formula (1.5.18) can reach all the results with a quick check. We are not going to expand the discussion on the details, but the readers may check by themselves.

## 5.F. *Non-Abelian Majorana modes in trivial superfluids*

So far we have focused on the topological superconductors which host Majorana modes in the vortex cores or boundary. Note that MZMs in SCs (e.g. bound at vortices) are topological defect modes, which correspond to nonlocal extrinsic deformations in the Hamiltonian of the topological system. For example, MZMs in the chiral $p_x + ip_y$ SC harbor at vortices which exhibit nonlocal phase windings of the SC order (a global deformation in the original uniform Hamiltonian). This feature tells that the MZMs at vortices are not intrinsic topological excitations, but extrinsic modes of a SC. In this regard, one may conjecture that the existence of MZMs is not uniquely corresponding to the bulk topology of a SC, and there might be much broader range of experimental systems which can host such exotic modes, besides those based on topologically nontrivial SCs. This conjecture was confirmed in a generic theorem shown in the recent works.[170,182]

### 5.F.1. *Chern-Simons invariant: a generic theorem*

Here we introduce the generic as shown in Ref. 170 that the existence of MZMs localized in the vortex cores does not rely on the bulk topology of a 2D SC. For a generic 2D normal system with $N$ FSs and gapped out by SC pairings, one can show that the existence of MZMs at the SC vortices is characterized by an emergent $\mathbb{Z}_2$ Chern-Simons invariant $\nu_3$:

$$\nu_3 = \sum_i^N n_i w_i \bmod 2, \tag{1.5.38}$$

$$w_i \equiv \frac{1}{2\pi} \oint_{\mathrm{FS}_i} \nabla_{\mathbf{k}} \arg \Delta_{\mathbf{Q}_i}(\mathbf{k}) \cdot d\mathbf{k},$$

where $\Delta_{\mathbf{Q}_i}(\mathbf{k})$ is the SC order projected onto the $i$-th FS and is generically momentum dependent, $w_i$ counts the phase winding of $\Delta_{\mathbf{Q}_i}(\mathbf{k})$ in the $\mathbf{k}$-space around the $i$-th FS loop, and $n_i$ denotes the integer vortex winding number (vorticity) attached to $\Delta_{\mathbf{Q}_i} \to \Delta_{\mathbf{Q}_i} e^{i n_i \theta(\mathbf{r})}$, namely the winding in the real space. At least a single MZM is protected when the index $\nu_3 = 1$ if there is no symmetry protection, even the bulk topology of the SC, characterized by Chern number, can be trivial.

*Proof.* We introduce the proof of the generic theorem given in Eq. (1.5.38) for the Chern-Simons invariant, which governs the existence of the MZM in a 2D SC. The essential idea for the present proof is similar to that done in proving the generic formula of Chern number for chiral SC, as discussed in the previous section. For a system with multiple normal bands and FSs, the superconducting pairings may occur within each FS (intra-FS pairings) and between different FSs (inter-FS pairings). The theorem (1.5.38) is not affected by inter-FS pairings. Thus for convenience, we consider first the generic SC Hamiltonian with only intra-FS pairings, given by

$$H = \sum_{\mathbf{k}} C_{\mathbf{k}}^\dagger \hat{H}_0 C_{\mathbf{k}} + \sum_{i,\alpha,\beta,\mathbf{k}} c_{\mathbf{Q}_i/2+\mathbf{k},\alpha}^\dagger \hat{\Delta}_{\mathbf{Q}_i}^{\alpha\beta} c_{\mathbf{Q}_i/2-\mathbf{k},\beta}^\dagger + \mathrm{h.c.}, \tag{1.5.39}$$

where $C_{\mathbf{k}} = (c_{\mathbf{k},\alpha}, c_{\mathbf{k},\beta}, \dots, c_{\mathbf{k},\gamma}, \dots)^{\mathrm{T}}$, with $\alpha$ incorporating the orbital and spin indices, the normal band Hamiltonian $\hat{H}_0(\mathbf{k})$ is considered to have $N$ FSs, and the pairing matrix element $\hat{\Delta}_{\mathbf{Q}_i}^{\alpha\beta} \propto \sum_{\mathbf{k}} \langle c_{\mathbf{Q}_i/2+\mathbf{k},\alpha} c_{\mathbf{Q}_i/2-\mathbf{k},\beta} \rangle$ regarding the $i$-th FS has a central-of-mass momentum $\mathbf{Q}_i$. Here for convenience we take that each FS is circular and centered at a momentum $\mathbf{Q}_i/2$. Similar to the proof of Chern number for chiral topological superfluids, as introduced in previous section, one can always continuously deform the FSs to be circular without changing topology of the system, as long as the bulk gap keeps open during the deformation. In general the SC order exhibits spatial modulation in the real space, rendering the PDW or FFLO state,[167,168] and bears the form $\hat{\Delta}(\mathbf{r}) = \sum_i \hat{\Delta}_{\mathbf{Q}_i} e^{i\mathbf{Q}_i \cdot \mathbf{r}}$. Note that each PDW component $\hat{\Delta}_{\mathbf{Q}_i}$ possesses a $U(1)$ symmetry, implying that each of them can be attached with a vortex of winding number $n_i$ independently, giving $\hat{\Delta}(\mathbf{r}) = \sum_i \hat{\Delta}_{\mathbf{Q}_i} e^{-in_i \theta(\mathbf{r}) + i\mathbf{Q}_i \cdot \mathbf{r}}$, with $\theta(\mathbf{r})$ being the vortex phase profile.

The Chern-Simons invariant $\nu_3$ is defined in 3D space, for which one parameterizes the Bogoliubov de Gennes (BdG) Hamiltonian by taking the phase $\phi \in [0, 2\pi)$ of the SC order $\hat{\Delta}_{\mathbf{Q}_i} e^{-in_i \phi}$ as a synthetic dimension of ring geometry $S^1$. Together with the 2D physical space, the bulk BdG Hamiltonian can then be written down in a synthetic 3D torus $T^3 = T^2 \times S^1$ spanned by $(\mathbf{k}, \phi)$. In the synethetic 3D space, the $\mathbb{Z}_2$ Chern-Simons invariant[171–173] is computed by

$$\nu_3 = -\frac{1}{4\pi^2} \int_{T^2 \times S^1} \mathcal{Q}_3 \bmod 2$$

$$\mathcal{Q}_3 = \text{Tr}\left[\mathcal{A}d\mathcal{A} - \frac{2i}{3}\mathcal{A}^3\right], \tag{1.5.40}$$

where the elements of one-form Berry connection are given by $\mathcal{A}_{\lambda\lambda'}(\mathbf{k}, \phi) = i\langle\psi_\lambda|d\psi_{\lambda'}\rangle$, with $|\psi_\lambda\rangle$ denoting the corresponding eigenvector of the BdG Hamiltonian, and the trace is performed on the filled bands.

A direct computation of the index $\nu_3$ for the generic case is not realistic. To simplify the study we again take the advantage that the topology of the system is unchanged under any kind of continuous deformation without closing bulk gap. For this we further adiabatically deform the Hamiltonian $H$ to a new form

$$H' \equiv H[\hat{\Delta}_{\mathbf{Q}_i} \rightarrow \hat{\Delta}_{\mathbf{Q}_i} \Omega_{\mathbf{Q}_i}(\mathbf{k})], \tag{1.5.41}$$

where $\Omega_{\mathbf{Q}_i}(\mathbf{k})$ is a positive real smooth truncation function with $\Omega_{\mathbf{Q}_i}(\mathbf{S}_i) = 1$ inside the orientable vector area $\mathbf{S}_i$ enclosed by the $i$-th FS loop centered at $\mathbf{Q}_i/2$, and decays to zero at a short distance beyond this area. Since the system remains fully gapped for the continuous deformation, the invariant $\nu_3$ can be evaluated over $H'$. Denoting by $\mathcal{F}_i$ the vector area with $\Omega_{\mathbf{Q}_i}(\mathbf{k}) \neq 0$, The invariant given in Eq. (1.5.40) can be reduced to the integral over the disjoint union $\bigsqcup_i \mathcal{F}_i \times S^1$, as shown in Ref. 170, which facilitates the further study.

While in general the Hamiltonian $\hat{H}_0$ incorporates multiple normal bands, one can consider the weak SC pairing regime, in which case only the states around each FS will be effectively paired up. The coupling between states from different FSs and that from different bands can be ignored due to the energy detuning. In this way, the BdG $H'$ further reduces to an effective one-band form in the eigen-basis $u_\mathbf{k}$ of $\hat{H}_0$. In particular, for the momentum $\mathbf{k} \in \mathcal{F}_i$ around a specific FS centered at momentum $\mathbf{Q}_i/2$, the effective BdG Hamiltonian takes the form

$$h_i(\mathbf{k}, \phi) = \begin{bmatrix} \epsilon_{\mathbf{Q}_i/2+\mathbf{k}} & \Delta_{\mathbf{Q}_i}(\mathbf{k})\Omega_{\mathbf{Q}_i} e^{-in_i\phi} \\ \Delta^*_{\mathbf{Q}_i}(\mathbf{k})\Omega_{\mathbf{Q}_i} e^{in_i\phi} & -\epsilon_{\mathbf{Q}_i/2-\mathbf{k}} \end{bmatrix} \qquad (1.5.42)$$

where $\Delta_{\mathbf{Q}_i}(\mathbf{k}) \equiv \langle u_{\mathbf{k}}|\hat{\Delta}_{\mathbf{Q}_i}|u^*_{-\mathbf{k}}\rangle$ is the pairing term projected onto the $i$-th FS. Note that $\Delta_{\mathbf{Q}_i}(\mathbf{k})$ has captured the original band topology. The eigenstates of $h_{\mathbf{Q}_i}$ take the form $|\psi_{\mathbf{k}\pm}\rangle = (\alpha_{\mathbf{k}\pm}u_{\mathbf{k}}, \beta_{\mathbf{k}\pm}u^*_{-\mathbf{k}})^{\mathrm{T}}$ (refer to also the proof of Chern number for chiral topological superfluid in the previous section). Then $\nu_3$ can be decomposed into $\nu_3 = \sum_i \nu_3^{(i)}$ ('mod 2' temporarily omitted), and

$$\nu_3^{(i)} = -\frac{1}{4\pi^2} \int_{\mathcal{F}_i \times S^1} [\mathcal{A}_\phi \nabla_{\mathbf{k}} \times \mathcal{A}_{\mathbf{k}} + \mathcal{A}_{\mathbf{k}} \times \nabla_{\mathbf{k}}\mathcal{A}_\phi] d\phi d^2\mathbf{k}$$

for each $\mathcal{F}_i$, where $\mathcal{A}_\phi = i\langle\psi_{\mathbf{k}-}|\partial_\phi|\psi_{\mathbf{k}-}\rangle$, and $\mathcal{A}_{\mathbf{k}} \equiv (\mathcal{A}_{k_x}, \mathcal{A}_{k_y}) = i\langle\psi_{\mathbf{k}-}|\nabla_{\mathbf{k}}|\psi_{\mathbf{k}-}\rangle$, with $\nabla_{\mathbf{k}} \equiv (\partial_{k_x}, \partial_{k_y})$. The above result can be further simplified by taking the weak pairing order limit $\Delta_{\mathbf{Q}_i} \to 0^+$, in which case the gap becomes infinitesimal at the FSs, and the contribution to $\nu_3$ will completely come from the FS states. It can be derived directly on $\mathcal{F}_i$ that $\mathcal{A}_\phi = -n_i\Theta_{\mathbf{S}_i}$ and $\mathcal{A}_{\mathbf{k}} = (1 - 2\Theta_{\mathbf{S}_i})\mathcal{A}_{0,\mathbf{k}} + \Theta_{\mathbf{S}_i}(\nabla_{\mathbf{k}} \arg \Delta_{\mathbf{Q}_i} + \mathbf{A}^i_d)$, where $\Theta_{\mathbf{S}_i}$ is a step function equal to 1 within $\mathbf{S}_i$ and 0 otherwise, $\mathcal{A}_{0,\mathbf{k}} \equiv iu^{\dagger}_{\mathbf{k}}\nabla_{\mathbf{k}}u_{\mathbf{k}}$ represents the Berry connection for the normal band, and $\mathbf{A}^i_d$ is the *defect gauge field* as a consequence of the multivalueness of $\arg \Delta_{\mathbf{Q}_i}$.[174] Substituting these results into the formula of $\nu_3$ yields

$$\nu_3 = \sum_i \frac{n_i}{2\pi} \int_{\mathcal{F}_i} \Theta_{\mathbf{S}_i} \nabla_{\mathbf{k}} \times (\nabla_{\mathbf{k}} \arg \Delta_{\mathbf{Q}_i} + \mathbf{A}^i_d) \cdot d^2\mathbf{k}. \qquad (1.5.43)$$

The above result is exactly the one given in Eq. (1.5.38) by observing that the curl of gradient of SC phase vanishes, while the contribution from the defect gauge field $\mathbf{A}^i_d$ renders the phase winding of SC order in the momentum space around FS loop.[170] This completes the proof. The theorem is still valid if the system has inter-FS pairings and connected FSs.[170] Particularly, in the one band case, the $\mathbb{Z}_2$ Chern Simons invariant readily reduces to the Hopf invariant.

The above result shows that the existence of MZMs at vortex cores is essentially protected not by the bulk topology of the 2D SC, but by an emerging Chern-Simons invariants $\nu_3$, implying that a non-Abelian MZM can exist in a trivial SC. A famous example can be obtained from a Rashba SO coupled semiconductor with Zeeman splitting and in the presence of an $s$-wave superconductivity,[79,156] as introduced in the previous section. To obtain a chiral topological SC the chemical potential has to lie within the Zeeman gap and cross the bulk band for once. According to the above theorem, even the chemical potential is above the Zeeman gap and crosses two FSs, MZMs can in principle be generated if the SC orders in the two FSs are independent and only one of them is attached with vortex.

### 5.F.2. *Majorana zero modes in 2D trivial superfluids*

*2D Dirac metal.* The theorem in (1.5.38) suggests that MZMs can exist in broader range of physical systems. Now we introduce a minimal scheme, which can be readily achieved based on the optical Raman lattice scheme,[88,166] for the realization of MZMs. The total Hamiltonian takes the form

$$H = H_0 + H_U,$$
$$H_0 = \sum_{\mathbf{k}} (c_{\mathbf{k},\uparrow}^\dagger, c_{\mathbf{k},\downarrow}^\dagger) \mathcal{H}_0 \begin{pmatrix} c_{\mathbf{k},\uparrow} \\ c_{\mathbf{k},\downarrow} \end{pmatrix}$$
$$\mathcal{H}_0 = (m_z - 2t_x \cos k_x - 2t_y \cos k_y)\sigma_z$$
$$+ 2t_{\mathrm{so}} \sin k_x \sigma_x - \mu,$$
$$H_U = -U \sum_i n_{i\uparrow} n_{i\downarrow}, \tag{1.5.44}$$

where the Hubbard interaction is attractive $U > 0$. As is known that the Hamiltonian $H_0$ describes a topological Dirac semimetal for $|m_z| < 2(t_x + t_y)$, with two Dirac points at $\mathbf{Q}_\pm = (0, \pm \cos^{-1}((m_z - 2t_y)/2t_x))$ and possesses non-trivial spin texture on the FSs (Fig. 1.5.6). The difference of the present case from the one in Eq. (1.5.34) is that here no inversion symmetry is broken and no gap opening at the Dirac points.

*Self-consistent phase diagram.* The superfluid (SC) states can be studied with the above model. Having multiple FSs around various Dirac cones, generically one shall consider both the inter-cone (BCS) and the intra-cone (PDW) pairing orders, described by $\Delta_{\mathbf{q}} = (U/N) \sum_{\mathbf{k}} \langle c_{\mathbf{q}/2+\mathbf{k},\uparrow} c_{\mathbf{q}/2-\mathbf{k},\downarrow} \rangle$, with $\mathbf{q} = 2\mathbf{Q}_\pm$ or 0 and $N$ is total number of lattice sites. Generally, the order parameter in real space takes the form

$$\Delta(\mathbf{r}) = \Delta_0 + \Delta_{2\mathbf{Q}_+} e^{2i\mathbf{Q}_+ \cdot \mathbf{r}} + \Delta_{2\mathbf{Q}_-} e^{2i\mathbf{Q}_- \cdot \mathbf{r}} \tag{1.5.45}$$

and the BCS and PDW orders may compete with each other. The intra-cone PDW order can fully gap the bulk while reducing the translation symmetry. On the other hand, owing to the different spin-momentum lock at the FSs

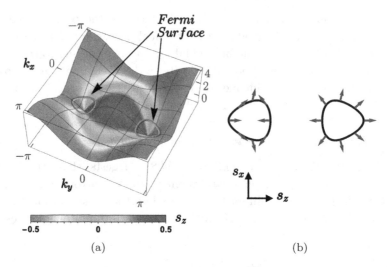

**Fig. 1.5.6.** An example of 2D SO coupled semimetal.[170] (a) The band structure of 2D topological Dirac metal with two Dirac points located at $\mathbf{Q}_\pm$, the gray thick loops around two Dirac points represent the Fermi surfaces, and the color represents the average value of the spin component $\langle s_z \rangle$. (b) Schematic of the spin orientations, shown by blue arrows, at the Fermi surfaces around the Dirac points. Parameters: $t_{x,y} = t_{so} = m_z = 1$, $\mu = 0.8$, and the corresponding Dirac node momenta $\mathbf{Q}_+ = -\mathbf{Q}_- = (0, 2\pi/3)$.

of the two Dirac cones [Fig. 1.5.6(b)], the inter-cone BCS pairing cannot fully gap out the bulk spectrum, and leaves four nodal points. These nodal points can be further gapped by charge density wave (CDW) orders

$$\rho_{\mathbf{q},\sigma} = (U/N) \sum_{\mathbf{k}} \langle c^\dagger_{\mathbf{k}-\mathbf{q}/2,\sigma} c_{\mathbf{k}+\mathbf{q}/2,\sigma} \rangle. \tag{1.5.46}$$

The BCS, PDW, and CDW orders may compete to dominate in different parameter regimes, and can be solved self-consistently.

The phase diagram are obtained by self-consistent calculation with proper parameters so that the Dirac points are located at $\mathbf{Q}_\pm = (0, \pm 2\pi/3)$, as shown in Fig. 1.5.7. The phase diagram is dominated by PDW order with $|\Delta_{2\mathbf{Q}_\pm}| \neq 0$, which appears only for finite $U$. With increasing chemical potential, the Dirac cone becomes less isotropic (Fig. 1.5.6) and the FSs are less well-nested. As a consequence, a narrow gapless region with nonzero PDW orders $|\Delta_{2\mathbf{Q}_\pm}| \neq 0$ is obtained for $\mu > 0.05$ [Fig. 1.5.7(a)], while the spectrum becomes fully gapped when $|\Delta_{2\mathbf{Q}_\pm}|$ increases exceeding some finite value. In the fully gapped region at larger $U$, one can readily check that the Chern number vanishes $\text{Ch}_1 = 0$ from the generic result shown in formula (1.5.18), and the bulk is trivial for the present (class D)

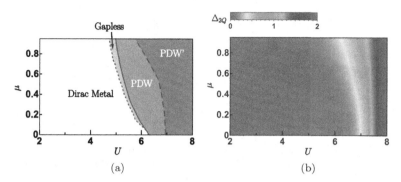

**Fig. 1.5.7.** Self-consistent result for the order parameters.[170] (a) Mean field phase diagram of the Dirac-Hubbard Hamiltonian versus attractive Hubbard interaction $U$ and chemical potential $\mu$. In the "Dirac metal" phase, all $\Delta_\mathbf{q} = 0$; in the narrow "Gapless" region, $\Delta_{2\mathbf{Q}_\pm}$ are finite but not strong enough to fully gap the system. In the "PDW" phase, the system is fully gapped by finite PDW. In the "PDW'" phase at large $U$, BCS and CDW orders of small magnitude also appear and accompany the PDW. (b) Magnitude of PDW order $\Delta_{2\mathbf{Q}_\pm}$. The parameters are the same as those in Fig. 1.5.6.

superfluid.[166] In the PDW' phase at large $U$, BCS and CDW orders of small magnitude also appear and accompany the PDW. They can be regarded as small perturbations to the PDW phase and hence the PDW' phase is topologically equivalent to the PDW phase.

*MZMs for the trivial superfluid.* Despite the topologically trivial superconducting state here, the system can host non-trivial MZM bound to vortices and protected by the Chern-Simons invariant shown in the present work. In general, the vortices proliferated to the PDW order read

$$\Delta(\mathbf{r}) = \Delta_{2\mathbf{Q}_+} e^{2i\mathbf{Q}_+ \cdot \mathbf{r} + in_+ \theta(\mathbf{r})} + \Delta_{2\mathbf{Q}_-} e^{2i\mathbf{Q}_- \cdot \mathbf{r} + in_- \theta(\mathbf{r})}$$
$$= 2\Delta_{2\mathbf{Q}_\pm} e^{i(n_+ + n_-)\theta(\mathbf{r})/2} \cos\left[2\mathbf{Q}_+ \cdot \mathbf{r} + (n_+ - n_-)\theta(\mathbf{r})/2\right].$$
$$(1.5.47)$$

The minimal allowed defect corresponds to a half-vortex, given by $n_+ + n_- = \pm 1$, while a full vortex is given by $n_+ + n_- = \pm 2$.[175,176] In particular, in Fig. 1.5.8(a,c) we consider the half vortex regime with two unit vortices of opposite vorticities $\pm 2\pi$ (i.e. $n_+ = \pm 1$) attached only to $\Delta_{2\mathbf{Q}_+}$ and located with a finite distance between each other in the real space. The real space BdG Hamiltonian with vortices is then numerically solved and the two lowest energy modes with finite-size energies $E = \pm 1.039 \times 10^{-4}$ are obtained [Fig. 1.5.8(c)]. Spatial wave function density $\sum_{s=\uparrow,\downarrow} |\psi_s(\mathbf{r})|^2$ for one of the solutions (the other is the same) is plotted in Fig. 1.5.8(a), showing that it is

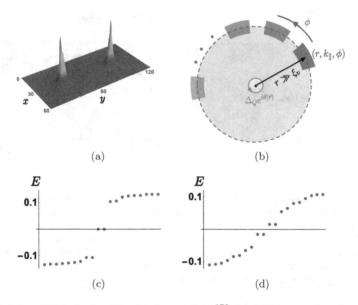

(a)                         (b)

(c)                         (d)

**Fig. 1.5.8.** MZMs in the 2D trivial superfluid.[170] (a) MZM wave function density $\sum_{s=\uparrow,\downarrow} |\psi_s(\mathbf{r})|^2$ computed in the Dirac-Hubbard model, with $\mu = 0.8$, $U = 5.5$, which gives the pairing order $\Delta_{2\mathbf{Q}_\pm} = 0.23$. System size: $N_x = N_y/2 = 60$. The vortices with opposite unit vorticities are located at $(30, 30)$ and $(30, 90)$ and the vortex field $e^{i\theta(\mathbf{r})}$ is attached only to $\Delta_{2\mathbf{Q}_+}$. (b) Schematic of the physical origin of MZMs at vortex cores. The vortex core can be viewed approximately as the open boundary of $r$-dimension in the 3D space spanned by $(r, k_\parallel, \phi)$. Energy spectrum for the half vortex (c) and full vortex (d) regime. The MZMs are obtained in the former case. Other parameters are the same as those in Fig. 1.5.6.

in the zero angular-momentum channel and well-localized at vortex cores, thus being a MZM. The robustness of MZMs against impurities can be shown straightforwardly.[170] The physical origin of the exsistence of MZMs can be viewed as a direct consequence of *bulk-boundary correspondence*, as illustrated in Fig. 1.5.8(b). Consider the region far away enough form the vortex core so that at each azimuthal angle $\phi$ we can find a microscopically large region with approximately constant SC phase $\theta$. This region can be thought of as a 2D system in $(r, k_\parallel; \phi)$ with fixed $\phi = \theta$, periodic boundary along $k_\parallel$ direction and open boundary along $r$ direction. Combining all such 2D systems with $\phi \in [0, 2\pi)$ yields an effective 3D space with periodic boundary with respect to $k_\parallel$ and $\phi$, while open boundary along $r$ axis due to the existence of vortex. With this picture when the parameterized 3D system has a nontrivial Chern-Simons invariant $\nu_3$, which is the case for half-vortex regime based on a direct numerical check, MZM is obtained as a boundary zero mode at the vortex core. In comparison, we have performed a similar calculation by attaching a full vortex with $n_+ + n_- = 2$ to $\Delta_{2\mathbf{Q}_\pm}$, which gives a null $\nu_3$. In Fig. 1.5.8(d), the corresponding low energy spectrum reveals that no zero mode but finite energy Andreev bound states are present in the system, consistent with the $\nu_3$ result.

### 5.F.3. *Chiral Majorana modes in 3D trivial superfluids*

The existence of non-Abelian Majorana modes in the trivial phase of a superconducting 2D Dirac semimetal can be generalized to the 3D case, as studied in.[170] In this case a 3D Weyl semimetal together with an attractive Hubbard interaction was considered, with the Hamiltonian given by

$$H = \sum_{\mathbf{p}} \psi_{\mathbf{p}}^\dagger h_{\mathbf{p}} \psi_{\mathbf{p}} - U \sum_i n_{i\uparrow} n_{i\downarrow},$$

$$h_{\mathbf{p}} = [m_z - 2t_0(\cos p_x + \cos p_y) - 2t_z^c \cos p_z]\sigma_z$$
$$+ 2t_{SO}(\sin p_x \sigma_x + \sin p_y \sigma_y) - \mu - 2t_z^s \sin p_z. \qquad (1.5.48)$$

Here the hopping terms along $z$ direction with $(t_z^c, t_z^s) = t_z(\cos \varphi_0, \sin \varphi_0)$, which break inversion symmetry of the Weyl semimetal unless $\varphi_0 = n\pi/2$ with integer $n$. For $m_z = 4t_0 + 2t_z^c \cos Q$ $(0 < Q < \pi)$, the Weyl semimetal has two nodal points of chiralities $\chi = \pm$ located at $\mathbf{Q}_\chi = (0, 0, \chi Q)$ and with energies $E_\pm = -\mu \mp 2t_z^s \sin Q$, respectively. It is convenient to choose $Q = \frac{2\pi}{3}$ and $t_{0,z} = t_{SO} = 1.0$ to facilitate further discussion. We note that the Weyl semimetal can be realized by generalizing the optical Raman lattice scheme to 3D regime, which is currently considered in both theory and experiment for ultracold atoms.

The SC phases can be induced with an attractive interaction $U > 0$. Similar to the results introduced in the previous sections for the 2D Dirac semimetal, the possible pairing orders are of two distinct types, namely the $s$-wave BCS and PDW phases, as sketched in Fig. 1.5.9(a). The former describes a uniform order $\Delta_0$ occurring between two different Weyl cones and with zero center-of-mass momentum of Cooper pairs, while the latter are spatially modulated orders $\Delta_{2\mathbf{Q}_\pm}$ occurring within each Weyl cone and the Cooper pairs have nonzero center-of-mass momentum. These pairing orders read $\Delta_{\mathbf{q}} = \frac{U}{2N} \sum_{\mathbf{k}} \langle c_{\mathbf{q}/2+\mathbf{k}\uparrow} c_{\mathbf{q}/2-\mathbf{k}\downarrow} \rangle$, with $\mathbf{q} = 0, 2\mathbf{Q}_\pm$, and $N$ being number of sites, and the interaction is decoupled into $\mathcal{H}_{MF} = \sum_{\mathbf{k},\mathbf{q}} \Delta_{\mathbf{q}} c_{\mathbf{q}/2+\mathbf{k}\uparrow}^\dagger c_{\mathbf{q}/2-\mathbf{k}\downarrow}^\dagger + \text{h.c}$, with $\mathbf{k}$ being summed over the entire Brillouin zone. Again the order parameter takes the following generic form in real space $\Delta(\mathbf{r}) = \Delta_0 + \Delta_{+2Q} e^{i2\mathbf{Q}_+ \cdot \mathbf{r}} + \Delta_{-2Q} e^{i2\mathbf{Q}_- \cdot \mathbf{r}}$. The

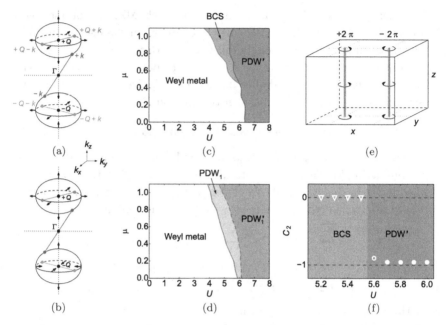

**Fig. 1.5.9.** Weyl metals, mean field phase diagrams, vortex line configuration, and 2nd Chern numbers.[177] (a,b) Schematic for FSs and pairing order of Weyl semimetals with (a) or without (b) inversion symmetry. The two ellipsoids centered at $\mathbf{Q}_\pm$ are the FSs. The spin orientations (black arrows) at the FS around $\mathbf{Q}_-$ is flipped when $0 < \mu < 2t_z^s \sin|Q_\pm|$. The colored lines represent the pairing order parameters $\Delta_\mathbf{q}$ with $\mathbf{q} = 0, 2\mathbf{Q}_\pm$. (c,d) Phase diagram for cases with (c) or without (d) inversion symmetry versus attractive Hubbard interaction $U$ and chemical potential $\mu$. (e) Configuration of two vortex lines with opposite vorticities. (f) The emergent 2nd Chern number $C_2$ for a 4D synthetic space generalized from the 3D physical system in the inversion symmetric case and with $\mu = 0.7$. Upon increasing $U$, the system undergoes a transition from BCS to PDW' phase. The circles (triangles) gives the 2nd Chern number in BCS (PDW') phase.

numerical simulation reveals different phase diagrams for the inversion-symmetric [Fig. 1.5.9(b)] and inversion symmetry broken [Fig. 1.5.9(d)] Weyl metals. For the inversion-symmetric Weyl metal with $\varphi_0 = 0$, a direct transition from Weyl metal phase to the PDW' phase, which has $|\Delta_{+2Q}| = |\Delta_{-2Q}| \gg |\Delta_0| \neq 0$, is obtained by increasing $U$ in the low chemical potential regime with $|\mu| < \mu_c$ and $\mu_c \sim 0.4$. The equality $|\Delta_{+2Q}| = |\Delta_{-2Q}|$ in the PDW' phase is a consequence of the inversion symmetry. However, when $\mu$ is tuned beyond the critical value $|\mu| > \mu_c$, the BCS phase with $\Delta_0 \neq 0$ and $\Delta_{\pm 2Q} = 0$ appears between the Weyl metal and PDW' phase. This result reflects that the pairings within each

Weyl Fermi surface dominates in relatively small $\mu$ regime. Moreover, if the inversion symmetry is broken, the BCS phase is suppressed, and the system enters from Weyl metal into $PDW_1$ state first and then $PDW_1'$ phase by increasing $U$, as shown in Fig. 1.5.9(d) with $\varphi_0 = \frac{3}{32}\pi$. The $PDW_1$ phase is characterized by $|\Delta_{+2Q}| \geq |\Delta_{-2Q}|$ for $\mu \geq 0$ and $\Delta_0 = 0$, while in the $PDW_1'$ phase a small $|\Delta_0|$ ($< |\Delta_{\pm 2Q}|$) also emerges.

*Vortex Line Modes for trivial superfluid phase.* Both BCS and PDW states have only the particle-hole symmetry, so they belong to class D according to the Altland-Zirnbauer symmetry classification.[123] Both BCS and PDW phases are $p_x + ip_y$ SCs stacked along $z$ direction and are topological in the 2D subspace, but are trivial in the 3D space. Namely, any invariant defined in the 3D space is zero for the present BCS and PDW phases.

While the phases are trivial for the 3D space, the nontrivial vortex line modes can be obtained. We consider first the PDW dominated phase, and take that $|\Delta_{+2Q}| = |\Delta_{-2Q}|$ and $\Delta_0 = 0$, since a small perturbative BCS order does not affect the results. Let a vortex line of winding $n$ be along $z$-axis and attached to $\Delta_{+2Q}$, so that $\Delta_{+2Q} = |\Delta_{+2Q}| \exp[in\phi(\mathbf{r})]$, with $\tan(\phi) = y/x$. In the low-energy limit the effective Hamiltonian can be obtained by linearizing the 3D Weyl cone Hamiltonian around $\mathbf{Q}_+$ point

$$H_{\text{eff}} = -\left( \sum_{j=x,y,z} iv_j \sigma_j \partial_j + \mu_{\text{eff}} \right) \tau_z$$
$$+ \bar{\Delta}(\rho)\cos(n\phi)\tau_x + \bar{\Delta}(\rho)\sin(n\phi)\tau_y, \qquad (1.5.49)$$

where $v_j$ is the Fermi velocity along $j$-th direction, $\mu_{\text{eff}}$ is the chemical potential measured from the Weyl node, and $\tau$'s are Pauli matrices for the Nambu space spanned by $\hat{f}(\mathbf{r}) = [c_\uparrow(\mathbf{r}), c_\downarrow(\mathbf{r}), c_\uparrow^\dagger(\mathbf{r}), -c_\uparrow^\dagger(\mathbf{r})]^T$. For the vortex line located at $x = y = 0$ along $z$-axis, one can choose cylindrical coordinate $(\rho, \phi, z)$, with $\bar{\Delta}(\rho \to 0) = 0$, $\bar{\Delta}(\rho \to \infty) = $ constant, and $n\phi$ the phase winding of the PDW component. For $n = \pm 1$ the Majorana in-gap modes can be obtained analytically and satisfy[170]

$$H_{\text{eff}}\gamma_z(\rho, \phi, k_z) = \mathcal{E}_{k_z}\gamma_z(\rho, \phi, k_z),$$
$$\mathcal{E}_{k_z} = \text{sgn}(nv_xv_y)v_zk_z, \qquad (1.5.50)$$

which implies that the vortex Majorana modes are chiral, with chirality $\chi_{M\ell} = \text{sgn}(nv_xv_yv_z)$ being related to $n$ and Weyl node chirality $\chi$. The Majorana operator reads $\hat{\gamma}_z(k_z) = \int d^2\mathbf{r}\, \gamma_z(\rho, \phi, k_z)\hat{f}(\rho, \phi, k_z)$, with $\hat{\gamma}_z(k_z) = \hat{\gamma}_z^\dagger(-k_z)$ for real Majorana states. This solution can be readily generalized to the case of a generic winding $n = \mathcal{N}$. Actually, such a vortex line is topologically equivalent to $|\mathcal{N}|$ vortex lines with unity winding $n = \text{sgn}(\mathcal{N})$. In the later case each line hosts a branch of chiral Majorana modes. Due to the chiral property putting the $|\mathcal{N}|$ branches of vortex modes together with couplings can at most deform their dispersions, but cannot annihilate them, yielding $|\mathcal{N}|$ chiral Majorana modes.

The chirality of Majorana vortex modes imply that these modes are gapless and traverse the bulk gap of the PDW phase. This property can be further confirmed by performing a full real-space numerical calculation based on the lattice model without linearity assumption. In particular, the inversion symmetric Weyl metal with $\mu = 0.7$ and $U = 5.8$ is considered. The self-consistent calculation for this regime reveals a PDW$'$ phase with $\Delta_{\pm 2Q} \approx 0.201$ and $\Delta_0 \approx -1.85 \times 10^{-2}$, and the system has a bulk gap $E_{\text{gap}} \approx 0.21$. A vortex line ($n = 1$) and anti-vortex line ($n = -1$) are considered along $z$-axis, separating from each other in $x$-$y$ plane, and are attached to one of $\Delta_{0,\pm 2Q}$, as sketched in Fig. 1.5.9(e). With this configuration appropriate periodic boundary condition can be applied in the numerical calculation. The local spectral function $A(x, y, k_z, E)$ can be obtained from the retarded Green's function $G^R(E)$ of the system

$$A = -\frac{1}{\pi} \sum_{s=\uparrow,\downarrow} \Im \langle x, y, k_z, s | G^R(E) | x, y, k_z, s \rangle, \qquad (1.5.51)$$

where $|x, y, k_z, s\rangle$ is the Bloch basis with momentum $k_z$. Computing $A(k_z, E)$ near the vortex core gives the energy spectra of the bulk and vortex line modes.

Figures 1.5.10(a) and (b) show the spectra measured from the vortex line ($n = 1$) and anti-vortex line ($n = -1$), respectively. In both cases the chiral Majorana modes traverse the bulk gap connecting the lower and upper bands. The chirality of these modes depends on the vortex line winding number, consistent with the previous analytic solution. This result is fundamentally different from that for a BCS phase, as shown in Fig. 1.5.10(c), where we compute the Majorana modes by attaching vortex and anti-vortex lines to $\Delta_0$ with $\Delta_0 = 0.25$ and $\Delta_{\pm 2Q} = 0$. Majorana zero-energy flat bands for the vortex lines are obtained. These Majorana zero modes are simply the vortex modes of $p_x + i p_y$ SCs with different momenta $k_z$.

Chiral gapless modes have to be protected by chiral topological invariants, e.g. the Chern numbers. However, it can be verified that for any 2D sub-plane incorporating $z$-axis the 1st Chern number is zero. Similar to the case in 2D superconducting Dirac semimetal system, an emerging

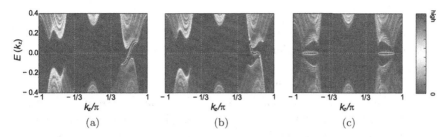

**Fig. 1.5.10.** Chiral Majorana modes as in-gap bound states to a vortex line.[177] (a–b) Chiral vortex line modes shown from spectral functions $A(k_z, E)$ of the inversion symmetric Weyl metal with PDW orders based on self consistent calculation for $\mu = 0.7$ and $U = 5.8$. The vortex line is attached to $\Delta_{+2Q}$, with winding number $n = +1$ for (a) and $n = -1$ for (b). (c) Spectral function $A(k_z, E)$ for the BCS dominated phase in the inversion symmetric Weyl metal with $\Delta_0 = 0.25$. Two segments of Majorana flat bands are obtained. Periodic boundary conditions are considered and the system size is $N_x/2 = N_y = 55$ and $N_z = 162$.

invariant defined in a higher-dimensional space shall provide the protection of the present chiral Majorana modes. Similarly, the SC order can be parameterized by its phase factor $\Delta_{\mathbf{q}} e^{in\theta}$, where the constant $\theta \in [0, 2\pi)$ forms a 1D periodic parameter space $S^1$. Together with the 3D lattice, we construct a 4D synthetic space $T^4 = T^3 \times S^1$ spanned by $\mathsf{p} = (\mathbf{p}, p_\theta)$ with $\mathbf{p} = (p_x, p_y, p_z)$ and $p_\theta = \theta$. In this synthetic 4D space we define the 2nd Chern number by $C_2 = \frac{1}{32\pi^2} \int_{T^4} d^4\mathsf{p}\, \epsilon_{ijk\ell} \mathrm{Tr}[\mathsf{F}_{ij}\mathsf{F}_{k\ell}] \in \mathbb{Z}$, where $\epsilon$ is the antisymmetric tensor and $\mathsf{F}_{ij}$ are the gauge field strengths calculated by diagonalizing Hamiltonian $H_{\mathrm{MF}}(\mathbf{p}, p_\theta)$ for every $\mathbf{p}$ and $p_\theta$. The second Chern number $C_2$ can be numerically computed for BCS and PDW phases, as shown numerically in Fig. 1.5.9(f). It is seen that $C_2 = 0$ for the BCS dominated phase, while $C_2 \approx -0.956$ for the PDW phase obtained with the same parameters except for $U$ as in Fig. 1.5.10(a). The deviation of $C_2$ from an integer is due to finite size effect. It was further shown and proposed that when considering the ring configuration of vortex lines, a 3D non-Abelian loop-loop braiding statistics can be obtained.[170] These results reveal a real physical system to explore the exotic 3D non-Abelian braiding statistics, and shall attract further studies based on realistic ultracold atom platforms.

## 1.6. Discussion and Outlook

In conclusion, in the present review we have pedagogically introduced the realization of spin-orbit (SO) coupling and topological quantum phases for ultracold atoms, and have focused on the latest progresses in theory and experiment, particularly for the SO coupling beyond one-dimension (1D) in optical lattices, the topological insulating states, and topological superfluid phases. We systematically discussed the optical Raman lattice schemes, with which the 1D and 2D Dirac types, 2D Rashba type, and 3D Weyl type SO couplings can be synthesized. Being of the high feasibility, the proposed SO couplings and topological quantum physics have motivated several experimental studies, with some of the proposals having been successfully realized. The topological superfluids and Majorana modes have also been discussed. After a brief introduction to the background, we showed a generic theorem for the chiral topological superfluid/superconductor phases, which provides a simple but generic approach to determine the topology of 2D chiral superfluids/superconductors. Moreover, the existence of non-Abelian Majorana modes at vortices (or vortex lines for 3D systems) is revisited and is found to be irrespective of the bulk topology of the superfluid/superconductor phases, and is protected by emergent topological numbers defined in the synthetic spaces with dimension higher than the physics system. As a direct consequence, the non-Abelian Majorana modes can exist in trivial superfluids/superconductors, with the minimal experimental schemes for the realization have been proposed and studied.

The realization of high-dimensional SO couplings has opened intriguing opportunities to explore novel quantum physics with ultracold atoms, ranging from traditional spintronic effects to the exotic topological physics. In particular, the recently proposed new optical Raman lattice scheme[149] exhibits high controllability in engineering various types of high-dimensional SO couplings, with the realized SO coupled quantum gases having long lifetime (up to several seconds depending on atom candidates). Especially, with such scheme a long-lived topological Bose gas has been achieved in experiment, confirming the high feasibility of the new proposal. With these progresses the research of the high-dimensional SO coupled quantum gases is developing into a mature topic, and the realization of high-dimensional SO couplings shall become routine studies in experiment in the near future. The next important issues in this direction include to investigate the quantum-far-from equilibrium dynamics with novel topology and the interacting topological quantum states for the SO coupled systems. For ultracold atoms, the nonequilibrium dynamics could be particularly useful to explore topological quantum physics, since the equilibrium ground state is not easy to reach due to the heating and loss in the system.

Nonequilibrium dynamics may bring up new physics beyond the achievability of equilibrium studies. The interacting topological quantum physics, e.g. the topological superfluids and fractional topological insulating states, are highly-sought-after but very challenging due to heating/loss in the real experiments. Note that the optical Raman lattice schemes, introduced in the present review, can be generically applied to any type of atom candidates, including the alkali earth and lanthanide atoms which have been confirmed to have much less heating in observing SO effects. It is therefore of high interests to investigate the novel interacting topological physics with high-dimensional SO coupled alkali earth and lanthanide quantum gases.

## References

1. R. Winkler, *Spin-Orbit Coupling in Two-Dimensional Electron and Hole Systems* (2003) (Spinger-Verlag, New York).
2. J. Sinova and A. H. MacDonald, Theory of spin-orbit effects in semiconductors. *Semiconductors and Semimetals*, **82**, 45–87 (2008).
3. N. Nagaosa, J. Sinova, S. Onoda, A. H. MacDonald, and N. P. Ong, Anomalous hall effect. *Reviews of Modern Physics*, **82**, 1539 (2010).
4. M. Z. Hasan and C. L. Kane, Topological insulators. *Rev. Mod. Phys.*, **82**, 3045–3067 (2010).
5. X.-L. Qi and S.-C. Zhang, Topological insulators and superconductors. *Rev. Mod. Phys.*, **83**, 1057–1110 (2011).
6. M. Sato and Y. Ando, Topological superconductors: a review. *Rep. Prog. Phys.*, **80** 076501 (2017).
7. S. Murakami, N. Nagaosa, and S. C. Zhang, Dissipationless quantum spin current at room temperature. *Science*, **301**, 1348–1351 (2003).
8. J. Sinova, D. Culcer, Q. Niu, N. A. Sinitsyn, T. Jungwirth, and A. H. MacDonald, Universal intrinsic spin Hall effect. *Physical Review Letters*, **92**, 126603 (2004).
9. K. V. Klitzing, G. Dorda, and M. Pepper, New method for high-accuracy determination of the fine-structure constant based on quantized Hall resistance. *Physical Review Letters*, **45**, 494 (1980).
10. D. C. Tsui, H. L. Stormer, and A. C. Gossard, Two-dimensional magneto-transport in the extreme quantum limit. *Physical Review Letters*, **48**, 1559 (1982).
11. B. A. Bernevig and T. L. Hughes, *Topological Insulators and Topological Superconductors* (2013) (Princeton University Press).
12. A. Bansil, H. Lin, and T. Das, Colloquium: Topological band theory. *Reviews of Modern Physics*, **88**, 021004 (2016).
13. L. Fu, Topological crystalline insulators. *Physical Review Letters*, **106**, 106802 (2011).

14. C.-K. Chiu, J. C. Teo, A. P. Schnyder, and S. Ryu, Classification of topological quantum matter with symmetries. *Reviews of Modern Physics*, **88**, 035005 (2016).
15. X. Wan, A. M. Turner, A. Vishwanath, and S. Y. Savrasov, Topological semimetal and Fermi-arc surface states in the electronic structure of pyrochlore iridates. *Phys. Rev. B*, **83**, 205101 (2011).
16. A. A. Burkov and L. Balents, Weyl semimetal in a topological insulator multilayer. *Phys. Rev. Lett.*, **107**, 127205 (2011).
17. S.-Y. Xu *et al.*, Discovery of a Weyl fermion semimetal and topological Fermi arcs. *Science*, **349**, 613 (2015).
18. B. Q. Lv, H. M. Weng, B. B. Fu, X. P. Wang, H. Miao, J. Ma, P. Richard, X. C. Huang, L. X. Zhao, G. F. Chen, Z. Fang, X. Dai, T. Qian, and H. Ding, Experimental discovery of Weyl semimetal TaAs. *Phys. Rev. X*, **5**, 031013 (2015).
19. F. Wilczek, Majorana returns. *Nat. Phys.*, **5**, 614 (2009).
20. J. Alicea, New directions in the pursuit of Majorana fermions in solid state systems. *Reports on Progress in Physics*, **75**, 076501 (2012).
21. M. Franz, Majorana's wires. *Nat. Nano.*, **8**, 149 (2013).
22. S. R. Elliott and M. Franz, Colloquium: Majorana fermions in nuclear, particle, and solid-state physics. *Rev. Mod. Phys.*, **87**, 137 (2015).
23. C. Nayak and F. Wilczek, $2n$-quasihole states realize $2n - 1$-dimensional spinor braiding statistics in paired quantum Hall states. *Nucl. Phys. B*, **479**, 529 (1996).
24. D. A. Ivanov, Non-abelian statistics of half-quantum vortices in $p$-wave superconductors. *Phys. Rev. Lett.*, **86**, 268 (2001).
25. S. Das Sarma, M. Freedman, and C. Nayak, Topologically protected qubits from a possible non-Abelian fractional quantum Hall state. *Phys. Rev. Lett.*, **94**, 166802 (2005).
26. J. Alicea, Y. Oreg, G. Refael, F. von Oppen, and M. P. A. Fisher, Non-Abelian statistics and topological quantum information processing in 1D wire networks. *Nat. Phys.*, **7**, 412 (2011).
27. X.-J. Liu, Chris L. M. Wong, and K. T. Law, Non-abelian majorana doublets in time-reversal-invariant topological superconductors. *Phys. Rev. X*, **4**, 021018 (2014).
28. A. Kitaev, *Annals of Physics*, **303**, 2 (2003).
29. C. Nayak, S. H. Simon, A. Stern, M. Freedman, and S. Das Sarma, *Rev. Mod. Phys.*, **80**, 1083 (2008).
30. J. K. Pachos, *Introduction to Topological Quantum Computation* (Cambridge University Press, 2012).
31. V. Mourik, K. Zuo, S. M. Frolov, S. R. Plissard, E. P. A. M. Bakkers, and L. P. Kouwenhoven, *Science*, **336**, 1003 (2012).
32. M. T. Deng, C. L. Yu, G. Y. Huang, M. Larsson, P. Caroff, and H. Q. Xu, *Nano Lett.*, **12**, 6414 (2012).
33. A. Das, Y. Ronen, Y. Most, Y. Oreg, M. Heiblum, and H. Shtrikman, *Nat. Phys.*, **8**, 887 (2012).

34. A. D. K. Finck, D. J. Van Harlingen, P. K. Mohseni, K. Jung, and X. Li, *Phys. Rev. Lett.*, **110**, 126406 (2013).
35. S. Nadj-Perge, I. K. Drozdov, J. Li, H. Chen, S. Jeon, J. Seo, A. H. MacDonald, B. A. Bernevig, and A. Yazdani, *Science*, **346**, 602 (2014).
36. M. Ruby, F. Pientka, Y. Peng, F. von Oppen, B. W. Heinrich, and K. J. Franke, *Physical Review Letters*, **115**, 197204 (2015).
37. R. Pawlak, M. Kisiel, J. Klinovaja, T. Meier, S. Kawai, T. Glatzel, D. Loss, and E. Meyer, *Npj Quantum Information*, **2**, 16035 (2016).
38. J.-P. Xu, M.-X. Wang, Z. L. Liu, J.-F. Ge, X. Yang, C. Liu, Z. A. Xu, D. Guan, C. L. Gao, D. Qian, Y. Liu, Q.-H. Wang, F.-C. Zhang, Q.-K. Xue, and J.-F. Jia, *Phys. Rev. Lett.*, **114**, 017001 (2015).
39. H.-H. Sun, K.-W. Zhang, L.-H. Hu, C. Li, G.-Y. Wang, H.-Y. Ma, Z.-A. Xu, C.-L. Gao, D.-D. Guan, Y.-Y. Li, C. Liu, D. Qian, Y. Zhou, L. Fu, S.-C. Li, F.-C. Zhang, and J.-F. Jia, *Phys. Rev. Lett.*, **116**, 257003 (2016).
40. Q. L. He *et al.*, Chiral Majorana fermion modes in a quantum anomalous Hall insulator superconductor structure. *Science*, **357**, 294–299 (2017).
41. C. J. Pethick and H. Smith, *Bose-Einstein Condensation in Dilute Gases* (Cambridge University Press, 2002).
42. I. Bloch, J. Dalibard, and W. Zwerger, Many-body physics with ultracold gases. *Rev. Mod. Phys.*, **80**, 885–964 (2007).
43. I. Bloch, J. Dalibard, and S. Nascimbène, Quantum simulations with ultracold quantum gases. *Nature Physics*, **8**, 267–276 (2012).
44. J. Dalibard, F. Gerbier, G. Juzeliūnas, and P. Öhberg, Artificial gauge potentials for neutral atoms. *Rev. Mod. Phys.*, **83**, 1523–1543 (2011).
45. N. Goldman, G. Juzeliūnas, P. Öhberg, and I. B. Spielman, Light-induced gauge fields for ultracold atoms. *Rep. Prog. Phys.*, **77**, 126401 (2014).
46. D. Jaksch and P. Zoller, Creation of effective magnetic fields in optical lattices: the Hofstadter butterfly for cold neutral atoms. *New J. Phys.* **5**, 56 (2003).
47. G. Juzeliūnas and P. Öhberg, Slow light in degenerate Fermi gases. *Phys. Rev. Lett.* **93**, 033602 (2004).
48. N. R. Cooper, Rapidly rotating atomic gases. *Adv. Phys.* **57**, 539–616 (2008).
49. A. L. Fetter, Rotating trapped Bose-Einstein condensates. *Rev. Mod. Phys.* **81**, 647–691 (2009).
50. X.-J. Liu, H. Jing, X. Liu, and M.-L. Ge, Generation of two-flavor vortex atom laser from a five-state medium. *Eur. Phys. J. D*, **37**, 261–265 (2006); quant-ph/0410096.
51. M. DeMarco and H. Pu, Angular spin-orbit coupling in cold atoms. *Phys. Rev. A*, **91**, 033630 (2015).
52. K. Sun, C. Qu, and C. Zhang, Spin orbital-angular-momentum coupling in Bose-Einstein condensates. *Phys. Rev. A*, **91**, 063627 (2015).
53. L. Chen, H. Pu, and Y. Zhang, Spin-orbit angular momentum coupling in a spin-1 Bose-Einstein condensate. *Phys. Rev. A*, **93**, 013629 (2016).
54. L. Jiang, Y. Xu, and C. Zhang, *Phys. Rev. A*, **94**, 043625 (2016).
55. X.-J. Liu, X. Liu, L. C. Kwek, and C. H. Oh, Optically induced spin-hall effect in atoms. *Phys. Rev. Lett.*, **98**, 026602 (2007).

56. S.-L. Zhu, H. Fu, C.-J. Wu, S.-C. Zhang, and L.-M. Duan, Spin hall effects for cold atoms in a light-induced Gauge potential. *Phys. Rev. Lett.*, **97**, 240401 (2006).

57. K. Osterloh, M. Baig, L. Santos, P. Zoller, and M. Lewenstein, Cold atoms in non-abelian Gauge potentials: From the hofstadter "moth" to lattice Gauge theory. *Phys. Rev. Lett.*, **95**, 010403 (2005).

58. J. Ruseckas, G. Juzeliūnas, P. Öhberg, and M. Fleischhauer, Non-abelian Gauge potentials for ultracold atoms with degenerate dark states. *Phys. Rev. Lett.* **95**, 010404 (2005).

59. X.-J. Liu, M. F. Borunda, X. Liu, and J. Sinova, Effect of induced spin-orbit coupling for atoms via laser fields. *Phys. Rev. Lett.* **102**, 046402 (2009).

60. J. Higbie and D. M. Stamper-Kurn, Periodically dressed Bose-Einstein condensate: a superfluid with an anisotropic and variable critical velocity. *Phys. Rev. Lett.*, **88**, 090401 (2002).

61. Y.-J. Lin, K. Jiménez-García, and I. B. Spielman, Spin-orbit-coupled Bose-Einstein condensates. *Nature*, **471**, 83–86 (2011).

62. R. A. Williams, L. J. LeBlanc, K. Jiménez-García, M. C. Beeler, A. R. Perry, W. D. Phillips, and I. B. Spielman, Synthetic partial waves in ultracold atomic collisions. *Science*, **335**, 314–317 (2012).

63. P. Wang, Z.-Q. Yu, Z. Fu, J. Miao, L. Huang, S. Chai, H. Zhai, and J. Zhang, Spin-orbit coupled degenerate fermi gases. *Phys. Rev. Lett.*, **109**, 095301 (2012).

64. L. W. Cheuk, A. T. Sommer, Z. Hadzibabic, T. Yefsah, W. S. Bakr, and M. W. Zwierlein, Spin-injection spectroscopy of a spin-orbit coupled Fermi gas. *Phys. Rev. Lett.*, **109**, 095302 (2012).

65. J.-Y. Zhang, S.-C. Ji, Z. Chen, L. Zhang, Z.-D. Du, B. Yan, G.-S. Pan, B. Zhao, Y. Deng, H. Zhai, S. Chen, and J.-W. Pan, Collective dipole oscillations of a spin-orbit coupled Bose–Einstein condensate. *Phys. Rev. Lett.*, **109**, 115301 (2012).

66. C. Qu, C. Hamner, M. Gong, C. Zhang, and P. Engels, Observation of Zitterbewegung in a spin-orbit-coupled Bose-Einstein condensate. *Phys. Rev. A*, **88**, 021604(R) (2013).

67. A. J. Olson, S.-J. Wang, R. J. Niffenegger, C.-H. Li, C. H. Greene, and Y. P. Chen, Tunable Landau-Zener transitions in a spin-orbit-coupled Bose-Einstein condensate. *Phys. Rev. A*, **90**, 013616 (2014).

68. Z. Fu, L. Huang, Z. Meng, P. Wang, L. Zhang, S. Zhang, H. Zhai, P. Zhang, and J. Zhang, Production of feshbach molecules induced by spin-orbit coupling in Fermi gases. *Nat. Phys.*, **10**, 110–115 (2014).

69. S.-C. Ji, J.-Y. Zhang, L. Zhang, Z.-D. Du, W. Zheng, Y.-J. Deng, H. Zhai, S. Chen, and J.-W. Pan, Experimental determination of the finite-temperature phase diagram of a spin-orbit coupled Bose gas. *Nat. Phys.*, **10**, 314–320 (2014).

70. S.-C. Ji, L. Zhang, X.-T. Xu, Z. Wu, Y. Deng, S. Chen, and J.-W. Pan, Softening of roton and phonon modes in a Bose-Einstein condensate with spin-orbit coupling. *Phys. Rev. Lett.*, **114**, 105301 (2015).

71. C. Hamner, Y. Zhang, M.-A. Khamehchi, M. J. Davis, and P. Engels, Spin-orbit-coupled Bose-Einstein condensates in a one-dimensional optical lattice. *Phys. Rev. Lett.*, **114**, 070401 (2015).
72. K. Jiménez-García, L. J. LeBlanc, R. A. Williams, M. C. Beeler, C. Qu, M. Gong, C. Zhang, and I. B. Spielman, Tunable spin-orbit coupling via strong driving in ultracold-atom systems. *Phys. Rev. Lett.* **114**, 125301 (2015).
73. N. Q. Burdick, Y. Tang, and B. L. Lev, Long-lived spin-orbit-coupled degenerate dipolar Fermi gas. *Phys. Rev. X*, **6**, 031022 (2016).
74. B. Song, C. He, S. Zhang, E. Hajiyev, W. Huang, X.-J. Liu, and G.-B. Jo, Spin-orbit-coupled two-electron Fermi gases of ytterbium atoms. *Phys. Rev. A*, **94**, 061604(R) (2016).
75. J. Li, W. Huang, B. Shteynas, S. Burchesky, F. Ç. Top, E. Su, J. Lee, A. O. Jamison, and W. Ketterle, Spin-orbit coupling and spin textures in optical superlattices. *Phys. Rev. Lett.*, **117**, 185301 (2016).
76. L. F. Livi, G. Cappellini, M. Diem, L. Franchi, C. Clivati, M. Frittelli, F. Levi, D. Calonico, J. Catani, M. Inguscio, and L. Fallani, Synthetic dimensions and spin-orbit coupling with an optical clock transition. *Phys. Rev. Lett.* **117**, 220401 (2016).
77. S. Kolkowitz, S. L. Bromley, T. Bothwell, M. L. Wall, G. E. Marti, A. P. Koller, X. Zhang, A. M. Rey, and J. Ye, Spin-orbit-coupled fermions in an optical lattice clock. *Nature*, **542**, 66–70 (2017).
78. C. Zhang, S. Tewari, R. M. Lutchyn, and S. Das Sarma, $p_x + ip_y$ superfluid from $s$-wave interactions of Fermionic cold atoms. *Phys. Rev. Lett.*, **101**, 160401 (2008).
79. M. Sato, Y. Takahashi, and S. Fujimoto, Non-abelian topological order in $s$-wave superfluids of ultracold Fermionic atoms. *Phys. Rev. Lett.*, **103**, 020401 (2009).
80. G. Juzeliūnas, J. Ruseckas, and Jean Dalibard, Generalized Rashba-Dresselhaus spin-orbit coupling for cold atoms. *Phys. Rev. A*, **81**, 053403 (2010).
81. D. L. Campbell, G. Juzeliūnas, and I. B. Spielman, Realistic Rashba and Dresselhaus spin-orbit coupling for neutral atoms. *Phys. Rev. A*, **84**, 025602 (2011).
82. J. D. Sau, R. Sensarma, S. Powell, I. B. Spielman, and S. Das Sarma, Chiral Rashba spin textures in ultracold Fermi gases. *Phys. Rev. B*, **83**, 140510(R) (2011).
83. B. M. Anderson, I. B. Spielman, and G. Juzeliūnas, Magnetically generated spin-orbit coupling for ultracold atoms. *Phys. Rev. Lett.*, **111**, 125301 (2013).
84. Z.-F. Xu, L. You, and M. Ueda, Atomic spin-orbit coupling synthesized with magnetic-field-gradient pulses. *Phys. Rev. A*, **87**, 063634 (2013).
85. X.-J. Liu, K. T. Law, and T. K. Ng, Realization of 2D spin-orbit interaction and exotic topological orders in cold atoms. *Phys. Rev. Lett.*, **112**, 086401 (2014); *ibid Phys. Rev. Lett.*, **113**, 059901 (2014).
86. B. M. Anderson, G. Juzeliūnas, V. M. Galitski, and I. B. Spielman, Synthetic 3D spin-orbit coupling. *Phys. Rev. Lett.*, **108**, 235301 (2013).

87. L. Huang, Z. Meng, P. Wang, P. Peng, S.-L. Zhang, L. Chen, D. Li, Q. Zhou, and J. Zhang, Experimental realization of two-dimensional synthetic spin-orbit coupling in ultracold Fermi gases. *Nature Physics*, **12**, 540–544 (2016).

88. Z. Wu, L. Zhang, W. Sun, X.-T. Xu, B.-Z. Wang, S.-C. Ji, Y. Deng, S. Chen, X.-J. Liu, and J.-W. Pan, Realization of two-dimensional spin-orbit coupling for Bose-Einstein condensates. *Science*, **354**, 83–88 (2016).

89. Y.-J. Lin, R. L. Compton, A. R. Perry, W. D. Phillips, J. V. Porto, and I. B. Spielman, Bose-Einstein condensate in a uniform light-induced vector potential. *Phys. Rev. Lett.*, **102**, 130401 (2009).

90. Y.-J. Lin, R. L. Compton, K. Jiménez-García, J. V. Porto, and I. B. Spielman, Synthetic magnetic fields for ultracold neutral atoms. *Nature*, **462**, 628–632 (2009).

91. Y.-J. Lin, R. L. Compton, K. Jiménez-García, W. D. Phillips, J. V. Porto, and I. B. Spielman, A synthetic electric force acting on neutral atoms. *Nature Physics* **7** 531–534 (2011).

92. H. Zhai, Degenerate quantum gases with spin-orbit coupling: a review. *Rep. Prog. Phys.* **78** 026001 (2015).

93. M. Aidelsburger, M. Atala, S. Nascimbène, S. Trotzky, Y.-A. Chen, and I. Bloch, Experimental realization of strong effective magnetic fields in an optical lattice. *Phys. Rev. Lett.*, **107**, 255301 (2011).

94. M. Aidelsburger, M. Atala, M. Lohse, J. T. Barreiro, B. Paredes, and I. Bloch, Realization of the Hofstadter Hamiltonian with ultracold atoms in optical lattices. *Phys. Rev. Lett.*, **111**, 185301 (2013).

95. H. Miyake, G. A. Siviloglou, C. J. Kennedy, W. C. Burton, and W. Ketterle, Realizing the Harper Hamiltonian with laser-assisted tunneling in optical lattices. *Phys. Rev. Lett.*, **111**, 185302 (2013).

96. M. V. Berry, Quantal phase factors accompanying adiabatic changes. *Proc. R. Soc. Lond. A*, **392**, 45–57 (1984).

97. F. Wilczek and A. Zee, Appearance of Gauge structure in simple dynamical systems. *Phys. Rev. Lett.*, **52**, 2111–2114 (1984).

98. X.-J. Liu, X. Liu, L.-C. Kwek, and C. H. Oh, Manipulating atomic states via optical orbital angular-momentum. *Front. Phys. China*, **3**, 113–125 (2008).

99. L. L. Wang, Q. Sun, W. M. Liu, G. Juzeliūnas, and A. C. Ji, Fulde-Ferrell-Larkin-Ovchinnikov state to topological superfluidity transition in bilayer spin-orbit-coupled degenerate Fermi gases. *Physical Review A*, **95**, 053628 (2017).

100. X.-J. Liu, Z.-X. Liu, and M. Cheng, Manipulating topological edge spins in a one-dimensional optical lattice. *Phys. Rev. Lett.*, **110**, 076401 (2013).

101. R. Grimm, M. Weidemüller, and Y. B. Ovchinnikov, Optical dipole traps for neutral atoms, *Adv. At. Mol. Opt. Phys.*, **42**, 95 (2000).

102. D. A. Steck, *Rubidium 87 D Line Data.* (available online, revision 2.1.5, 2015).

103. T. G. Tiecke, *Feshbach Resonances in Ultracold Mixtures of the Fermionic Quantum Gases* $^6$Li *and* $^{40}$K. (PhD thesis, University of Amsterdam, 2009).

104. W. Zheng, Z.-Q. Yu, X. Cui, and H. Zhai, Properties of Bose gases with the Raman-induced spin-orbit coupling. *J. Phys. B: At. Mol. Opt. Phys.*, **46** 134007 (2013).

105. Y. Li, L. P. Pitaevskii, and S. Stringari, Quantum tricriticality and phase transitions in spin-orbit coupled Bose-Einstein condensates. *Phys. Rev. Lett.*, **108**, 225301 (2012).

106. C. Wang, C. Gao, C.-M. Jian, and H. Zhai, Spin-orbit coupled spinor Bose-Einstein condensates. *Phys. Rev. Lett.*, **105**, 160403 (2010).

107. C.-J. Wu, I. Mondragon-Shem, and X.-F. Zhou, Unconventional states of bosons with the synthetic spin-orbit coupling. *Chin. Phys. Lett.*, **28**, 097102 (2011).

108. T.-L. Ho and S. Zhang, Bose-Einstein condensates with spin-orbit interaction. *Phys. Rev. Lett.*, **107**, 150403 (2011).

109. Y. Li, G. I. Martone, L. P. Pitaevskii, and S. Stringari, Superstripes and the excitation spectrum of a spin-orbit-coupled Bose-Einstein condensate. *Phys. Rev. Lett.*, **110**, 235302 (2013).

110. G. I. Martone, Y. Li, L. P. Pitaevskii, and S. Stringari, Anisotropic dynamics of a spin-orbit-coupled Bose-Einstein condensate. *Phys. Rev. A*, **86**, 063621 (2012).

111. J. P. Vyasanakere, S. Zhang, and V. B. Shenoy, BCS-BEC crossover induced by a synthetic non-abelian gauge field. *Phys. Rev. B*, **84**, 014512 (2011).

112. H. Hu, L. Jiang, X.-J. Liu, and H. Pu, Probing anisotropic superfluidity in atomic fermi gases with rashba spin-orbit coupling. *Phys. Rev. Lett.*, **107**, 195304 (2011).

113. Z.-Q. Yu and H. Zhai, Spin-orbit coupled Fermi gases across a Feshbach resonance. *Phys. Rev. Lett.*, **107**, 195305 (2011).

114. L. He and X.-G. Huang, BCS-BEC crossover in 2D Fermi gases with rashba spin-orbit coupling. *Phys. Rev. Lett.*, **108**, 145302 (2012).

115. R. A. Williams, M. C. Beeler, L. J. LeBlanc, K. Jiménez-García, and I. B. Spielman, Raman-induced interactions in a single- component Fermi gas near an *s*-wave Feshbach resonance. *Phys. Rev. Lett.*, **111**, 095301 (2013).

116. L. Huang, P. Wang, P. Peng, Z. Meng, L. Chen, P. Zhang, and J. Zhang, Dissociation of Feshbach molecules via spin-orbit coupling in ultracold Fermi gases. *Phys. Rev. A*, **91**, 041604(R) (2015).

117. N. Goldman, I. Satija, P. Nikolic, A. Bermudez, M. A. Martin-Delgado, M. Lewenstein, and I. B. Spielman, Realistic time-reversal invariant topological insulators with neutral atoms. *Phys. Rev. Lett.*, **105**, 255302 (2010).

118. S.-L. Zhu, L.-B. Shao, Z. D. Wang, and L.-M. Duan, Probing non-abelian statistics of majorana fermions in ultracold atomic superfluid. *Phys. Rev. Lett.*, **106**, 100404 (2011).

119. K. Seo, L. Han, and C. A. R. Sá de Melo, Emergence of majorana and dirac particles in ultracold fermions via tunable interactions, spin-orbit effects, and zeeman fields. *Phys. Rev. Lett.*, **109**, 105303 (2012).

120. X. Cui and W. Yi, Universal borromean binding in spin-orbit-coupled ultracold fermi gases. *Phys. Rev. X*, **4**, 031026 (2014).

121. Y. Xu and C. Zhang, Berezinskii-Kosterlitz-Thouless phase transition in 2D spin-orbit-coupled fulde-ferrell superfluids. *Phys. Rev. Lett.*, **114**, 110401 (2015).
122. Note that the original scheme proposed in Ref. 100 considered a red-detuned optical lattice, while the blue-detuned scheme, as introduced here, has been presented afterward by one of the authors (X.J.L) in lots of conferences and workshops since 2013. The blue-detuned optical Raman lattcie scheme was later proposed to realize 2D SO coupling[85]
123. A. Altland and M. R. Zirnbauer, Nonstandard symmetry classes in mesoscopic normal-superconducting hybrid structures. *Phys. Rev. B*, **55**, 1142–1161 (1997).
124. X. Zhou, J. S. Pan, Z. X. Liu, W. Zhang, W. Yi, G. Chen, and S. Jia, Symmetry-protected topological states for interacting fermions in Alkaline-Earth-like atoms. *Physical Review Letters*, **119**, 185701 (2017).
125. T. Giamarchi, *Quantum Physics in One Dimension* (Oxford University Press, New York, 2004).
126. J. J. He, J. Wu, T. P. Choy, X. J. Liu, Y. Tanaka, and K. T. Law, Correlated spin currents generated by resonant-crossed Andreev reflections in topological superconductors. *Nat. Comm.*, **5**, 3232 (2014).
127. J.-S. Pan, X.-J. Liu, W. Zhang, W. Yi, and G.-C. Guo, Topological superradiant states in a degenerate Fermi gas. *Phys. Rev. Lett.*, **115**, 045303 (2015).
128. D. Yu, J.-S. Pan, X.-J. Liu, W. Zhang, and W. Yi, Topological superradiant state in Fermi gases with cavity induced spin-orbit coupling. *Front. Phys.*, **13**, 136701 (2018).
129. H. Chen, X.-J. Liu, and X. C. Xie, Hidden nonsymmorphic symmetry in optical lattices with one-dimensional spin-orbit coupling. *Phys. Rev. A*, **93**, 053610 (2016).
130. L.-J. Lang, S.-L. Zhang, and Q. Zhou, Nodal Brillouin-zone boundary from folding a Chern insulator. *Phys. Rev. A*, **95**, 053615 (2017).
131. B. Song, L. Zhang, C. He, T. F. J. Poon, E. Hajiyev, S. Zhang, X.-J. Liu, and G.-B. Jo, Observation of a symmetry-protected topological phase with ultracold fermions. *Science Advances*, **4**, eeao4748 (2018) arXiv:1706.00768.
132. J. Wu, J. Liu, and X.-J. Liu, Topological spin texture in a quantum anomalous hall insulator. *Phys. Rev. Lett.*, **113**, 136403 (2014).
133. C.-Z. Chang *et al.*, Experimental observation of the quantum anomalous hall effect in a magnetic topological insulator. *Science*, **340**, 167–170 (2013).
134. X. J. Liu, X. Liu, C. Wu, and J. Sinova, Quantum anomalous Hall effect with cold atoms trapped in a square lattice. *Physical Review A*, **81**, 033622 (2010).
135. J. S. Douglas and K. Burnett, Imaging of quantum Hall states in ultracold atomic gases. *Physical Review A*, **84**, 053608 (2011).
136. A. Zamora, G. Szirmai, and M. Lewenstein, Layered quantum Hall insulators with ultracold atoms. *Physical Review A*, **84**, 053620 (2011).
137. N. Goldman, J. Beugnon, and F. Gerbier, Detecting chiral edge states in the Hofstadter optical lattice. *Physical Review Letters*, **108**, 255303 (2012).

138. M. Mancini, G. Pagano, G. Cappellini, L. Livi, M. Rider, J. Catani, C. Sias, P. Zoller, M. Inguscio, M. Dalmonte, and L. Fallani, Observation of chiral edge states with neutral fermions in synthetic Hall ribbons. *Science*, **349**, 1510–1513 (2015).
139. B. K. Stuhl, H.-I. Lu, L. M. Aycock, D. Genkina, and I. B. Spielman, Visualizing edge states with an atomic Bose gas in the quantum Hall regime. *Science*, **349**, 1514–1518 (2015).
140. E. Alba, X. Fernandez-Gonzalvo, J. Mur-Petit, J. K. Pachos, and J. J. Garcia-Ripoll, Seeing topological order in time-of-flight measurements. *Physical Review Letters*, **107**, 235301 (2011).
141. H. M. Price and N. R. Cooper, Mapping the Berry curvature from semiclassical dynamics in optical lattices. *Physical Review A*, **85**, 033620 (2012).
142. G. Jotzu, M. Messer, Rémi Desbuquois, M. Lebrat, T. Uehlinger, D. Greif, and T. Esslinger, Experimental realization of the topological Haldane model with ultracold fermions. *Nature*, **515**, 237–240 (2014).
143. L. Duca, T. Li, M. Reitter, I. Bloch, M. Schleier-Smith, and U. Schneider, An Aharonov-Bohm interferometer for determining Bloch band topology. *Science*, **347**, 288–292 (2015).
144. M. Aidelsburger, M. Lohse, C. Schweizer, M. Atala, J. T. Barreiro, S. Nascimbène, N. R. Cooper, I. Bloch, and N. Goldman, Measuring the Chern number of Hofstadter bands with ultracold bosonic atoms. *Nature Physics*, **11**, 162–166 (2015).
145. N. Fläschner, B. S. Rem, M. Tarnowski, D. Vogel, D.-S. Lühmann, K. Sengstock, and C. Weitenberg, Experimental reconstruction of the Berry curvature in a Floquet Bloch band. *Science*, **352**, 1091–1094 (2016).
146. X.-J. Liu, K. T. Law, T. K. Ng, and P. A. Lee, Detecting topological phases in cold atoms. *Phys. Rev. Lett.*, **111**, 120402 (2013).
147. L. Fu and C. L. Kane, Topological insulators with inversion symmetry. *Phys. Rev. B*, **76**, 045302 (2007).
148. J.-S. Pan, W. Zhang, W. Yi, and G.-C. Guo, *Phys. Rev. A*, **94**, 043619 (2016).
149. B.-Z. Wang, Y.-H. Lu, W. Sun, S. Chen, Y, Deng, and X.-J. Liu, Dirac-, Rashba-, and Weyl-type spin-orbit couplings: Toward experimental realization in ultracold atoms. *Phys. Rev. A*, **97**, 011605(R) (2018).
150. W. Sun, B.-Z. Wang, X.-T. Xu, C.-R. Yi, L. Zhang, Z. Wu, Y. Deng, X.-J. Liu, S. Chen, J.-W. Pan, Long-lived 2D spin-orbit coupled topological bose gas. arXiv:1710.00717.
151. E. Majorana, A symmetric theory of electrons and positrons. *Nuovo Cimento*, **5**, 171 (1937) [*Soryushiron Kenkyu Electronics*, **63**, 149–162 (1981).]
152. A. Y. Kitaev, Unpaired Majorana fermions in quantum wires. *Physics-Uspekhi*, **44**, 131 (2001).
153. N. Read and D. Green, Paired states of fermions in two dimensions with breaking of parity and time-reversal symmetries and the fractional quantum Hall effect. *Phys. Rev. B* **61**, 10267 (2000).

154. G. Moore and N. Read, Nonabelions in the fractional quantum Hall effect. *Nucl. Phys. B*, **360**, 362 (1991).
155. L. Fu and C. L. Kane, Superconducting proximity effect and Majorana fermions at the surface of a topological insulator. *Phys. Rev. Lett.*, **100**, 096407 (2008).
156. J. D. Sau, R. M. Lutchyn, S. Tewari, and S. Das Sarma, Generic new platform for topological quantum computation using semiconductor heterostructures. *Phys. Rev. Lett.* **104**, 040502 (2010).
157. R. M. Lutchyn, J. D. Sau, and S. Das Sarma, Majorana fermions and a topological phase transition in semiconductor-superconductor heterostructures. *Phys. Rev. Lett.*, **105**, 077001 (2010).
158. Y. Oreg, G. Refael, and F. von Oppen, Helical liquids and Majorana bound states in quantum wires. *Phys. Rev. Lett.*, **105**, 177002 (2010).
159. T.-P. Choy, J. M. Edge, A. R. Akhmerov, and C.W. J. Beenakker, Majorana fermions emerging from magnetic nanoparticles on a superconductor without spin-orbit coupling. *Phys. Rev. B*, **84**, 195442 (2011).
160. S. Nadj-Perge, I. K. Drozdov, B. A. Bernevig, and A. Yazdani, Proposal for realizing Majorana fermions in chains of magnetic atoms on a superconductor. *Phys. Rev. B*, **88**, 020407 (2013).
161. S.-L. Zhu, L. B. Shao, Z. D. Wang, and L.-M. Duan, Probing non-Abelian statistics of Majorana fermions in ultracold atomic superfluid. *Phys. Rev. Lett.*, **106**, 100404 (2011).
162. C. Qu, Z. Zheng, M. Gong, Y. Xu, L. Mao, X. Zou, G. Guo, and C. Zhang, Topological superfluids with finite-momentum pairing and Majorana fermions. *Nat. Commun.*, **4**, 2710 (2013).
163. W. Zhang and W. Yi, Topological Fulde Ferrell Larkin Ovchinnikov states in spin-orbit-coupled Fermi gases. *Nat. Commun.*, **4**, 2711 (2013).
164. Y. Cao, S.-H. Zou, X.-J. Liu, S. Yi, G.-L. Long, and H. Hu, Gapless topological Fulde-Ferrell superfluidity in spin-orbit coupled Fermi gases. *Phys. Rev. Lett.*, **113**, 115302 (2014).
165. C. Chin, R. Grimm, P. Julienne, and E. Tiesinga. Feshbach resonances in ultracold gases. *Reviews of Modern Physics*, **82**, 1225 (2010).
166. T.-F. J. Poon and X.-J. Liu, From a semimetal to a chiral Fulde-Ferrell superfluid. *Phys. Rev. B*, **97**, 020501(R) (2018).
167. P. Fulde and R. A. Ferrell, Superconductivity in a strong spin-exchange field. *Phys. Rev.*, **135**, A550 (1964).
168. A. I. Larkin and Y. N. Ovchinnikov, Inhomogeneous state of superconductors (Production of superconducting state in ferromagnet with Fermi surfaces, examining Green function). *Zh. Eksp. Teor. Fiz.*, **47**, 1136 (1964) [*Sov. Phys. JETP*, **20**, 762 (1965)].
169. S. M. Young and C. L. Kane, Dirac semimetals in two dimensions. *Phys. Rev. Lett.*, **115**, 126803 (2015).
170. C. Chan, L. Zhang, T. F. J. Poon, Y.-P. He, Y.-Q.Wang, and X.-J. Liu, Generic theory for Majorana zero modes in 2D superconductors. *Phys. Rev. Lett.*, **119**, 047001 (2017).
171. L. S. Pontryagin, *Mat. Sb.*, **9**, 331 (1941).

172. J. C. Y. Teo and C. L. Kane, *Phys. Rev. B*, **82**, 115120 (2010).
173. J. E. Moore, Y. Ran, and X.-G. Wen, *Phys. Rev. Lett.*, 101, 186805 (2008).
174. H. Kleinert, *Multivalued Fields in Condensed Matter, Electromagnetism, and Gravitation* (World Scientific, Singapore, 2008).
175. D. F. Agterberg and H. Tsunetsugu, Dislocations and vortices in pair-density-wave superconductors. *Nat. Phys.*, **4**, 639 (2008).
176. A.-C. Ji, W. M. Liu, J. L. Song, and F. Zhou, Dynamical creation of fractionalized vortices and vortex lattices. *Phys. Rev. Lett.*, **101**, 010402 (2008).
177. C. Chan and X. -J. Liu, Non-Abelian Majorana modes protected by an emergent second Chern number. *Physical Review Letters*, **118**, 207002 (2017).
178. Robert-Jan Slager, Andrej Mesaros, Vladimir Juricic, and Jan Zaanen, *Nature Physics*, **9**, 98 (2013).
179. Jorrit Kruthoff, Jan de Boer, Jasper van Wezel, Charles L. Kane, and Robert-Jan Slager, *Phys. Rev. X*, 7, 041069 (2017).
180. F. Setiawan, K. Sengupta, I. B. Spielman, and Jay D. Sau, *Phys. Rev. Lett.*, **115**, 190401 (2015).
181. Jami J. Kinnunen, Jildou E. Baarsma, Jani-Petri Martikainen and Päivi Törmä, *Rep. Prog. Phys.*, **81**, 046401 (2018).
182. Z. Yan, R. Bi, and Z. Wang, *Phys. Rev. Lett.*, **118**, 147003 (2017).

## Chapter 2

## Quasi-low Dimensional Fermi Gases with Spin-orbit Coupling

Ren Zhang*

*School of Science, Xi'an Jiaotong University,
Shanxi, 710049, China*

We review recent progress on the spin-orbit coupled Fermi gas in low-dimensions. To show the basic method of tackling such system, we present two examples of one- and two-dimensional fermi gas. We firstly solve a two-body problem for spin-orbit coupled Fermi gas confined in quasi-low dimensional trapping potential. By analyzing the two-body bound state, we find that the population of the excited states in the tightly-confined direction would be significant when the two-body binding energy becomes comparable to or exceeds the energy gap by confinement. To study the impact of these excited modes of the trapping potentials, one needs to construct an effective Hamiltonian in the form of a two-channel model, where the dressed molecules in the closed channel consist of the conventional Feshbach molecules as well as the excited state's occupation in the confined direction. The effective Hamiltonian provides a proper description for a quasi-low dimensional Fermi gas with spin-orbit coupling. In particular, the stability region of the quasi-two dimensional topological superfluid phase is increased. Confinement induced resonance in quasi-one dimensional spin-orbit coupled quantum gas is also investigated. High spin atoms and alkaline-earth atoms with spin-orbit coupling are briefly introduced as well.

### 2.1. Overview

Recently, the experimental realization of spin-orbit coupling (SOC) in cold atomic gas has stimulated intense theoretical interests on simulation various quantum phase in such systems, such as topological phase, exotic superfluid

---

*Correspondence addressed to: renzhang@xjtu.edu.cn.

phase. Meanwhile, the property of the low-dimensional system can be significantly different from that of the three-dimensional system. Furthermore, there exist some exact analytical solutions to the low-dimensional system, such as Bethe-ansatz, Bosonization, *et al.*, and some reliable numerical methods to tackle the low-dimensional system have been developed, such as DMRG, which make the low-dimensional system quite interesting. In this chapter, we focus on the low-dimensional spin-orbit coupling quantum gas.

In cold atomic gas, the low-dimension system is realized by manipulating the interaction between atom and light. The scalar potential felt by atoms in the presence of laser plays an important role to generate the trapping potential. To produce the two-dimensional system, we can enhance the laser strength in $z$-axis such that the rotational degree of freedom in the $z$-axis is frozen. The low energy physics of such system is same as that of the two-dimensional system. Then such system can be *viewed* as a two-dimensional system.[3,4] Similar approach can be applied to generate the one-dimensional system. However, it should be noticed that the low-dimensional system is *quasi*-low-dimensional, not really low-dimensional. To illustrate this statement, let us consider two atoms in a tube. As the distance between these two atoms is much larger the cross-section radius of the tube, these two atoms seems to be located on a line. However, when these two atoms collide, the distance between them is of the order of van der Waals radius which is much smaller than the cross-section radius and the interaction between atoms in a tube is same as that in 3D configuration. Thus we state that atom is in the quasi-one dimension. To capture the physics in quasi-low-dimensional system, we have to solve a confined system with 3D interaction, which is a tough task. Fortunately, what we are interested in is the low energy physics, which is dominated by the two-body interaction and enable us to develop an effective Hamiltonian to capture the low-energy physics.[5-7] If the bound state with large binding energy is involved, it has been shown that the occupation of trapping potential excited state can be significant, and the degrees of freedom along the confined dimension can lead to measurable many-body effects on the BEC side of the Feshbach resonance,[8-11,14] and the desired effective Hamiltonian should incorporate such effect. This effective Hamiltonian enables us to study the interplay between atom-atom interaction, spin-orbit coupling and dimensionality.

As we know, the density of state (DoS) of the low-dimensional system is quite different from that of the 3D system. At low energy limit, DoS is finite or diverge for a two-dimensional and one-dimensional system, respectively. SOC can also enhance the low energy DoS. To investigate the enhancement of DoS of the low-dimensional system, we can start with a two-body problem. It is found that enhancement of DoS facilitates the appearance of the bound state, which implies that paring of atoms can be easier and the transition temperature of superfluid would be higher. In 3D two-component fermi gas, Z.-Q. Yu *et al.* found that $T_c$ can be enhanced to a sizable fraction of Fermi temperature.[13,29] In two-dimensions, the similar effect has been discovered as well.[12,15] It is known that since the phase fluctuation the long-range superfluid order is destroyed in two dimensions at finite temperature. The phase transition in two-dimensions is the Berezinskii-Kosterlitz-Thouless (BKT) transition, which is characterized by the presence of unbinding of the topological defect excitation. The ground state properties and BKT transition temperature have been investigated.[1,12] It is found that in two-dimensional spin-orbit coupled fermi gas, the finite temperature is necessary for the observation of topological Majorana fermions. P. Zou *et al.* recently predicted the existence of peculiar Majorana solitons, which host two Majorana fermions and feature a phase jump of $\pi$ across the soliton in low-dimensional spin-orbit coupled fermi gas.[2] Low-dimensional spin-orbit coupled fermi gas has enabled us to study the topological phase, such as topological insulator (TI) or topological superfluid (TSF). Here we would like to figure out that the meaning of *topological* here is different from that in topological defect excitation. The topological superfluid is characterized by the nonzero Chern number band structure. The phase structure for the two-dimensional spin-orbit coupled fermi gas have obtained theoretically.[24–28] The typical phase diagram are shown in Fig. 2.1.1. Nonabelian topological order in spin-orbit coupled low-dimensional fermi gas

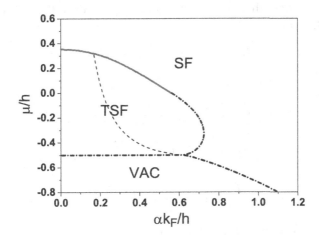

**Fig. 2.1.1.** The phase diagram of 2D spin-orbit coupled fermi gas with spin-orbit coupling.[26]

have been proposed.[35,36] Two-dimensional spin-orbit coupled fermi gas is sign-problem free, which enable us to simulate such system using Monte Carlo method. Equation of state, the momentum distributions, the pseudospin correlations, and the pair wave functions have been obtained accurately.[44]

## 2.2. Quasi-two Dimensional Fermi Gases with Spin-orbit Coupling

As we have mentioned in the above section, quasi-2D Fermi gas with spin-orbit coupling (SOC) and strong $s$-wave interaction is an ideal system for the realization of topological superfluid (TSF) state. In this section, we show the concrete procedure of deriving the effective Hamiltonian for the spin-orbit coupled Fermi gas. Firstly, we present how to solve the two-body problem with spin-orbit coupling. Illuminated by the two-body physics, we could construct the effective Hamiltonian which captures the low-energy physics of two-body process and available for investigating many-body effect. In particular, the stability region of the topological superfluid phase is increased in the presence of spin-orbit coupling, which is inspiring to explore topological phase experimentally.

### 2.A. *Two-body bound state in quasi-2D confinement*

Let us consider a spin-1/2 Fermi gas confined in a one-dimensional harmonic potential with trapping frequency $\omega_z$.[15] In the presence of SOC, the system can be described by a conventional two-channel field theory around a Feshbach resonance[16,17]

$$H = H_0 + H_{\text{soc}} + H_{\text{bf}} + H_{\text{int}}. \tag{2.2.1}$$

The terms in the Hamiltonian are

$$
\begin{aligned}
H_0 &= \sum_{\sigma=\uparrow,\downarrow} \int d^3 r \, \psi_\sigma^\dagger \left( -\frac{\hbar^2 \nabla^2}{2m_f} + \frac{1}{2} m_f \omega_z^2 z^2 \right) \psi_\sigma \\
&\quad + \int d^3 r \, \phi^\dagger \left( -\frac{\hbar^2 \nabla^2}{4m_f} + m_f \omega_z^2 z^2 + \bar{\nu}_b \right) \phi, \\
H_{\text{bf}} &= \bar{g}_b \int d^3 r \left( \phi^\dagger \psi_\downarrow \psi_\uparrow + \text{H.C.} \right) \\
H_{\text{int}} &= \bar{U}_b \int d^3 r \, \psi_\uparrow^\dagger \psi_\downarrow^\dagger \psi_\downarrow \psi_\uparrow,
\end{aligned}
\tag{2.2.2}
$$

where $\psi_\sigma(\boldsymbol{r})$ is the annihilate operator of the fermion located at $\boldsymbol{r}$ with spin index $\sigma = (\uparrow, \downarrow)$, $\phi(\boldsymbol{r})$ is annihilate operator of the bosonic molecular located at $\boldsymbol{r}$, $\bar{\nu}_b$ is the bare detuning between closed channel and open channel, $\bar{g}_b$ is the bare atom-molecule interaction strength, $\bar{U}_b$ is the bare background scattering amplitude, and H.C. denotes Hermitian conjugate. The relation between the bare scattering parameters (with subscript $p$) and the physical ones (with subscript $p$) are the standard renormalization relations[22]

$$U_c^{-1} = -\int \frac{d^3\boldsymbol{k}}{(2\pi^3)} \frac{1}{2\bar{\epsilon}_{\boldsymbol{k}}}, \quad \Gamma^{-1} = 1 + \frac{\bar{U}_p}{U_c},$$

$$\bar{U}_p = \Gamma^{-1}\bar{U}_b, \quad \bar{g}_p = \Gamma^{-1}\bar{g}_b, \quad \bar{\nu}_p = \bar{\nu}_b + \Gamma \frac{\bar{g}_p^2}{U_c}. \quad (2.2.3)$$

Here, $\bar{\epsilon}_{\boldsymbol{k}} = \hbar^2 k^2/(2m_f)$ is the dispersion relation for fermion with mass $m_f$ and 3D momentum $\boldsymbol{k}$, and the integral is taken in three dimensions with an explicit energy cutoff $E_c$, so $U_c^{-1} = \sqrt{E_c}/2^{3/2}\pi$. The physical parameters can be obtained from the scattering measurements with

$$\bar{U}_p = \frac{4\pi\hbar^2 a_{\mathrm{bg}}}{m_f}, \quad \bar{g}_p = \sqrt{\frac{4\pi\hbar^2 \mu_{\mathrm{co}} W |a_{\mathrm{bg}}|}{m_f}}, \quad \bar{\nu}_p = \mu_{\mathrm{co}}(B - B_0), \quad (2.2.4)$$

where $a_{\mathrm{bg}}$ is the 3D background scattering length, $W$ is the resonance width, $\mu_{\mathrm{co}}$ is the difference in magnetic moments between the closed and open channels, and $B_0$ is the resonance position. It is pointed out that this model is renormalizable in the sense that the physical result is independent on the artificial energy cutoff $E_c$.

The second term in the Hamiltonian Eq. (2.2.1) represents the Rashba type spin-orbit coupling (SOC),

$$H_{\mathrm{soc}} = -i\hbar\bar{\lambda} \int d^3\boldsymbol{r}\bar{\psi}^\dagger \left(\sigma_x \partial_x + \sigma_y \partial_y\right) \bar{\psi}. \quad (2.2.5)$$

The parameter $\bar{\lambda}$ denotes the SOC constant, $\sigma_{i=x,y}$ are the Pauli matrices, and $\bar{\psi} = (\psi_\uparrow, \psi_\downarrow)^T$. This SOC term spoils the conservation of spin such that spin is no longer a good quantum number. However, the single-particle Hamiltonian can be diagonalized in the helicity basis where each helix corresponds to particles with in-plane spin parallel or antiparallel to the in-plane momentum.

To discuss the two-body bound state, we notice that the center-of-mass (CoM) degree of freedom along the axial direction can be separated from the relative coordinate under a harmonic potential since it is not affected by the interaction nor the SOC. Thus, the CoM degree of freedom are assumed to be in the ground harmonic mode along the $z$-direction. In this CoM frame, the terms in Hamiltonian Eq. (2.2.1) can be rewritten in a second quantized form by expanding the field operators $\psi_\sigma$ and $\phi$ in terms of harmonic oscillators along the $z$-direction and plane waves in the $x$-$y$ plane, which leads to

$$H_0 = \sum_{m,\mathbf{k},\sigma} (\varepsilon_m + \epsilon_\mathbf{k}) c^\dagger_{m,\mathbf{k},\sigma} c_{m,\mathbf{k},\sigma} + \sum_{\ell,\mathbf{q}} (\nu_b + \varepsilon_\ell + \epsilon_\mathbf{q}/2) b^\dagger_{\ell,\mathbf{q}} b_{\ell,\mathbf{q}},$$

$$H_\mathrm{soc} = \lambda \sum_{m,\mathbf{k}} \left[ (k_x - i k_y) c^\dagger_{m,\mathbf{k},\uparrow} c_{m,\mathbf{k},\downarrow} + \mathrm{H.C.} \right],$$

$$H_\mathrm{bf} = g_b \sum_{m,n,\ell,\mathbf{k},\mathbf{q}} \gamma_{mn\ell} \left( b^\dagger_{\ell,\mathbf{q}} c_{m,-\mathbf{k}+\mathbf{q}/2,\downarrow} c_{n,\mathbf{k}+\mathbf{q}/2,\uparrow} + \mathrm{H.C.} \right)$$

$$H_\mathrm{int} = U_b \sum_{m,n,\mathbf{k},m',n',\mathbf{k}',\mathbf{q}} \gamma^{m'n'}_{mn} c^\dagger_{m,\mathbf{k}+\mathbf{q}/2,\uparrow} c^\dagger_{n,-\mathbf{k}+\mathbf{q}/2,\downarrow}$$

$$\times c_{n',-\mathbf{k}'+\mathbf{q}/2,\downarrow} c_{m',\mathbf{k}'+\mathbf{q}/2,\uparrow}. \tag{2.2.6}$$

Here, we have taken the energy unit as the $z$-direction trapping energy $\hbar\omega_z$, and the length unit as its characteristic length $a_t = \sqrt{\hbar/(m_f\omega_z)}$. The corresponding dimensionless parameters are then defined as $g_b = \bar{g}_b a_t^{-3/2}/(\hbar\omega_z)$, $U_b = \bar{U}_b a_t^{-3}/(\hbar\omega_z)$, $\nu_b = \bar{\nu}_b/(\hbar\omega_z)$, and $\lambda = \bar{\lambda} a_t^{-1}/(\hbar\omega_z)$. The bosonic field $b_{\ell,\mathbf{q}}$ represents the molecular state with axial harmonic mode $\ell$ and 2D transverse momentum $\mathbf{q}$, which is also dimensionless in the unit of $a_t^{-1}$. The fermionic field $c_{m,\mathbf{k},\sigma}$ represents the atomic state with axial harmonic mode $m$ and 2D momentum $\mathbf{k}$, and is characterized with axial mode energy $\varepsilon_m = m + 1/2$ and plane wave energy $\epsilon_\mathbf{k} = (k_x^2 + k_y^2)/2$ in the dimensionless form. The factors appearing in $H_\mathrm{bf}$ and $H_\mathrm{int}$ are defined as the overlap of harmonic oscillators

$$\gamma_{mn\ell} = 2^{1/4} \int dz \Phi_m(z) \Phi_n(z) \Phi_\ell(\sqrt{2}z),$$

$$\gamma^{m'n'}_{mn} = \int dz \Phi_m(z) \Phi_n(z) \Phi_{m'}(z) \Phi_{n'}(z) = \sum_\ell \gamma_{mn\ell} \gamma^*_{m'n'\ell}. \tag{2.2.7}$$

Here, $\Phi_n$ is the wavefunction of the $n^\mathrm{th}$ harmonic oscillator

$$\Phi_n(x) = \frac{e^{-x^2/2}}{\pi^{1/4}\sqrt{2^n n!}} H_n(x) \tag{2.2.8}$$

with $H_n(x)$ the $n^\mathrm{th}$ Hermite polynomial.

When treating the few-body problem, since the dimension of Hilbert space is finite, it is possible to write down the general wave function. Here, the general two-body state involving atoms and molecule can be written as the following ansatz

$$|\Psi\rangle_{\ell,\mathbf{q}} = \left( \beta_{\ell,\mathbf{q}} b^{\dagger}_{\ell,\mathbf{q}} + \sum_{m,n,\mathbf{k}}' \sum_{\sigma,\sigma'} \eta^{\sigma\sigma'}_{\ell,m,n,\mathbf{k},\mathbf{q}} c^{\dagger}_{m,\mathbf{k}+\mathbf{q}/2,\sigma} c^{\dagger}_{n,-\mathbf{k}+\mathbf{q}/2,\sigma'} \right) |0\rangle,$$

(2.2.9)

where $\sum'_{m,n,\mathbf{k}}$ indicates summation over mode $(m,n)$ and transverse momentum $\mathbf{k}$ for $k_y > 0$. The summation over spin runs over all four combinations of $(\sigma,\sigma')$. The coefficients $\beta_{\ell,\mathbf{q}}$ and $\eta^{\sigma\sigma'}_{\ell,m,n,\mathbf{k},\mathbf{q}}$ are determined by solving the Schrödinger's equation $H|\Psi\rangle_{\ell,\mathbf{q}} = E_{\ell,\mathbf{q}}|\Psi\rangle_{\ell,\mathbf{q}}$ under the normalization relation.

Under the quasi-2D condition, the $z$ confinement is much greater than the Fermi energy $E_F$ and temperature $k_B T$. In this case, the population of two-body states with nonzero CoM axial mode $\ell > 0$ is neglectable, and the system's properties will be dominated by two-body physics in the $\ell = 0$ subspace. As a consequence, we are allowed to focus on two-body states with zero axial mode $\ell = 0$, and obtain the equations determining the two-body bound state energy and the corresponding coefficients $\beta$ and $\eta$'s

$$S_p(E_{\mathbf{q}}) = \left[ U_p - \frac{g_p^2}{\nu_p + \epsilon_{\mathbf{q}}/2 - E_{\mathbf{q}}} \right]^{-1},$$

(2.2.10)

$$\beta_{\mathbf{q}} = \left[ 1 - Z_p^2(E_{\mathbf{q}}) \frac{\partial S_p(E_{\mathbf{q}})}{\partial E_{\mathbf{q}}} \right]^{-1/2},$$

(2.2.11)

$$|\eta^{\uparrow\uparrow}_{m,n,\mathbf{k},\mathbf{q}}|^2 = |\eta^{\downarrow\downarrow}_{m,n,\mathbf{k},\mathbf{q}}|^2$$

$$= \frac{\lambda^2 |\eta^{\mathrm{s}}_{m,n,\mathbf{k},\mathbf{q}}|^2}{\mathcal{E}^2_{m,n,\mathbf{k},\mathbf{q}}} \left[ \left( k_x - \frac{\lambda^2 q_y(-k_x q_y + k_y q_x)}{\mathcal{E}^2_{m,n,\mathbf{k},\mathbf{q}} - \lambda^2 q^2} \right)^2 + (x \leftrightarrow y) \right],$$

(2.2.12)

$$\eta^{\uparrow\downarrow}_{m,n,\mathbf{k},\mathbf{q}} = \frac{1}{2} \left( \eta^{\mathrm{s}}_{m,n,\mathbf{k},\mathbf{q}} + \eta^{\mathrm{t}}_{m,n,\mathbf{k},\mathbf{q}} \right),$$

(2.2.13)

$$\eta^{\downarrow\uparrow}_{m,n,\mathbf{k},\mathbf{q}} = \frac{1}{2} \left( \eta^{\mathrm{t}}_{m,n,\mathbf{k},\mathbf{q}} - \eta^{\mathrm{s}}_{m,n,\mathbf{k},\mathbf{q}} \right),$$

(2.2.14)

where the functions are defined as

$$S_p(E_{\mathbf{q}}) \equiv \sum_{m,n,\mathbf{k}} \gamma^2_{mn0} \left[ \mathcal{P}^{-1}_{m,n,\mathbf{k},\mathbf{q}} + \frac{1}{2\epsilon_{\mathbf{k}}} \right],$$

(2.2.15)

$$Z_p(E_\mathbf{q}) \equiv g_p - \frac{U_p}{g_p} \left(\nu_p + \epsilon_\mathbf{q}/2 - E_\mathbf{q}\right), \tag{2.2.16}$$

$$\mathcal{P}_{m,n,\mathbf{k},\mathbf{q}} \equiv \mathcal{E}_{m,n,\mathbf{k},\mathbf{q}} - \frac{4\lambda^2 k^2}{\mathcal{E}_{m,n,\mathbf{k},\mathbf{q}}} - \frac{4\lambda^4 \left(k_x q_y - k_y q_x\right)^2}{\mathcal{E}_{m,n,\mathbf{k},\mathbf{q}} \left(\mathcal{E}_{m,n,\mathbf{k},\mathbf{q}}^2 - \lambda^2 q^2\right)}, \tag{2.2.17}$$

$$\eta_{m,n,\mathbf{k},\mathbf{q}}^\mathrm{s} = \frac{2\gamma_{mn0} Z_p}{\mathcal{P}_{m,n,\mathbf{k},\mathbf{q}}} \beta_\mathbf{q}, \tag{2.2.18}$$

$$\eta_{m,n,\mathbf{k},\mathbf{q}}^\mathrm{t} = \frac{2i\lambda^2 \left(-k_x q_y + k_y q_x\right)}{\mathcal{E}_{m,n,\mathbf{k},\mathbf{q}} - \lambda^2 q^2} \eta_{m,n,\mathbf{k},\mathbf{q}}^\mathrm{s} \tag{2.2.19}$$

with $\mathcal{E}_{m,n,\mathbf{k},\mathbf{q}} = E_\mathbf{q} - k^2 - 1 - m - n - q^2/4$. Here, we drop the subscript $\ell = 0$ in $E_\mathbf{q}$, $\beta_\mathbf{q}$ and $\eta$'s to simplify notation. The dimensionless physical parameters are related to the bare ones via the same renormalization relations as in Eq. (2.2.3). Considering the parity of wave function, the coefficients $\gamma_{mn0} = 0$ for $m + n$ is odd, and

$$\gamma_{mn0} = \frac{(-1)^{(m-n)/2}}{(2\pi^3)^{1/4}\sqrt{m!n!}} \Gamma\left(\frac{m+n+1}{2}\right) \tag{2.2.20}$$

for $m + n$ is even, where $\Gamma(x)$ is the Euler Gamma function.

The energy of two-body bound state is determined by solving Eq. (2.2.10), and the corresponding eigenstate can be extracted from Eqs. (2.2.11–2.2.14). As typical examples, we focus on the s-wave wide Feshbach resonances around $B_0 = 202$ G for $^{40}$K and $B_0 = 834$ G for $^6$Li. The scattering parameters are taken to be the same as the cases without SOC, with $W = 8$ G, $a_{bg} = 174a_B$, $\mu_{\mathrm{co}} = 1.68\mu_B$ for $^{40}$K,[18–20] and $W = 300$ G, $a_{bg} = -1405a_B$, $\mu_{\mathrm{co}} = 2\mu_B$ for $^6$Li.[21] With a typical trapping frequency $\omega_z = 2\pi \times 62$ kHz,[3] the dimensionless physical parameters are then given by $g_p = 23$ (272), $U_p = 1.7$ (−5.5) for $^{40}$K ($^6$Li). We notice that in the presence of SOC, the open and closed channels involved in the Feshbach resonance will be altered since the spin conservation is spoiled. This modification can quantitatively change the resonance position as well as the scattering parameters. However, as long as we express the physical quantities in terms of the 3D scattering length $a_s$, the results would remain valid due to the requirement of universality near resonance.

We first consider two-body bound states with zero transverse CoM momentum $q = 0$. We show in Fig. 2.2.1 the eigenstate energy $E_{q=0}$ of two-body bound states for (a) $^{40}$K and (b) $^6$Li. For both cases, a bound state is always present around the resonance point. This result is a combined effect of axial confinement and Rashba SOC, which both drive the

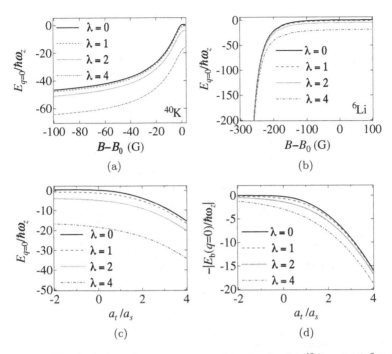

**Fig. 2.2.1.** Two-body bound state energy vs detuning for (a) $^{40}$K and (b) $^6$Li with transverse CoM momentum $q = 0$. (c) When plotted as functions of $a_t/a_s$, results for $^{40}$K and $^6$Li are almost indistinguishable, showing universal behavior around unitarity. (d) The presence of SOC tends to enhance the two-body binding energy, which is the difference between bound state energy and open-channel threshold.

system to an effective 2D model in the low-energy limit. The bound state energy can also be illustrated as a function of the 3D scattering length, as shown in Fig. 2.2.1(c). In this plot, results for $^{40}$K and $^6$Li are not distinguishable around the resonance point, indicating universal behavior at unitarity. This is consistent with case of absence of SOC.[8,23]

In order to discuss the SOC effect, we show in Fig. 2.2.1(d) the binding energy of the bound state as the difference between the eigenstate energy and the open channel threshold $-|E_b(q = 0)| = E_{q=0} - E_{th}$. In the presence of SOC, the open channel consists of two colliding particles residing on the ground state of the lower helicity band. Thus, the threshold $E_{th} = -\lambda^2 + 1/2$, where $1/2$ denotes the zero-point energy of the relative degree of freedom along the confined direction. From Fig. 2.2.1(d), we notice that SOC tends to enhance the two-body binding energy throughout the entire BCS-BEC crossover region. As we have mentioned in the above section, the

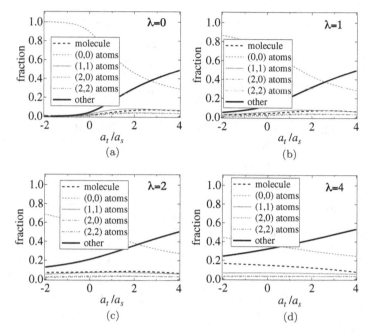

**Fig. 2.2.2.** Population fraction of the two-body bound state in different axial harmonic levels for $^{40}$K. Excited axial states are significantly populated around unitarity and on the BEC side of the resonance, as a direct consequence of the increasing two-body binding energy. The presence of SOC tends to enhance the binding energy, hence makes the population in excited states more prominent and may become notable even on the BCS side of the resonance.

underlying reason is SOC enhances the single particle density of states in the low-energy limit.

In Fig. 2.2.2, we show the molecular fraction $|\beta|^2$ and the population distribution in the axial harmonic levels within the two-body bound state for $^{40}$K. The population distribution in the axial $(m, n)$ mode are defined as $P_{m,n} = \sum_{\sigma\sigma'} \sum'_{\mathbf{k}} |\eta^{\sigma\sigma'}_{m,n,\mathbf{k},q=0}|^2$ which is a combination of all four spin states. From these results, it is clear that many axial excited states are occupied. The population fraction still goes down as the energy of the modes goes up, but the convergence is poor and there are so many excited harmonics that the total population in the excited levels becomes significant at unitarity and eventually dominates on the BEC side of the resonance. This result is a direct consequence of the two-body binding energy approaching or exceeding the axial confinement, which can populate fermions to axial excited states. As SOC tends to enhance the two-body binding energy, this effect will be more eminent with stronger SOC, such that the population in excited harmonics can become dominant even on the BCS of the resonance.

Up to now, we have discussed two-body bound state in a two-component Fermi gas with Rashba SOC confined in quasi two-dimensions. We find that the axial excited harmonic modes will be significantly occupied as the dimer's binding energy becomes comparable to or exceeds the axial confinement energy gap. Compared to the case without SOC, this effect is more significant in the presence of SOC as the two-body binding energy is increased due to the enhancement of low energy density of state. As a direct consequence, the population of axial excited modes can play an important role even in the BCS regime if the SOC strength is large enough. In this case, these higher excited states must be taken into account for a correct description of the underlying system. Next, we present a 2D effective model which incorporates the effect of axial excited states into a so-called *dressed molecule* state. This model can describe the quasi-2D, but truly 3D Fermi gas in the low-energy limit.

## 2.B. *Dressed molecule and effective 2D Hamiltonian*

The original quasi-2D Hamiltonian Eq. (2.2.6) contains three types of degree-of-freedom (DoF), including fermions in the axial ground state, fermions in axial excited states, and Feshbach molecules. The desired effective 2D Hamiltonian should be with respect to these three types of DoF. To this end, it is important to realize that these three types of DoF can be categorized into different length and energy scales. For fermions in the axial ground state, the corresponding length and energy scales are the inter-particle separation $d$ and the Fermi energy $E_F$, respectively. For axial excited state fermions, the corresponding length and energy scales are $a_t$ and $\hbar\omega_z$. The Feshbach molecular state is related to the short-range details of the atom-atom interaction potential, which has length scale $R_e$ and energy scale $V_e$. In ultra-cold Fermi gases with quasi-2D confinement, these three length and energy scales are usually well separated, satisfying $d \gg a_t \gg R_e$ and $E_F \ll \hbar\omega_z \ll V_e$. This observation is crucial for us to write down an effective Hamiltonian to describe physics in the long-wavelength and low-energy limit.

To capture the high-energy DoF including fermions in axial excited states, we combine them and the Feshbach molecule to define a *dressed* molecular state, which can be viewed as the Feshbach molecule dressed by excited fermions.[8,9] It is pointed that the dressed molecule is structure-less since all short-range details associated with excited fermions and the Feshbach molecule become irrelevant in the low-energy regime. With this dressed molecule and the fermions in the axial ground state, we write down an effective 2D Hamiltonian in the form of a two-channel model,

$$H_{\text{eff}} = \sum_{\mathbf{k},\sigma} \epsilon_{\mathbf{k}} a^{\dagger}_{\mathbf{k},\sigma} a_{\mathbf{k},\sigma} + \sum_{\mathbf{q}} (\delta_b + \epsilon_{\mathbf{q}}/2) \, d^{\dagger}_{\mathbf{q}} d_{\mathbf{q}}$$

$$+ \lambda' \sum_{\mathbf{k}} \left[ (k_x - ik_y) a^{\dagger}_{\mathbf{k},\uparrow} a_{\mathbf{k},\downarrow} + \text{H.C.} \right]$$

$$+ \alpha_b \sum_{\mathbf{k},\mathbf{q}} \left( d^{\dagger}_{\mathbf{q}} a_{\mathbf{k}+\mathbf{q}/2,\uparrow} a_{-\mathbf{k}+\mathbf{q}/2,\downarrow} + \text{H.C.} \right)$$

$$+ V_b \sum_{\mathbf{k},\mathbf{k}',\mathbf{q}} a^{\dagger}_{\mathbf{k}+\mathbf{q}/2,\uparrow} a^{\dagger}_{-\mathbf{k}+\mathbf{q}/2,\downarrow} a_{-\mathbf{k}'+\mathbf{q}/2,\downarrow} a_{\mathbf{k}'+\mathbf{q}/2,\uparrow}.$$

$$(2.2.21)$$

Here, $a^{\dagger}_{\mathbf{k},\sigma} (a_{\mathbf{k},\sigma})$ are fermionic creation (annihilation) operator in two dimensions, $\epsilon_{\mathbf{k}} = k^2/2$ is the corresponding dispersion relation, $d^{\dagger}_0$ and $d_0$ denote the dressed molecular operators, and $\lambda'$ is the Rashba SOC strength. The three bare scattering parameters $\delta_b$, $\alpha_b$, and $V_b$ can be linked to physical ones via a 2D renormalization relation

$$V_c^{-1} = - \int \frac{d^2\mathbf{k}}{(4\pi^2)} \frac{1}{2\epsilon_{\mathbf{k}} + 1}, \quad \Omega^{-1} = 1 + \frac{V_p}{V_c},$$

$$V_p = \Omega^{-1} V_b, \quad \alpha_p = \Omega^{-1} \alpha_b, \quad \delta_p = \delta_b + \Omega \frac{\alpha_p^2}{V_c}. \quad (2.2.22)$$

To determine the parameters in the effective Hamiltonian, we need to check the boundary condition that should be respected. First, we require that the effective theory needs to reproduce the single-particle dispersion in the low-energy limit. Second, when the system is off resonance which means the population of both the Feshbach molecule and excited fermions are negligible, the effective theory can describe a 2D system with all fermions staying in the axial ground state. This observation indicates that the background interaction $V_b$ in the effective 2D theory should be related to the 3D background interaction $U_b$ by integrating out the harmonic ground state wavefunction along the axial direction.

Across Feshbach resonance, we further require the effective 2D model to reproduce the correct two-body physics which dominate the three-body or four body physics. For the quasi-2D system, the ground state is a two-body bound state with axial CoM mode $\ell = 0$ and transverse CoM momentum $q = 0$, even in the BCS regime. Therefore, it is reasonable to require that the effective 2D theory should give the same two-body binding energy $|E_b(\ell = 0, q = 0)|$ as the original Hamiltonian. Besides, since the dressed molecule is a combination of the Feshbach molecule and axial excited fermions, we also require the population of dressed molecule to equal the population of Feshbach molecule plus that of the dimers formed with axial excited fermions.

The general two-body wave function within the effective 2D model can be written as

$$|\Phi\rangle_{\mathbf{q}} = \left( \beta d_{\mathbf{q}}^\dagger + \sum_{\mathbf{k}}{}' \sum_{\sigma,\sigma'=(\uparrow,\downarrow)} \eta_{\mathbf{k}}^{\sigma\sigma'} a_{\mathbf{k}+\mathbf{q}/2,\sigma}^\dagger a_{-\mathbf{k}+\mathbf{q}/2,\sigma'}^\dagger \right) |0\rangle, \tag{2.2.23}$$

where $\sum_{\mathbf{k}}'$ indicates summation over 2D momentum with $k_y > 0$. For the case of $q = 0$, the Schrödinger's equation $H_{\text{eff}}|\Phi\rangle_{q=0} = E_b|\Phi\rangle_{q=0}$ leads to

$$\left[ V_p - \frac{\alpha_p^2}{\delta_p - E_b} \right]^{-1} = \sqrt{2\pi}\sigma_p(E_b), \tag{2.2.24}$$

where the function $\sigma_p$ is defined as

$$\sigma_p(E_b) = \int \frac{d^2\mathbf{k}}{(2\pi)^{5/2}} \left[ \frac{1}{E_b - 2\epsilon_{\mathbf{k}} - \frac{4(\lambda')^2 k^2}{E_b - 2\epsilon_{\mathbf{k}}}} + \frac{1}{1 + 2\epsilon_{\mathbf{k}}} \right]$$
$$- b \frac{\pi + 2\tan^{-1}\left( \frac{b+2E_b}{\sqrt{-b(4E_b+b)}} \right)}{8\pi\sqrt{2\pi}\sqrt{-b(4E_b+b)}} + \frac{\ln(-E_b)}{4\pi\sqrt{2\pi}} \tag{2.2.25}$$

and $b \equiv 4(\lambda')^2$ is adopted to simplify notation.

By matching single- and two-body physics shown here with that given above, we can fix the parameters in the effective 2D Hamiltonian

$$\lambda' = \lambda,$$
$$V_p^{-1} = \sqrt{2\pi}\left( U_p^{-1} - C_p \right),$$
$$\delta_p = E_b - \frac{\sigma_p(E_b)}{\partial\mathcal{P}(E_b)/\partial E_b} \left( 1 - \frac{\sigma_p(E_b)}{U_p^{-1} - C_p} \right),$$
$$\alpha_p^2 = \frac{1}{\sqrt{2\pi}\partial\mathcal{P}(E_b)/\partial E_b} \left( 1 - \frac{\sigma_p(E_b)}{U_p^{-1} - C_p} \right)^2. \tag{2.2.26}$$

Here, the parameters are defined as

$$C_p = S_p(E_b^{\text{inf}} + 1/2) - \sigma_p(E_b^{\text{inf}}),$$
$$\mathcal{P}(E_b) = \left[ 1/U_p^{\text{eff}} - S_p(E_b + 1/2) + \sigma_p(E_b) \right],$$
$$U_p^{\text{eff}} = U_p - g_p^2/(\nu_p - E_b), \tag{2.2.27}$$

with $E_b^{\text{inf}}$ denoting the two-body binding energy in quasi two dimensions for $\nu_p \to \infty$. Notice that the eigenenergy of the two-body bound state as determined by Eq. (2.2.10) is shifted from the binding energy by the zero-point energy along the strongly confined axial direction.

Equations (2.2.26), together with the renormalization relation Eq. (2.2.22), determine the parameters in the effective 2D Hamiltonian (2.2.21) as functions of two-body binding energy $E_b$ In Fig. 2.2.3, we show the parameters $\delta_p$ and $\alpha_p$ for $^{40}$K and $^6$Li across resonance for various SOC strengths, using the same 3D parameters as in Fig. 2.2.1. Due to the difference in sign of their individual background interaction, these effective parameters behave qualitatively differently for $^{40}$K and $^6$Li. For future use, the effective interaction $V_p^{\text{eff}} \equiv V_p - \alpha_p^2/(\delta_p - E_b)$ is also shown.

Up to now, by matching single- and two-body physics with the original quasi-2D system, we have introduced an effective 2D Hamiltonian . This effective theory is derived by grouping the high-energy DoF of axial excited fermions and Feshbach molecules to define a dressed molecular state, while keeping the low-energy axial ground fermions to capture the correct low-energy physics of the corresponding many-body system.

### 2.C.  Many-body calculations with the effective two-channel model

To demonstrate the significance of the dress molecules, we now proceed with the many-body calculations using the effective two-channel model. For the sake of simplicity, we consider only the possibility of BCS superfluid and focus on the zero CoM momentum pairing state with $q = 0$. The dimensionless many-body Hamiltonian can be written as

$$H_{\text{eff}} - \mu \left( N_\uparrow + N_\downarrow \right) - h \left( N_\uparrow - N_\downarrow \right)$$
$$= \sum_{\mathbf{k},\sigma} (\epsilon_{\mathbf{k}} - \mu_\sigma) \, a_{\mathbf{k},\sigma}^\dagger a_{\mathbf{k},\sigma} + (\delta_b - 2\mu) d_0^\dagger d_0$$

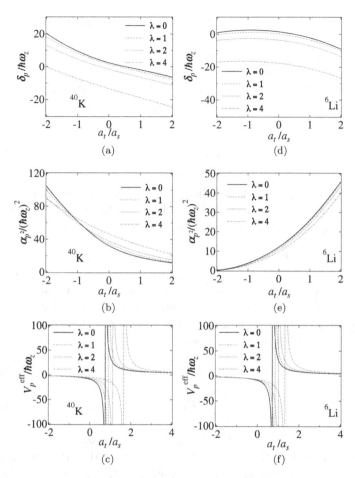

**Fig. 2.2.3.** Parameters in the effective 2D Hamiltonian as functions of $a_t/a_s$. The energy independent background scattering amplitude $V_p$ for different SOC constants are: 3.134 ($\lambda = 0$), 3.147 ($\lambda = 1$), 3.186 ($\lambda = 2$), 3.341 ($\lambda = 4$) for $^{40}$K, and $-1.801$ ($\lambda = 0$), $-1.828$ ($\lambda = 1$), $-1.914$ ($\lambda = 2$), $-2.155$ ($\lambda = 4$) for $^6$Li.

$$+\lambda \sum_{\mathbf{k}} \left[ (k_x - ik_y)a^\dagger_{\mathbf{k},\uparrow}a_{\mathbf{k},\downarrow} + (k_x + ik_y)a^\dagger_{\mathbf{k},\downarrow}a_{\mathbf{k},\uparrow} \right]$$

$$+\alpha_b \sum_{\mathbf{k}} \left( d^\dagger_0 a_{\mathbf{k},\uparrow}a_{-\mathbf{k},\downarrow} + \text{H.C.} \right) + V_b \sum_{\mathbf{k},\mathbf{k}'} a^\dagger_{\mathbf{k},\uparrow}a^\dagger_{-\mathbf{k},\downarrow}a_{-\mathbf{k}',\downarrow}a_{\mathbf{k}',\uparrow},$$

$$(2.2.28)$$

where $N_\uparrow$ ($N_\downarrow$) is the total number of particles in the corresponding spin state, and the bare parameters are fixed by the renormalization relation

Eq. (2.2.22) and Eqs. (2.2.26). The harmonic potential in the $x$-$y$ plane are assumed to be slow-varying enough. We take the local density approximation (LDA) in the transverse direction such that $\mu(\tilde{\mathbf{r}}) = \mu(0) - V(\tilde{\mathbf{r}})$, where $V(\tilde{\mathbf{r}})$ is the dimensionless external trapping potential, $\mu(\tilde{\mathbf{r}})$ is the local chemical potential, $h$ is the effective Zeeman field. Akin to the mean-field theory present in Ref.,[26] the zero-temperature thermodynamic potential can be written as

$$\Omega = -\frac{|\Delta|^2}{V_p^{\text{eff}}} + \frac{1}{2} \sum_{\mathbf{k},s=\pm} (\xi_s - E_{\mathbf{k},s}), \qquad (2.2.29)$$

where the order parameter for the two-channel model are defined as $\Delta = \alpha_p \langle d_0 \rangle + V_p \sum_{\mathbf{k}} \langle a_{-\mathbf{k},\downarrow} a_{\mathbf{k},\uparrow} \rangle$, and $V_p^{\text{eff}} \equiv V_p - \alpha_p^2/(\delta_p - E_b)$. The ground state energy for quasi-particles in the presence of SOC are given as[26]

$$E_{\mathbf{k},\pm} = \sqrt{\xi_{\mathbf{k}}^2 + \lambda^2 k^2 + |\Delta|^2 + h^2 \pm 2E_0}, \qquad (2.2.30)$$

where $E_0 = \sqrt{h^2 \left(\xi_{\mathbf{k}}^2 + |\Delta|^2\right) + \lambda^2 \xi_{\mathbf{k}}^2 k^2}$ and $\xi_{\mathbf{k}} = \epsilon_{\mathbf{k}} - \mu(\tilde{\mathbf{r}})$. Minimizing the thermodynamic potential, we can obtain the stable ground state.

The local chemical potentials can be determined self-consistently from the dimensionless number equations

$$\tilde{N} = \int d^2\tilde{\mathbf{r}}[\tilde{n}_\uparrow(\tilde{\mathbf{r}}) + \tilde{n}_\downarrow(\tilde{\mathbf{r}})], \qquad (2.2.31)$$

$$P = \frac{1}{\tilde{N}} \int d^2\tilde{\mathbf{r}}[\tilde{n}_\uparrow(\tilde{\mathbf{r}}) - \tilde{n}_\downarrow(\tilde{\mathbf{r}}))], \qquad (2.2.32)$$

where $\tilde{n}_\uparrow = -(\partial\Omega/\partial\mu + \partial\Omega/\partial h)/2$, $\tilde{n}_\downarrow = -(\partial\Omega/\partial\mu - \partial\Omega/\partial h)/2$, $\tilde{N} = N\omega_x\omega_y/\omega_z^2$, with $N$ the total particle number in the trap. Solving these equations self-consistently while minimizing the local thermodynamic potential, we get density distributions of the gas in a typical quasi-2D trapping potential. The population of the dressed molecules are given as[23]

$$\tilde{n}_b(\tilde{\mathbf{r}}) = \langle d_0^\dagger d_0 \rangle = |\Delta(\tilde{\mathbf{r}})|^2 \left[\alpha_p - \frac{U_p(\delta_p - 2\mu(\tilde{\mathbf{r}}))}{\alpha_p}\right]^{-2}. \qquad (2.2.33)$$

The order parameter characterizing the topological superfluid (TSF) state is the same as that of the superfluid (SF) state, since the two states are with respect to the same symmetry and are not separated by a spontaneous symmetry breaking. It has been shown that when the Zeeman field $h$ crosses $\sqrt{\mu^2 + \Delta^2}$ from below, an excitation gap closes and then opens up again, while the pairing order parameter remains finite.[26,31] The system then undergoes a topological phase transition from an SF state to a

TSF state,[31] where the winding number (Chern's number) becomes non-zero and Majorana zero modes can be stabilized at the center of vortex excitations.[31] Hence, the TSF phase can be identified in the trap where $h > \sqrt{\mu^2(\tilde{\mathbf{r}}) + |\Delta(\tilde{\mathbf{r}})|^2}$.[26,31]

We now investigate the typical density distributions of a polarized quasi-2D Fermi gas with SOC. To see the effect of dressed molecules in the closed channel, we compare results by the effective two-channel model with those by a single-channel model.[26] In the context of quasi-2D gases, the single-channel model can be understood as an approximation in which the population of all excited levels in the axial direction are neglected. One may then integrate out the lowest axial mode and relate the two-body binding energy $|E_b|$ appearing in the renormalization condition in the single-channel model to the 3D scattering length $a_s$ via $|E_b| \approx 0.915\pi^{-1} \exp\left(\sqrt{2\pi}a_t/a_s\right)\hbar\omega_z$.[5–7] In Fig. 2.2.4 we present the typical density distributions by these two models. It is clear that the dressed molecules affect properties of the trapped gas in two different ways. First, the density distribution can be significantly modified by the inclusion of dressed molecules as shown in Fig. 2.2.4(a). Second, in the presence of dressed molecules, the in-trap phase structure can be different dramatically. The first-order phase boundaries between the conventional superfluid phase (SF) at the center and the topological superfluid phase (TSF) at the edge are shifted. Notably, on the BCS side of the Feshbach resonance, the dressed molecules can qualitatively alter the in-trap phase structure by inducing an additional TSF phase which is absent in a single-channel calculation as shown in Fig. 2.2.4(b).

## 2.D. *Summary for this section*

In the present section, we have shown the low energy physics of a quasi-two-dimensional Fermi gas confined in a strong axial harmonic potential with Rashba spin-orbit coupling and population imbalance. We analyze the two-body bound state in such a system at first, and find that the axial excited harmonic levels can be significantly populated as the binding energy becomes comparable to or exceeds the axial confinement. The presence of Rashba spin-orbit coupling increases the density of states in the low-energy limit, hence causes an enhancement of two-body binding energy, and eventually excites more particles to the high-lying axial harmonic modes. This observation implies that the degrees of freedom along the third dimension would be crucial for a satisfactory description of the quasi-two-dimensional, but truly three-dimensional system.

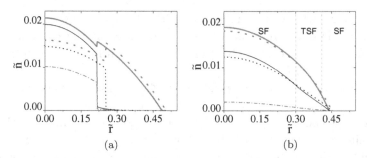

**Fig. 2.2.4.** Density distribution for spin-up (bold, red) and spin-down (thin, black) atoms in the single-channel (dotted) and the effective two-channel (solid) models, with parameters: (a) $a_t/a_s \sim 0.34$, $\lambda \sim 0.14$, $\omega_x/\omega_z = 10^{-3}$, $N \sim 10^4$, $P \sim 0.50$, $2N_b/N \sim 13\%$; (b) $a_t/a_s \sim -0.40$, $\lambda \sim 0.22$, $\omega_x/\omega_z = 10^{-3}$, $N \sim 10^4$, $P \sim 0.27$, $2N_b/N \sim 5\%$. For the effective two-channel calculation, the percentage of atoms in the closed channel $2N_b/N = 2\int d^2r n_b(\mathbf{r})/N$, where the density distribution of the closed channel atoms $2n_b$ are shown with the dash-dotted line (green). The first-order phase boundaries in (a) between the SF and TSF phases are indicated by the abrupt jump in the density distributions. The vertical dashed lines in (b) indicate the second-order phase boundaries between the SF and TSF phases in the two-channel model.

In order to incorporate the effects of these axial excited modes, we introduce a dressed molecule degree of freedom and construct an effective two-dimensional Hamiltonian in the form of a two-channel model. The parameters in this effective model are fixed by matching single- and two-body physics with the original Hamiltonian. Specifically, we require the effective Hamiltonian to give the correct single-particle dispersion, background scattering amplitude, two-body binding energy, and the fraction of Feshbach molecules plus dimers formed with axial excited fermions. Therefore, the effective model can mimic the low-energy physics as the original Hamiltonian provided that the chemical potential is not too far away from one half of the two-body binding energy. This condition usually holds for ultra-cold dilute Fermi gases in quasi-2D confinement.

Many-body properties of a quasi-2D Fermi gas using this effective Hamiltonian are studied as well. We conclude that the inclusion of dressed molecules is crucial for the investigation of the underlying system. Specifically, we find that the stability region of the topological superfluid phase is increased as a result of the occupation of axial excited modes. Results present in this section are helpful for the experimental search for the topological superfluid phase in ultra-cold Fermi gases, and have interesting implications for 3D or quasi-low-dimensional polarized Fermi gases in general.[30]

## 2.3. Quasi-one-dimensional Fermi Gases with Spin-orbit Coupling

In this section, we turn to investigate quasi-one dimensional fermi system with in-line SOC and transverse Zeeman field.[37] Similar to quasi-two dimensions, we solve the two-body problem at the outset. Then with the insight by the two-body physics, we construct the effective 1D Hamiltonians for different energy regimes. For energy below the threshold, we match solutions of two-body bound states and obtain the effective Hamiltonian in a two-channel model form. This 1D model is suitable for the investigation on pairing phases in quasi-1D Fermi gases with SOC. For energy around threshold, we match the zero-energy scattering amplitude and obtain the effective Hamiltonian in a single-channel form. We further analyze the effect of SOC and Zeeman field on the position of confinement-induced resonances (CIR),[38] which can be used to create 1D systems in the strongly-interacting limit. We note that with the present technology, such a system is experimentally achievable by implementing anisotropic traps or 2D optical lattices together with NIST-type SOC.

### 3.A. *Two-body problems with NIST-type SOC in quasi-1D trapping potential*

We consider two spin-1/2 fermionic atoms with one-dimensional SOC in a quasi-1D configuration. The free Hamiltonian of this two-body system is given by

$$\hat{H}_0 = -\frac{\hbar^2}{2m}\left(\nabla_{\mathbf{r}_1}^2 + \nabla_{\mathbf{r}_2}^2\right) + U(\mathbf{r}_1) + U(\mathbf{r}_2) + \hat{H}_{\text{SOC}}^1 + \hat{H}_{\text{SOC}}^2, \qquad (2.3.1)$$

where $m$ is the atomic mass, $\mathbf{r}_{j=1,2} \equiv (x_j, y_j, z_j)$ denotes the spatial position of the $j$-th atom, and $U(\mathbf{r}_j) \equiv m\omega^2(y_j^2 + z_j^2)/2$ is the quasi-1D confinement within the radial $y$-$z$ plane with trapping frequency $\omega$. The SOC term takes the form

$$\hat{H}_{\text{SOC}}^j = \lambda p_x^{(j)} \sigma_x^{(j)} + h\sigma_z^{(j)} + h_x \sigma_x^{(j)}, \qquad (2.3.2)$$

where $p_x^{(j)}$ is the linear momentum along the $x$-axis of the $j$-th atom, $\sigma_i^{(j)}$ are Pauli matrices, $\lambda$ is the intensity of in-line SOC, $h$ and $h_x$ are the effective Zeeman fields along the transverse and axial directions, respectively. Such type of SOC has been successfully implemented in Fermi systems using the NIST scheme, where the transverse Zeeman field $h$ and in-line field $h_x$ correspond to the Rabi frequency of Raman lasers $\Omega$ and the two-photon detuning $\delta$, respectively. It is pointed out that the Zeeman field along the transverse direction is necessary.

Due to the presence of SOC, the center-of-mass (CoM) and relative degrees of freedom along the SOC direction of the two-particle system is not separable. In fact, by representing the quantum state of the relative degree of freedom by a spinor wave function

$$|\psi(\mathbf{r})\rangle = \psi_{\uparrow\uparrow}(\mathbf{r})|\uparrow\rangle_1|\uparrow\rangle_2 + \psi_{\uparrow\downarrow}(\mathbf{r})|\uparrow\rangle_1|\downarrow\rangle_2$$
$$+ \psi_{\downarrow\uparrow}(\mathbf{r})|\downarrow\rangle_1|\uparrow\rangle_2 + \psi_{\downarrow\downarrow}(\mathbf{r})|\downarrow\rangle_1|\downarrow\rangle_2, \qquad (2.3.3)$$

the relative motion along the $x$-axis depends on the corresponding CoM momentum $Q_x$. Thus the Hamiltonian can be expressed as

$$\hat{H}_{\text{rel}} = -\sum_{s=x,y,z}\frac{\partial^2}{\partial s^2} + \frac{1}{4}\sum_{s=y,z}(s^2-1)$$
$$+ \lambda\sum_{j=1,2}\left[\frac{Q_x}{2} + (-1)^j k_x\right]\sigma_x^{(j)} + h\sum_{j=1,2}\sigma_z^{(j)}. \qquad (2.3.4)$$

where $k_x$ is the relative momentum along the $x$-axis. Here, we adopt the natural unit with $\hbar = m = 1$, and set $\omega = 1$ as the energy unit. We also drop the zero-point energy of the transverse motion along the $y$ and $z$ directions for simplicity, since the relative and CoM degrees of freedom remain separable along these axes. The interaction effect between the two particles can be analyzed by implementing the Bethe-Peierls boundary condition in three dimensions

$$\psi \propto (1/r - 1/a_s), \quad r \to 0. \qquad (2.3.5)$$

The validity of this boundary condition can be understood by noticing the separation of energy or length scales. In fact, The length scale associated with the inter-particle interacting potential $R_e$ is in the scale of $10^{-9}$m. As a comparison, the characteristic length scales associated with the transverse trap $a_\perp$ and the spin-orbit coupling $a_\lambda \equiv 1/\lambda$ are both in the scale of $10^{-6}$m, which are several orders of magnitude larger than $R_e$. Thus, the presence of a transverse trap and SOC will not alter the divergence behavior of scattering wave functions. For sufficiently low energy, i.e. $kR_e \ll 1$, where $k$ is the relative wave vector between two particles, the scattering is dominated by $s$-wave contribution, and the behavior of the wave function at the inter-particle distance $r \gg R_e$ is determined by the 3D $s$-wave scattering length $a_s$, which is related to the scattering phase shift as $\delta_s = -\arctan(ka_s)$. Hence, to obtain the correct expression for the wave function at distance

$r \gg R_e$, which is the regime we are interested in, we need to solve the Shrödinger equation for the relative motion Eq. (2.3.4) under the boundary condition Eq. (2.3.5).

To investigate the two-body scattering process in the presence of SOC, we first define the single-particle spin state for the $j$-th atom

$$(t_x \sigma_x^{(j)} + t_z \sigma_z^{(j)})|\alpha_j, \mathbf{t}\rangle \equiv \alpha_j |t| |\alpha, \mathbf{t}\rangle, \qquad (2.3.6)$$

where $\mathbf{t} = (t_x, t_z)$ is a two-dimensional vector in the $x$-$z$ plane, and $\alpha_j = \pm 1$ denotes the helicity index. From now on, we focus on the zero CoM momentum case with $Q_x = 0$ as a particular example, and note that the same procedure can be easily extended to cases with finite $Q_x$. The two-particle spin state can be defined as

$$|\alpha(k_x)\rangle = |\alpha_1, (\lambda k_x, h)\rangle_1 |\alpha_2, (-\lambda k_x, h)\rangle_2 \qquad (2.3.7)$$

with $\alpha \equiv (\alpha_1, \alpha_2)$ the shorthand notation. The incident wave function with the proper symmetry thus can be expressed in the following form

$$|\psi_c^{(0)}(\mathbf{r})\rangle = \frac{e^{ik_x x}}{2\sqrt{\pi}} \left[ |\alpha(k_x)\rangle - |\alpha'(-k_x)\rangle \right] \phi_0(y)\phi_0(z), \qquad (2.3.8)$$

where $\phi_0$ is the ground state of a one-dimensional harmonic oscillator, and $\alpha' \equiv (\alpha_2, \alpha_1)$. In the rest of this section, we denote the scattering channel by $c = (\alpha, k_x)$. The eigen-energy corresponding to the incident wave function can be obtained by a straightforward calculation, leading to

$$\varepsilon = \varepsilon_c + (m + n), \qquad (2.3.9)$$

where $\varepsilon_c = k_x^2 + (\alpha_1 + \alpha_2)\sqrt{\lambda^2 k_x^2 + h^2}$. One should notice that in the presence of SOC, the threshold energy is shifted from zero to a nonzero value

$$\varepsilon_{\text{th}} = \begin{cases} -2h, & \lambda^2 < h; \\ -\lambda^2 - h^2/\lambda^2, & \lambda^2 \geq h. \end{cases} \qquad (2.3.10)$$

As the scattering energy is low enough with $\varepsilon_c - \varepsilon_{\text{th}} \ll 1/R_e^2$, the wave function of the scattering state can be expressed as[39-41]

$$|\psi_c(\mathbf{r})\rangle \approx |\psi_c^{(0)}(\mathbf{r})\rangle + \frac{A(c)}{\phi_0(0)^2} G_0(\varepsilon_c; \mathbf{r}, \mathbf{0})|0, 0\rangle \qquad (2.3.11)$$

in the asymptotic region with $r \gtrsim R_e$. Here, $A(c)$ is a coefficient to be determined, $G_0(\varepsilon_b; \mathbf{r}, \mathbf{r}')$ is the Green's function associated with the free

Hamiltonian which describes the relative motion of the two colliding atoms

$$G_0(\mathcal{E}; \mathbf{r}, \mathbf{r}') = \frac{1}{\mathcal{E} + i0^+ - \hat{H}_{\text{rel}}} \delta(\mathbf{r} - \mathbf{r}'), \tag{2.3.12}$$

and $|0, 0\rangle$ represents a spin singlet state

$$|0, 0\rangle = \frac{1}{\sqrt{2}} (|\uparrow\rangle_1 |\downarrow\rangle_2 - |\downarrow\rangle_1 |\uparrow\rangle_2). \tag{2.3.13}$$

Similarly, when the energy $\varepsilon_b$ of the bound state is close enough to the threshold with $\varepsilon_{\text{th}} - \varepsilon_b \ll 1/R_e^2$, the wave function $|\psi_b(\mathbf{r})\rangle$ of the two-body bound state can be approximated as

$$|\psi_b(\mathbf{r})\rangle \approx B G_0(\varepsilon_b; \mathbf{r}, \mathbf{0})|0, 0\rangle \tag{2.3.14}$$

in the region of $r \gtrsim R_e$, where $B$ is the normalization constant. The coefficients $A(c)$ and $B$ in Eqs. (2.3.11) and (2.3.14) can be derived by implementing the Bethe-Peierls boundary condition (2.3.5). Specifically, by using the identity

$$\delta(\mathbf{r} - \mathbf{r}') = \int_{-\infty}^{\infty} dk_x \frac{e^{ik_x(x-x')}}{2\pi} \left( \sum_\alpha |\alpha(k_x)\rangle\langle\alpha(k_x)| \right)$$

$$\times \left( \sum_{m,n} \phi_m^*(y')\phi_m(y)\phi_n^*(z')\phi_n(z) \right) \tag{2.3.15}$$

and the Schrödinger equation

$$\hat{H}_{\text{rel}} e^{ik_x x} \phi_m(y)\phi_n(z)|\alpha(k_x)\rangle$$
$$= (\varepsilon_c + m + n) e^{ik_x x} \phi_m(y)\phi_n(z)|\alpha(k_x)\rangle \tag{2.3.16}$$

with $\phi_m$ the $m^{\text{th}}$ eigenstate of a one-dimensional harmonic oscillator, the behavior of the Green's function $G_0(\mathcal{E}; \mathbf{r}, \mathbf{0})$ at $r \to 0$ can be obtained as

$$\langle 0, 0|G_0(\mathcal{E}; \mathbf{r}, \mathbf{0})|0, 0\rangle = \langle 0, 0|g(\mathcal{E}; \mathbf{r}, \mathbf{0})|0, 0\rangle + \mathcal{S}(\mathcal{E}, \mathbf{r}). \tag{2.3.17}$$

The functions in the equation above are defined as

$$\langle 0, 0|g(\mathcal{E}; \mathbf{r}, \mathbf{0})|0, 0\rangle = \sum_{m,n} \int dk_x \frac{e^{ik_x x}}{2\pi} \frac{\phi_m^*(0)\phi_m(y)\phi_n^*(0)\phi_n(z)}{\mathcal{E} + i0^+ - (k_x^2 + m + n)}, \tag{2.3.18}$$

$$\mathcal{S}(\mathcal{E}, \mathbf{r}) = \sum_{\alpha',m,n} \int_{-\infty}^{\infty} dk_x' \frac{e^{ik_x' x}}{2\pi} |\langle 0, 0|\alpha'(k_x')\rangle|^2 \phi_m^*(0)\phi_m(y)\phi_n^*(0)\phi_n(z)$$

$$\times \left[ \frac{1}{\mathcal{E} + i0^+ - (\varepsilon_{c'} + m + n)} - \frac{1}{\mathcal{E} + i0^+ - (k_x'^2 + m + n)} \right]. \tag{2.3.19}$$

The two terms on the right-hand-side of Eq. (2.3.17) can be understood as follows. The first term is the Green's function associated with the relative Hamiltonian as SOC is absent, while the second term is the contribution induced by the SOC. By substituting Eqs. (2.3.17-2.3.18) into Eqs. (2.3.11) and (2.3.14), the scattering and bound states can be solved under the Bethe-Peierls boundary condition Eq. (2.3.5), respectively.

### 3.B. *Bound states and the two-channel effective Hamiltonian*

In this section, we focus on the energy regime below threshold and investigate the two-body bound state with eigen-energy $\varepsilon_b$. From Eq. (2.3.18), the behavior of $\langle 0, 0|g(\varepsilon_b; \mathbf{r}, \mathbf{0})|0, 0\rangle$ at vanishing $r$ reads

$$\lim_{r \to 0^+} \langle 0, 0|g(\varepsilon_b; \mathbf{r}, \mathbf{0})|0, 0\rangle = -\frac{1}{2\pi\sqrt{-\varepsilon_b}} - \mathcal{F}(\varepsilon_b), \qquad (2.3.20)$$

where $\mathcal{F}(\varepsilon_b)$ is defined as,

$$\mathcal{F}(\varepsilon_b) = \frac{1}{2^{3/2}\pi} \lim_{x \to 0^+} \sum_{s=1}^{\infty} \frac{e^{-\sqrt{s-\varepsilon_b/2}\sqrt{2x}}}{\sqrt{s - \varepsilon_b/2}}. \qquad (2.3.21)$$

Here, we set the value of $\phi_0(0)$ in Eqs. (2.3.18) and (2.3.18) to be real and positive, leading to the following expressions for $|\phi_{2m}(0)|^2 = \Gamma(m + 1/2)/[\pi\Gamma(m+1)]$. Besides, we also utilize the identity

$$\sum_{n=0}^{s} \frac{\Gamma(s-n+1/2)\Gamma(n+1/2)}{\Gamma(s-n+1)\Gamma(n+1)} = \pi, \quad \text{for } s = 1, 2, 3... \qquad (2.3.22)$$

to obtain the final form of Eq. (2.3.21).

The summation over $s$ in Eq. (2.3.21) includes contribution from all transverse excited states, and it cannot be interchanged with the limit of $x \to 0^+$ since the summation is not uniformly convergent. By using the relation

$$\int_0^{\infty} \frac{\exp(-\sqrt{s}\xi)}{\sqrt{s}} ds = \sum_{s=1}^{\infty} \int_{s-1}^{s} \frac{\exp(-\sqrt{s'}\xi)}{\sqrt{s'}} ds' = \frac{2}{\xi}, \qquad (2.3.23)$$

we can finally obtain the behavior of $\langle 0, 0|g(\varepsilon_b; \mathbf{r}, \mathbf{0})|0, 0\rangle$ at vanishing $r$

$$\lim_{r \to 0^+} \langle 0, 0|g(\varepsilon_b; \mathbf{r}, \mathbf{0})|0, 0\rangle = \frac{-1}{2\pi}\left[\frac{1}{r} - \mathcal{C} + \frac{\overline{\mathcal{L}}(-\varepsilon_b/2)}{\sqrt{2}} + \frac{1}{\sqrt{-\varepsilon_b}}\right], \qquad (2.3.24)$$

where the function $\overline{\mathcal{L}}(\varepsilon)$ is defined as

$$\overline{\mathcal{L}}(\varepsilon) = \sum_{n=1}^{\infty} (-1)^n \frac{\zeta(n+1/2)(2n-1)!!\varepsilon^n}{2^n n!} = \zeta\left[\frac{1}{2}, 1+\varepsilon\right] + \sqrt{2}\mathcal{C}$$

(2.3.25)

with $\mathcal{C} \approx 1.0326$ and $\zeta(s, a)$ is the Hurwitz Zeta function.

By using the identity Eq. (2.3.22), we can evaluate the contribution to the Green's function induced by the presence of SOC, leading to

$$\mathcal{S}_b(\varepsilon_b) \equiv \lim_{r \to 0^+} \mathcal{S}(\varepsilon_b, \mathbf{r}) = \frac{1}{2\pi} \sum_{s=0}^{\infty} \frac{b}{E_s b + d}$$
$$\times \left\{ \frac{\sqrt{2\beta - 2E_s - b}\,[2d - b(\beta - E_s)]}{4E_s b + b^2 + 4d} - \sqrt{-E_s} \right\},$$

(2.3.26)

where $\beta = \sqrt{E_s^2 - d}$, $E_s = \varepsilon_b - 2s$, $b = 4\lambda^2$ and $d = 4h^2$. By substituting Eqs. (2.3.24) and (2.3.26) into the bound state wave function Eq. (2.3.14) and the Bethe-Peierls boundary condition Eq. (2.3.5), one can easily reach

$$\frac{1}{a_s} = -\left( \frac{\zeta[1/2, 1 - \varepsilon_b/2]}{\sqrt{2}} + \frac{1}{\sqrt{-\varepsilon_b}} \right) + 2\pi \mathcal{S}_b(\varepsilon_b). \quad (2.3.27)$$

Notice that the effect of SOC is represented in the last term on the right-hand-side of the equation above. In fact, as the intensity of SOC goes to zero with $\lambda \to 0$, the function $\mathcal{S}_b$ vanishes and we retrieve the equation for two-body bound states in the absence of SOC.[38,42]

With the knowledge of the two-body bound states we have discussed above, next we derive the 1D effective Hamiltonian, which is valid within the energy regime around the two-body bound state energy $\varepsilon_b$. This effective Hamiltonian takes a two-channel model form, where the open channel describes fermions residing on the transverse ground state, and the closed channel is constructed by dressed molecules similar to the case of quasi-two dimensions. In the presence of SOC and effective Zeeman field, the effective Hamiltonian can be written as

$$\hat{H}_{\text{eff}}^{2c} = \hat{H}_{\text{eff}}^0 + \hat{H}_{\text{eff}}^{cc} + \hat{U}_{\text{eff}}, \quad (2.3.28)$$

where the free Hamiltonian and the atom-atom interaction are represented as,

$$\hat{H}_{\text{eff}}^0 = \sum_{k,\sigma} \epsilon_k a_{k\sigma}^\dagger a_{k\sigma} + \lambda' \sum_k k(a_{k\uparrow}^\dagger a_{k\downarrow} + a_{k\downarrow}^\dagger a_{k\uparrow})$$

$$+ h' \sum_k (a_{k\uparrow}^\dagger a_{k\uparrow} - a_{k\downarrow}^\dagger a_{k\downarrow}). \tag{2.3.29}$$

$$\hat{U}_{\text{eff}} = \frac{V_b}{L} \sum_{k,k'} a_{k\uparrow}^\dagger a_{-k\downarrow}^\dagger a_{-k'\downarrow} a_{k'\uparrow}. \tag{2.3.30}$$

Here, $a_{k\sigma}^\dagger$ ($a_{k\sigma}$) is the creation (annihilation) operator for an atom with 1D momentum $k$ and spin $\sigma$, and $L$ is the quantization length of the system. The threshold energy is chosen as $-1$ to match the zero-point energy of the two transverse directions in the original quasi-1D configuration. The term involving the closed channel can be written as

$$\hat{H}_{\text{eff}}^{\text{cc}} = \delta_b d_0^\dagger d_0 + \frac{\alpha_b}{L^{1/2}} \sum_k \left( a_{k\uparrow}^\dagger a_{-k\downarrow}^\dagger d_0 + \text{H.C.} \right), \tag{2.3.31}$$

where $d_0^\dagger$ and $d_0$ are the creation and annihilation operators for dressed molecules, respectively. Here, $\delta_b$ is the detuning between open and closed channels, $\alpha_b$ denotes the atom-molecule interaction strength.

The three bare parameters $V_b$, $\delta_b$ and $\alpha_b$ are related to the physical ones via a 1D renormalization relations

$$V_c^{-1} = -\int \frac{dk}{(4\pi)} \frac{1}{\epsilon_k + 1}, \quad \Omega^{-1} = 1 + \frac{V_p}{V_c}$$

$$V_p = \Omega^{-1} V_b, \quad \alpha_p = \Omega^{-1} \alpha_b, \quad \delta_p = \delta_b + \Omega \frac{\alpha_p^2}{V_c}. \tag{2.3.32}$$

Following the scheme shown in the above section, we can reach

$$\left[ V_b - \frac{\alpha_b^2}{\delta_b - \varepsilon_b} \right]^{-1} = 2\pi \sigma_p(\varepsilon_b), \tag{2.3.33}$$

where the function $\sigma(\varepsilon_b)$ takes the form,

$$\sigma(\varepsilon_b) = \frac{1}{2^{5/2}\pi} - \frac{\sqrt{\varepsilon_b^2 - d} - \varepsilon_b}{4\pi \sqrt{2\sqrt{\varepsilon_b^2 - d} - 2\varepsilon_b - b\sqrt{\varepsilon_b^2 - d}}}. \tag{2.3.34}$$

The physical parameters in the effective Hamiltonian can be fixed by matching single- and two-body physics with the original quasi-1D Hamiltonian. Specifically, we require the effective Hamiltonian to give the same single-particle dispersion, two-body bound state energy, and the population of atoms in the transverse ground state as the original model. Under these conditions, we get

$$\lambda' = \lambda,$$

$$h' = h,$$

$$V_p^{-1} = 2\pi(U_p^{-1} - C_p),$$

$$\delta_p = \varepsilon_b - \frac{\sigma_p(\varepsilon_b)}{\partial \mathcal{P}(\varepsilon_b)/\partial \varepsilon_b} \left[ 1 - \frac{\sigma_p(\varepsilon_b)}{U_p^{-1} - C_p} \right], \qquad (2.3.35)$$

$$\alpha_p^2 = \frac{1}{2\pi \partial \mathcal{P}(\varepsilon_b)/\partial \varepsilon_b} \left[ 1 - \frac{\sigma_p(\varepsilon_b)}{U_p^{-1} - C_p} \right]^2.$$

Here, the parameters are defined as

$$C_p = \mathcal{S}_p(\varepsilon_b^{\text{inf}}) - \sigma_p(\varepsilon_b^{\text{inf}}),$$

$$\mathcal{P}(\varepsilon_b) = 1/V_{\text{eff}} - \mathcal{S}_p(\varepsilon_b) + \sigma_p(\varepsilon_p), \qquad (2.3.36)$$

$$V_{\text{eff}} = U_p - \frac{g_p^2}{\nu_p - E},$$

and $\varepsilon_b^{\text{inf}}$ denotes the two-body bound state energy in quasi-one dimension for $\nu_p \to \infty$. In Fig. 2.3.1, we plot the parameters $\delta_p$, $\alpha_p$ and $V_{\text{eff}} \equiv V_p - \alpha_p^2/(\delta_p - \varepsilon_b)$ for $^{40}$K and $^6$Li by tuning through a wide Feshbach resonance. The scattering parameters are same as the quasi-two dimensions. The qualitative difference of $\delta_p$ and $\alpha_p$ is mainly due to the difference in sign of their individual background interaction.

### 3.C. Scattering states and the single-channel effective Hamiltonian

From now on, we focus on the energy regime above the threshold, and derive a 1D effective model by analyzing the two-body scattering state. In this case, since a two-body bound state is absent, the transverse trapping potential remains the largest energy scale such that the population of all transverse excited states are negligible as varying the 3D scattering length. Thus, these high energy degrees of freedom can be dropped out when analyzing low-energy physics, and a single-channel form is sufficient for a 1D effective Hamiltonian.

From Eq. (2.3.18), the behavior of $\langle 0, 0 | g(\varepsilon_c; \mathbf{r}, \mathbf{0}) | 0, 0 \rangle$ at vanishing $r$ associated with a two-body scattering state with energy $\varepsilon_c$ reads

$$\lim_{r \to 0^+} \langle 0, 0 | g(\varepsilon_c; \mathbf{r}, \mathbf{0}) | 0, 0 \rangle = \frac{1}{2\pi i \sqrt{\varepsilon_c}} - \mathcal{F}(\varepsilon_c), \qquad (2.3.37)$$

where the function $\mathcal{F}$ is defined as in Eq. (2.3.21). Following a similar approach as outlined in the previous section, we obtain the contribution to

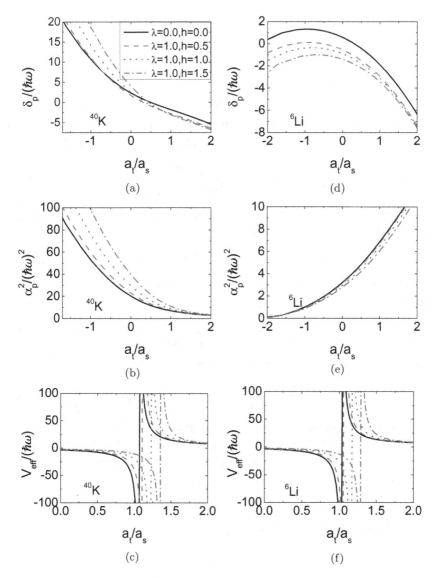

**Fig. 2.3.1.** Variation of parameters $\delta_p$, $\alpha_p$ and $V_{\text{eff}}$ in 1D effective Hamiltonian as functions of $a_t/a_s$.

the Green's function induced by the presence of SOC,

$$S_s(\varepsilon_c) = \frac{1}{2\pi^2} \sum_{s=0}^{\infty} \int_0^{\infty} dx \, \frac{b\sqrt{x}}{bx+d}$$

$$\times \left[ \frac{a_s + i0^+ - x}{(a_s + i0^+ - x)^2 - (bx+d)} - \frac{1}{a_s + i0^+ - x} \right]. \qquad (2.3.38)$$

By substituting Eqs. (2.3.37), (2.3.38), and (2.3.11) into the Bethe-Peierls boundary condition Eq. (2.3.5), we can determine the coefficient $A(c)$

$$A(c) = \frac{-2\pi^{3/2} \langle 0, 0 | \psi_c^{(0)}(\mathbf{0}) \rangle}{-i\varepsilon_c^{-1/2} - (a_s{}^{-1} - C) - \overline{\mathcal{L}}(-\varepsilon_c/2)/\sqrt{2} + 2\pi S_s(\varepsilon_c)} \qquad (2.3.39)$$

The scattering amplitude $f$ between the incident state $|\psi_c^{(0)}\rangle$ and the energy conserved output state $|\psi_{c'}^{(0)}\rangle$ is defined as

$$f(c' \leftarrow c) = -2\pi^2 \langle \psi_{c'}^{(0)}(\mathbf{0}) | 0, 0 \rangle A(c). \qquad (2.3.40)$$

In the low-energy limit $\varepsilon_c \to \varepsilon_{th}$, this 1D scattering amplitude can be expressed as

$$f(c' \leftarrow c) = -\frac{1}{1 + i\sqrt{\varepsilon_c} a_{1D}}, \qquad (2.3.41)$$

where the 1D scattering length takes the form

$$a_{1D} = -\frac{a_t^2}{a_s} \left\{ 1 - \frac{a_s}{a_t} \left[ 2\pi S_s(\varepsilon_c) - \frac{\zeta(1/2, 1 - \varepsilon_c/2)}{\sqrt{2}} \right] \right\}. \qquad (2.3.42)$$

Here, we have brought back the unit of length to give a complete expression.

To derive a 1D effective Hamiltonian, we notice that the 1D scattering length, and hence the solution of two-body scattering states, can be reproduced by considering a 1D pseudo-potential

$$U_{1D}(x) = g_{1D}\delta(x), \qquad (2.3.43)$$

where the coupling strength is

$$g_{1D}(x) = -\frac{\hbar^2}{ma_{1D}}. \qquad (2.3.44)$$

By further matching the single-particle dispersion, we obtain the 1D effective Hamiltonian in a single-channel model form,

$$\hat{H}_{\text{eff}}^{\text{sc}} = -\frac{\partial^2}{\partial x^2} + \lambda \sum_{j=1,2} \left[ \frac{qx}{2} + (-1)^j k_x \right] \sigma_x^{(j)} + h \sum_{j=1,2} \sigma_z^{(j)} + U_{1D}(x).$$

$$(2.3.45)$$

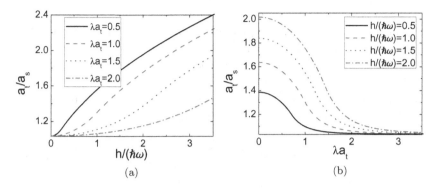

**Fig. 2.3.2.** Variation of CIR position $a_t/a_s$ as functions of (a) Zeeman field intensity $h$ and (b) SOC strength $\lambda$.

### 3.D. *Confinement-induced resonance*

The confinement-induced resonance is defined as a 1D resonance where the scattering amplitude $f = -1$, indicating a complete reflection with $a_{1D} = 0$. Thus, the position for the CIR to take place can be derived from Eq. (2.3.42), leading to

$$\frac{a_t}{a_s} = 2\pi \mathcal{S}_s(\varepsilon_{\text{th}}) - \frac{\zeta\left[1/2, 1 - \varepsilon_{\text{th}}/2\right]}{\sqrt{2}}. \qquad (2.3.46)$$

In Fig. 2.3.2, we show the position of CIR by varying the intensities of effective Zeeman field $h$ and SOC $\lambda$. For a given SOC strength, $a_t/a_s$ increases monotonically with the effective Zeeman field, as shown in Fig. 2.3.2(a). In the zero-field limit, since the SOC corresponds to an Abelian gauge field and can be dropped out from the problem via a unitary transformation, the position of CIR reduces to the known result of $a_t/a_s = \zeta(1/2, 1) \approx 1.0326$ for the case without SOC.[38] For a given effective Zeeman field, $a_t/a_s$ decreases monotonically with increasing SOC intensity, as shown in Fig. 2.3.2(b). Notice that in the large SOC limit, the effect of Zeeman field becomes negligible, such that the position of CIR takes the same value of $a_t/a_s = \zeta(1/2, 1)$ as in the zero-field limit. On the other hand, as the SOC strength tends zero, the position of CIR approaches to a limiting value, which is different from the result of the zero-SOC case, and depends on the intensity of the effective Zeeman field. In fact, since SOC mixes the two spin states, the two-body threshold energy will present an abrupt change in the presence of an infinitesimally small SOC, leading to a finite shift of the position where CIR takes place.

In the case without spin-orbit coupling, CIR can be understood as a Feshbach resonance which happens when the continuum threshold of the open channel, which corresponds to the transverse ground state, degenerates with the two-body bound state energy of the closed channel, which consists all transverse excited states.[42,43] In the presence of SOC, the two-body bound state energy $\varepsilon_e$ within the closed channel can be derived using the same approach as in Sec. 3.B, leading to

$$\frac{1}{a_s} = 2\pi \mathcal{S}_e(\varepsilon_e) - \frac{\zeta\left[1/2, 1 - \varepsilon_e/2\right]}{\sqrt{2}}, \qquad (2.3.47)$$

where the function $\mathcal{S}_e$ takes the form

$$\mathcal{S}_e(\varepsilon_e) = \frac{1}{2\pi} \sum_{s=1}^{\infty} \frac{b}{E_s b + d}$$

$$\times \left\{ \frac{\sqrt{2\gamma - 2E_s - b}\left[2d - b\left(\gamma - E_s\right)\right]}{4E_s b + b^2 + 4d} - \sqrt{-E_s} \right\}$$

$$(2.3.48)$$

with $\gamma = \sqrt{E_s^2 - d}$, $E_s = \varepsilon_e - 2s$, $b = 4\lambda^2$ and $d = 4h^2$. In Fig. 2.3.3, we plot the solution of $\varepsilon_e$ as a function of $a_t/a_s$ for a various combination of SOC intensity and the strength of the effective Zeeman field. From these results, we have confirmed that the crossing point of $\varepsilon_e$ and the open channel threshold $\varepsilon_{\text{th}}$ (denoted by dotted lines in Fig. 2.3.3) exactly coincides with

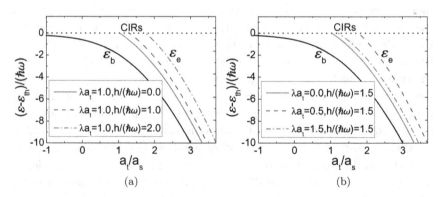

**Fig. 2.3.3.** Two-body bound state energies $\varepsilon_b$ and $\varepsilon_e$ as functions of $a_t/a_s$ by varying (a) the effective Zeeman field with a given SOC intensity and (b) the SOC intensity with a given effective Zeeman field. The crossing points of $\varepsilon_e$ and open channel threshold $\varepsilon_{\text{th}}$ (dotted) indicate the position of CIR to take place.

the position of CIR. This observation indicates that the understanding of CIR as a Feshbach resonance is still valid in the presence of SOC.

### 3.E. *Summary for this section*

In this section, we have shown the solution of the two-body problem of spin-1/2 fermionic atoms confined in a quasi-one-dimensional harmonic trap with NIST-type spin-orbit coupling, and derive one-dimensional effective Hamiltonians for all energy regimes. For energy regime close to the two-body bound state energy, since the transverse excited states can be significantly populated at unitarity of even on the BCS side of a Feshbach resonance, the 1D effective model takes a two-channel model form. The parameters in this effective Hamiltonian are fixed by matching single- and two-body physics, This effective Hamiltonian can be used to investigate pairing physics within such a system where the fermionic chemical potential is close to one half of the two-body bound state energy.

For energy slightly above the continuum threshold, the population of transverse excited states remains negligible such that these degrees of freedom can be safely ignored when discussing low-energy behavior of the system. Thus, we write down the 1D effective model in a single-model form. This model is useful when analyzing the scattering processes or low-energy physics on the upper repulsive branch of a Feshbach resonance. We also discuss the effect of SOC on the position of confinement-induced resonances, where the system undergoes a complete reflection. Furthermore, we show that in the presence of NIST-type SOC, CIR can still be understood as a Feshbach resonance where the continuum threshold of the transverse ground state degenerates with the bound state energy of the closed channel consisting of all transverse excited states. Considering the experimental realization of quasi-one-dimensionality and SOC in fermionic systems, the shift of CIR position induced by SOC can be detected using present technology at attainable temperatures.

### 2.4. Outlook

So far, spin-orbit coupled cold atomic gas is not restricted to alkali atoms anymore. SOC has also been realized in high spin atoms (Dy).[45] Dysprosium is characterized by large electronic orbital angular momentum and large magnetic moment which leads to the long range, anisotropic dipole-dipole interaction. The realization of spin-orbit coupled dysprosium gas enables

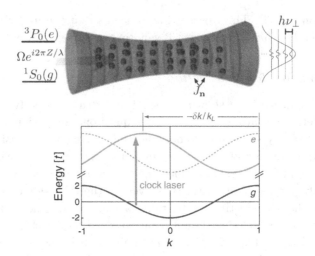

**Fig. 2.4.1.** (Top): Experimental setup for the realization of spin-orbit coupling in alkaline-earth atoms. A clock laser along the Z direction couples the ground state $^1S_0$ and clock state $^3P_0$ of alkaline-earth atoms trapped in an optical lattice with magic wavelength $\lambda_m$. (Bottom): Band structure for ground state(blue line) and clock state (green dashed line) without spin-orbit coupling. Shifting the band of clock state (green solid line) by $-\delta k$ represent the clock transition between the two states with a momentum-dependent energy.

people to study the interplay between long-range, anisotropic interaction, SOC and dimensionality.

Another interesting system is spin-orbit coupled alkaline-earth (like)(Sr,Yb) atomic gas.[46–48] Besides the ground state $^1S_0$, alkaline-earth (like) atoms have a long-lived excited state $^3P_0$ which also named as clock state, and the energy gap between clock $^3P_0$ and ground state $^1S_0$ is of the order of electronic volt. Ground state $^1S_0$ and clock state $^3P_0$ can be coupled by a clock laser, meanwhile the momentum of clock laser can be transferred to atom, which, in essence, is spin-orbit coupling. Here spin denotes the internal state of atom $^1S_0$ and $^3P_0$ and orbit denotes the spatial motion of atoms. In 1D optical lattice as shown in Fig.2.4.1, the Hamiltonian can be written as

$$H = -\sum_{n,q} \mathbf{B}_{nq} \cdot \hat{\mathbf{S}}_{nq} \tag{2.4.1}$$

where $\mathbf{n}$ denotes the transverse mode, $q$ is the quasi momentum, $\mathbf{B}_{nq} = [\Omega_n, 0, \Delta E_n(q, \phi)]$, $\Delta E_n(q, \phi) = 2J_n(\cos(q) - \cos(q + \phi))$ with $\Omega_n$ the Rabi frequency and $\phi$ the phase difference between adjacent lattice and $J_n$ the tunneling rate. The realized dispersion is shown in Fig.2.4.1, which shows the effective magnetic field felt by spin is momentum dependent, i.e., spin and orbit are coupled. We would like to stress that compared with the Raman scheme in alkali atoms, the advantage of such scheme to realize spin-orbit coupling is that benefiting from the ultranarrow linewidth of the clock state, the heating problem can be circumvented, which is favorable to the exploration of many-body effects. The interaction between clock state and ground state can be tuned by the recently realized orbital Feshbach resonance (OFR).[49–51] Since the nuclear spin is decoupled from the electronic orbital momentum and spin momentum, the interaction between alkaline earth atoms is respect to the SU($N$) symmetry. Here $N$ refers to the component number of nuclear spin. These ingredients may provide us flexible scheme to simulate exotic phase in low-dimensional alkaline-earth atoms. If we assume the $^3P_0$ as an impurity and $^1S_0$ as the itinerant electron, we can simulate the Kondo problem with spin-orbit coupling.[52,53]

# References

1. M. Gong, G. Chen, S. Jia, and C. Zhang. *Phys. Rev. Lett.*, **109**, 105302 (2012).
2. P. Zou, J. Brand, X.-J. Liu, and Hui Hu. *Phys. Rev. Lett.*, **117**, 225302 (2016).
3. M. Köhl, H. Moritz, T. Stoferle, K. Gunter, T. Esslinger. *Phys. Rev. Lett.*, **94**, 080403 (2005).
4. P. Dyke, E. D. Kuhnle, S. Whitlock, H. Hu, M. Mark, S. Hoinka, M. Lingham, P. Hannaford, C. J. Vale. *Phys. Rev. Lett.*, **106**, 105304 (2011).
5. M. Randeria, J. M. Duan, L. Y. Shieh. *Phys. Rev. B*, **41**, 327 (1990).
6. D. S. Petrov, G. V. Shlyapnikov. *Phys. Rev. A*, **64**, 012706 (2001).
7. J. Tempere, M. Wouters, J. T. Devreese. *Phys. Rev. B*, **75**, 184526 (2007).
8. J. P. Kestner, L.-M. Duan. *Phys. Rev. A*, **74**, 053606 (2006).
9. J. P. Kestner, L.-M. Duan. *Phys. Rev. A*, **76**, 063610 (2007).
10. W. Zhang, G.-D. Lin, L.-M. Duan. *Phys. Rev. A*, **77**, 063613 (2008).
11. W. Zhang, G.-D. Lin, L.-M. Duan. *Phys. Rev. A*, **78**, 043617 (2008).
12. L. He and X. -G. Huang. *Phys. Rev. Lett.*, **108**, 145302 (2012).
13. Z.-Q. Yu and H. Zhai. *Phys. Rev. Lett.*, **107**, 195305 (2011).
14. V. Pietilä, D. Pekker, Y. Nishida, E. Demler. *Phys. Rev. A*, **85**, 023621 (2012).
15. R. Zhang, F. Wu, J. Tang, G. Guo, W. Yi, and W. Zhang. *Phys. Rev. A*, **87** 033629 (2013).
16. M. Holland, S. J. J. M. F. Kokkelmans, M. L. Chiofalo, R. Walser. *Phys. Rev. Lett.*, **87**, 120406 (2001).
17. E. Timmermans, P. Tommasini, M. Hussein, A. Kerman. *Phys. Rep.*, **315**, 199 (1999).

18. C. A. Regal, M. Greiner, D. S. Jin. *Phys. Rev. Lett.*, **92**, 040403 (2004).
19. M. W. Zwierlein, C. A. Stan, C. H. Schunck, S. M. F. Raupach, A. J. Kerman, W. Ketterle. *Phys. Rev. Lett.*, **92**, 120403 (2004).
20. C. Chin, M. Bartenstein, A. Altmeyer, S. Riedl, S. Jochim, J. Hecker Denschlag, R. Grimm. *Science*, **305**, 1128 (2004).
21. M. Bartenstein, A. Altmeyer, S. Riedl, R. Geursen, S. Jochim, C. Chin, J. Hecker Denschlag, R. Grimm, A. Simoni, E. Tiesinga, C. J. Williams, P. S. Julienne. *Phys. Rev. Lett.*, **94**, 103201 (2005).
22. Q. Chen, J. Stajic, S. Tan, K. Levin. *Phys. Rep.*, **412**, 1 (2005).
23. W. Yi, L.-M. Duan. *Phys. Rev. A*, **73**, 063607 (2006).
24. M. Iskin, A. L. Subasi. *Phys. Rev. Lett.*, **107**, 050402 (2011).
25. W. Yi, G.-C. Guo. *Phys. Rev. A*, **84**, 031608(R) (2011).
26. J. Zhou, W. Zhang, W. Yi. *Phys. Rev. A*, **84**, 063603 (2011).
27. X. Yang and S. Wan. *Phys. Rev. A*, **85**, 023233 (2012).
28. L. Han, C. A. R. Sá de Melo. *Phys. Rev. A*, **85**, 011606(R) (2012).
29. J. P. Vyasanakere and V. B. S. *Phys. Rev. B*, **83**, 094515 (2011).
30. F. Wu, R. Zhang, T.-S. Deng, W. Zhang, W. Yi, and G.-C. Guo. *Phys. Rev. A*, **89**, 063610 (2014)
31. J. D. Sau, R. M. Lutchyn, S. Tewari, S. Das Sarma. *Phys. Rev. Lett.*, **104**, 040502 (2010).
32. P. Fulde, R. A. Ferrell. *Phys. Rev.*, **135**, A550 (1964).
33. A. I. Larkin, Y. N. Ovchinnikov. *Sov. Phys. JETP*, **20**, 762 (1965).
34. W. V. Liu, F. Wilczek. *Phys. Rev. Lett.*, **90**, 047002 (2003).
35. C. Zhang, S. Tewari, R. M. Lutchyn, S. Das Sarma. *Phys. Rev. Lett.*, **101**, 160401 (2008).
36. M. Sato, Y. Takahashi, S. Fujimoto. *Phys. Rev. Lett.*, **103**, 020401 (2009).
37. R. Zhang and W. Zhang. *Phys. Rev. A*, **88**, 053605 (2013).
38. M. Olshanii. *Phys, Rev. Lett.*, **81**, 938 (1998).
39. P. Zhang, L. Zhang, and W. Zhang. *Phys. Rev. A*, **86**, 042707 (2012).
40. J. R. Taylor. *Scattering Theory* (Wiley, New York, 1972).
41. D. S. Petrov and G. V. Shlyapnikov. *Phys. Rev. A*, **64**, 012706 (2001).
42. W. Zhang, P. Zhang. *Phys. Rev. A*, **83**, 053615 (2011).
43. T. Bergeman, M. G. Moore, and M. Olshanii. *Phys. Rev. Lett.*, **91**, 163201 (2003).
44. H. Shi, P. Rosenberg, S. Chiesa, and S. Zhang. *Phys. Rev. Lett.*, **117**, 040401 (2016).
45. N. Q. Burdick, Y. Tang, and B. L. Lev. *Phys. Rev. X*, **6**, 031022 (2016).
46. M. L. Wall, A. P. Koller, S. Li, X. Zhang, N. R. Cooper, J. Ye, and A. M. Rey. *Phys. Rev. Lett.*, **116**, 035301 (2016).
47. L. F. Livi, G. Cappellini, M. Diem, L. Franchi, C. Clivati, M. Frittelli, F. Levi, D. Calonico, J. Catani, M. Inguscio, and L. Fallani. *Phys. Rev. Lett.*, **117**, 220401 (2016).
48. S. Kolkowitz, S. L. Bromley, T. Bothwell, M. L. Wall, G. E. Marti, A. P. Koller, X. Zhang, A. M. Rey, and J. Ye. *Nature*, **542**, 66 (2017).
49. R. Zhang, Y. Cheng, H. Zhai, and P. Zhang. *Phys. Rev. Lett.*, **115**, 135301 (2015).

50. M. Höfer, L. Riegger, F. Scazza, C. Hofrichter, D. R. Fernandes, M. M. Parish, J. Levinsen, I. Bloch, and S. Fölling. *Phys. Rev. Lett.*, **115**, 265302 (2015).
51. G. Pagano, M. Mancini, G. Cappellini, L. Livi, C. Sias, J. Catani, M. Inguscio, and L. Fallani. *Phys. Rev. Lett.*, **115**, 265301 (2015).
52. R. Zhang, D. Zhang, Y. Cheng, W. Chen, P. Zhang, and H. Zhai. *Phys. Rev. A*, **93**, 043601 (2016).
53. L. Isaev, J. Schachenmayer, and A. M. Rey. *Phys. Rev. Lett.*, **117**, 135302 (2016).

# Chapter 3

# Rashba-spin-orbit Coupling in Interacting Fermi Gases

Jayantha P. Vyasanakere*

*School of Liberal Studies, Azim Premji University,
Bengaluru, 560100, India*

The possibility of employing cold atomic gases as emulators of condensed matter Hamiltonians has got boosted up by the birth and growth of the field of synthetic spin-orbit coupling. This being the case in the context of both bosonic and fermionic systems, here we discuss interacting Fermi gas in three spatial dimensions in presence of uniform synthetic non-Abelian gauge fields that induce a generalized Rashba-spin-orbit coupling (RSOC). We show that in presence of a class of RSOC, however small, a two-particle bound state exists even for a vanishingly small attraction. We discover that the fermion system can evolve to a Bose-Einstein condensate of a novel boson called the *rashbon*, whose properties are determined solely by RSOC and not by the interaction between the fermions or the fermion density. Most interestingly, via a study of collective excitations of the superfluid state, the rashbon-rashbon interaction is shown to be independent of the constituent fermion-fermion interaction. By constructing a fluctuation theory, we demonstrate that RSOC enhances the transition temperature of a weak Fermi superfluid to the order of Fermi temperature.

## 3.1. Introduction

Understanding the basic laws of nature alone cannot be sufficient to understand the nature itself, owing to the complexity of the real world. The art of extracting the information of interest regarding different substances in a unified fashion broadly defines the motto of condensed matter physics. The field has met with a tremendous amount of success — the Fermi liquid theory of metals, the Bardeen-Cooper-Schrieffer (BCS) theory of

---

*Correspondence addressed to: jayanth.vyasanakere@apu.edu.in

superconductivity — just to name a few. The main recent interests in quantum condensed matter research include high temperature superconductivity,[1] quantum computation[2] and topological aspects in quantum matter.[3–5]

The field of cold atoms has emerged as a remarkable tool in this process, engineering Hamiltonians of interest and achieving unprecedented control over parameters.[6–12] The landmark successes in cold atoms include experimental realization of Bose Einstein condensation (BEC) with alkali atoms,[13,14] for which the theory had to wait for around eight decades, experimental determination of the phase diagram of the Bose Hubbard model[15,16] and experimental realization of BCS-BEC crossover of fermions.[17,18] Apart from condensed matter physics, high energy physics[19] and astrophysics[20] are also beneficiaries of cold atoms.

Although cold atoms provide an ideal platform to work with, in practice however, there are several hurdles to overcome. One important among them is regarding achieving cold temperature itself that corresponds to quantum degeneracy regime, especially in presence of multiple lasers.[21,22] Another major roadblock is to emulate systems with orbital magnetism such as quantum Hall systems in charge-neutral cold atoms. One possibility to achieve an artificial magnetic field is through rotation.[23] This idea exploits the fact that the Coriolis force experienced in the rotating frame is very similar in form to the Lorentz force experienced by a charge moving in a magnetic field. This method has met with some success.[9] However, main issues in this approach are to maintain the stability of the rotational setup and to achieve high strength and nonhomogeneous magnetic fields. This serves as a motivation to look for synthetic gauge fields to emulate electromagnetic response.

A charged particle taken along a closed loop in a magnetic field experiences a phase, depending on the magnetic flux enclosed by the path. The key idea in synthetic gauge fields is the following. A *neutral* atom species which has a suitable hyperfine manifold is chosen. An arrangement of lasers (electromagnetic fields) appropriately couples different hyperfine states. The resultant set of states are called dressed states. The ground state sector now can be made to pick up a geometric phase when moved adiabatically in a cleverly engineered spatial variation of electromagnetic fields. The effective Hamiltonian for the ground state then mimics the presence of a U(1) gauge field.[24] Generation of a synthetic U(1) gauge field has been achieved and vortices were injected into a BEC of $^{87}$Rb atoms using the same method.[25] In this approach of usefully exploiting internal degrees of freedom of atoms, what would happen when the ground state of the dressed manifold is doubly degenerate? It turns out that this will give rise to a SU(2) (non-Abelian) gauge field. The idea of generating Abelian and non-Abelian synthetic gauge fields has attracted different experimental groups and has seen achievements both in the context of bosons[25–29] and fermions.[30–33] The field of cold atoms is also abundant with many theoretical proposals[34–38] for realizing synthetic gauge fields including the ones which do not employ laser lights.[39,40]

While a uniform Abelian gauge field is merely equivalent to a Galilean transformation, a uniform non-Abelian gauge field induces a generalized Rashba-spin-orbit coupling (RSOC) and is known to nurture interesting physics in case of bosons.[41–45] Here we review the effects of a uniform RSOC on a system of spin-$\frac{1}{2}$ fermions interacting via a contact attraction in three dimensional continuum that also forms the subject of our theoretical investigation.[46–51] We include as much details as possible, also focusing on the mathematically simplest special cases that can make the physics most transparent. Wide range of research directions has got opened up involving fermions in RSOC. Anisotropic superfluidity,[52] zero-temperature BCS-BEC crossover in the presence of Zeeman fields[53,54] are studied. Dresselhaus like spin-orbit interaction[55,56] and RSOC in lower dimensions and lattices have been investigated.[57–60] The reader is urged to also look at other reviews on this fast-paced field.[61,62] Certain aspects of physics of spin-orbit coupled fermions were reported earlier[63,64] before our independent discovery.

The exact form of RSOC and the interaction considered here, along with the classic results in absence of RSOC, will be described in Sec. 3.2. The rich physics of the two-body problem in presence of RSOC, both at zero and finite center of mass momentum, that sets the platform for the intuitions and calculations of future sections is dealt in Sec. 3.3. Section 3.4 describes the noninteracting Fermi gas in presence of RSOC, which also forms the foundation for novel many-body physics uncovered in the following sections. The effect of RSOC on the interacting many-body system is considered at mean field level in Sec. 3.5 where BCS to BEC crossover induced by RSOC and the rashbon condensate will be described. The emergent excitations and the rashbon-rashbon interaction will be studied in Sec. 3.6. The fluctuation theory, at a Gaussian level and beyond will be discussed in Sec. 3.7. We conclude with Sec. 3.8 providing future directions for further investigation.

## 3.2. Preliminaries

In this section we describe how the non-Abelian gauge fields induce RSOC on the motion of fermions; special types of gauge field configurations are discussed; the interaction term is introduced; we discuss the symmetries and the scales. Finally, a discussion on interacting fermions in absence of RSOC is included as a background.

### 2.A. *Rashba-spin-orbit coupling*

The Hamiltonian of spin-$\frac{1}{2}$ fermions moving in continuum of three spatial dimensions in presence of a spatially uniform and a temporally static SU(2) non-Abelian gauge field is[a]

$$\mathcal{H}_{GF} = \int \Delta r \, \Psi^\dagger(r) \left[ \frac{1}{2}(p_i 1 - A_i^\mu \tau^\mu)(p_i 1 - A_i^\nu \tau^\nu) \right] \Psi(r), \quad (3.2.1)$$

where $\Psi(r) = \{\psi_\sigma(r)\}, \sigma = \uparrow, \downarrow$ is a two component spinor field, $p_i$ is the $i$-component of the momentum operator, $\tau^\mu$ are Pauli spin operators ($i, \mu = x, y, z$), $1$ is the SU(2) identity, $A_i^\mu$ describe a uniform gauge field.

Motivated by experiments, we consider $A_i^\mu$ of the type $A_i^\mu = \lambda_i \delta_i^\mu$ which leads to a generalized Rashba Hamiltonian describing (in general) an anisotropic RSOC[b]

$$\mathcal{H}_R = \int \Delta r \Psi^\dagger(r) \left( \frac{p^2}{2} 1 - p_\lambda \cdot \tau \right) \Psi(r), \quad (3.2.2)$$

where, $p_\lambda = \lambda_x p_x e_x + \lambda_y p_y e_y + \lambda_z p_z e_z$; $\lambda \equiv \lambda \hat{\lambda} = \lambda_x e_x + \lambda_y e_y + \lambda_z e_z$ defines precisely the generalized RSOC, which we call a gauge field configuration (GFC). Here, $\lambda = \sqrt{\lambda_x^2 + \lambda_y^2 + \lambda_z^2}$ is the strength of RSOC. RSOC of these kinds produce a linear–momentum dependent magnetic field for the fermions.

The GFC where two of the couplings vanish (say $\lambda_x = \lambda_y = 0$) while only one is nonzero ($\lambda_z = \lambda$), called 'extreme prolate' GFC in the literature, has been experimentally realized.[27,30,31] GFCs of special interest are the 'oblate' GFCs where two of the couplings are equal and greater than the third. An extreme case of an oblate GFC is called 'extreme

---

[a]We work in units in which $\hbar = h/2\pi$, where $h$ is the Planck constant, fermion mass ($m$) and Boltzmann constant ($k_B$) are unity.
[b]The term containing $\tau^2$ is a constant and has no effect on the physics. Hence it is dropped.

oblate' GFC in the literature where the third component vanishes (say $\lambda_x = \lambda_y = \lambda/\sqrt{2}, \lambda_z = 0$). This GFC is also experimentally realized.[32] The most symmetric GFC is the one where all three couplings are equal $(\lambda_x = \lambda_y = \lambda_z = \lambda/\sqrt{3})$. Routes for experimental realization of this GFC, called 'spherical' GFC in the literature, is suggested.[65] The relative magnitudes of the components of $\lambda$ has a qualitative effect on the physics of the system. Here, we keep the formulation aspect general to a generic GFC, while, describing the results, we focus largely on the spherical GFC through which most of the interesting physics with RSOC is captured.

## 2.B. *Interaction*

Cold atom systems are not only ultra-cold, but they are also ultra-dilute. This is in the sense that the average interparticle spacing is too large for the particles to sample the detailed nature of the actual interaction potential. Hence the interaction, which can be tuned by a Feshbach resonance, between fermions can be described by a contact attraction model in the singlet channel[66] as

$$\mathcal{H}_v = \frac{v}{2} \int \Delta r \, S^\dagger(r) S(r) = v \int \Delta r \, \psi_\uparrow^\dagger(r) \psi_\downarrow^\dagger(r) \psi_\downarrow(r) \psi_\uparrow(r), \qquad (3.2.3)$$

where $S^\dagger(r) = \frac{1}{\sqrt{2}} \left( \psi_\uparrow^\dagger(r) \psi_\downarrow^\dagger(r) - \psi_\downarrow^\dagger(r) \psi_\uparrow^\dagger(r) \right) = \sqrt{2} \, \psi_\uparrow^\dagger(r) \psi_\downarrow^\dagger(r)$ is the singlet creation operator, and $v$ is the bare contact interaction. Expressing in Fourier (momentum) space, $\psi_\sigma^\dagger(r) = \frac{1}{\sqrt{V}} \sum_k e^{-ik \cdot r} C_{k\sigma}^\dagger \, ; S^\dagger(r) = \frac{1}{V} \sum_q e^{-iq \cdot r} \, S^\dagger(q)$, where $V$ is the volume and $C^\dagger$ is a fermion creation operator in momentum space. Hence,

$$S^\dagger(q) = \frac{1}{\sqrt{2}} \sum_k \left( C_{\frac{q}{2}+k\,\uparrow}^\dagger \, C_{\frac{q}{2}-k\,\downarrow}^\dagger - C_{\frac{q}{2}+k\,\downarrow}^\dagger \, C_{\frac{q}{2}-k\,\uparrow}^\dagger \right)$$

$$= \sqrt{2} \sum_k C_{\frac{q}{2}+k\,\uparrow}^\dagger \, C_{\frac{q}{2}-k\,\downarrow}^\dagger. \qquad (3.2.4)$$

With all these,

$$\mathcal{H}_v = \frac{v}{2V} \sum_q S^\dagger(q) S(q) = \frac{v}{V} \sum_{q\,k\,k'} C_{\frac{q}{2}+k\,\uparrow}^\dagger \, C_{\frac{q}{2}-k\,\downarrow}^\dagger \, C_{\frac{q}{2}-k'\,\downarrow} \, C_{\frac{q}{2}+k'\,\uparrow}. \qquad (3.2.5)$$

To describe in words, the contact interaction is a two-body operator and scatters an incoming pair with center of mass (COM) momentum $q$ and any relative momentum $(k')$ to the same COM momentum and any other relative momentum $(k)$ with equal amplitude.

**Fig. 3.2.1.** Variation of attraction strength as described by the scattering length $a_\mathrm{s}$.

The theory described by the Hamiltonian $\mathcal{H} = \mathcal{H}_R + \mathcal{H}_v$ requires an ultraviolet momentum cut-off $\Lambda$ given by $\Lambda = \frac{1}{V}\sum_k \frac{1}{k^2}$. The bare contact interaction parameter $v$ is $\Lambda$-dependent and satisfies the regularization relation

$$\frac{1}{v} + \frac{1}{V}\sum_k \frac{1}{k^2} = \frac{1}{4\pi a_\mathrm{s}} \tag{3.2.6}$$

where $a_\mathrm{s}$ is the s-wave scattering length in the *absence* of RSOC.[67] The condition $1/a_\mathrm{s} = 0$ is called resonance. The attraction is quantified in terms of the negative inverse of the scattering length and this variation is as shown in Fig. 3.2.1.

## 2.C. *The system, symmetries and scales*

Now the full Hamiltonian that describes our system is given by

$$\mathcal{H} = \mathcal{H}_R + \mathcal{H}_v. \tag{3.2.7}$$

In describing many-body system we will be working in the grand canonical ensemble, where the volume $(V)$, temperature $(T)$ and chemical potential $(\mu)$ are held fixed. We absorb the term $-\mu\mathcal{N}$ into the Hamiltonian (3.2.7) while working with the many-body system, where $\mathcal{N} = \int \Delta r\, \Psi^\dagger(r)\Psi(r)$ is the total number operator.

The Hamiltonian $\mathcal{H}$ generically has three symmetries — global translation, global phase and time reversal. Extra symmetries occur in special GFCs like extreme prolate, extreme oblate and spherical. Note that in presence of RSOC, there is no Galilean invariance, the consequences of which will be discussed in Sec. 3.D. The system also does not possess spatial inversion symmetry due to RSOC.

While $a_\mathrm{s}$ introduces a length scale, $\lambda$ introduces a momentum scale. The finite density $\rho$ of fermions also introduces an energy scale which is the Fermi energy $E_\mathrm{F}$ (along with Fermi wave vector $k_\mathrm{F}$) in absence of RSOC: $E_\mathrm{F} = \frac{k_\mathrm{F}^2}{2} = \frac{1}{2}\left(3\pi^2\rho\right)^{2/3}$.

## 2.D. *Interacting fermions in free vacuum*

We refer to the absence of RSOC as "free vacuum". The physics of inter-acting fermions in free vacuum which are of relevance[66,68–78] is summarized here.

### 2.D.1. *Two-body problem*

In one and two spatial dimensions, a system of two attracting fermions admits a bound state for arbitrarily small attractions. However, in three dimensional free vacuum a critical strength of attraction, characterized by the critical scattering length $a_{sc}$ is needed to form a bound state of two fermions. $1/a_{sc} = 0$ and the two-body binding energy is given by

$$E_{\mathrm{b}} = \begin{cases} 1/a_{\mathrm{s}}^2 & \text{if } a_{\mathrm{s}} \geq 0, \\ \text{no bound state} & \text{if } a_{\mathrm{s}} < 0. \end{cases} \tag{3.2.8}$$

Also, the bound state wave function is a singlet owing to the interaction being in singlet channel and the kinetic energy being uninfluenced by the spin structure.

### 2.D.2. *BCS-BEC crossover*

Moving to the many-body system, when the attraction is weak ($a_{\mathrm{s}} < 0, |k_{\mathrm{F}}a_{\mathrm{s}}| \ll 1$), the ground state is a superfluid with large overlapping Cooper pairs, which means that the extent of the pair wave function is much large compared to the average inter-particle spacing. This regime is well described by the BCS theory. The chemical potential in this state is essentially unchanged from that of the noninteracting system, *i.e.*, $\mu \approx E_{\mathrm{F}}$. The excitation gap $\Delta_0$, in this regime is exponentially small and is given by $\frac{\Delta_0}{E_{\mathrm{F}}} \approx \frac{8}{e^2} \exp\left(-\frac{\pi}{2k_{\mathrm{F}}|a_{\mathrm{s}}|}\right)$.

As the attraction increases, the pairs get more and more strongly attached. When the attraction is extremely large ($a_{\mathrm{s}} > 0, |k_{\mathrm{F}}a_{\mathrm{s}}| \ll 1$), the pairs get tightly bound. Since they are made up of *two* spin$-\frac{1}{2}$ fermions they assume a bosonic identity. Hence, the ground state can be looked upon as a BEC of such bosons. The chemical potential in this regime is determined by the binding energy of the two-body problem, *i.e.*, $\mu \sim -E_{\mathrm{b}}/2$ because, when 2 new particles are added to such a system, they form a bound state with binding energy $E_{\mathrm{b}}$ below the threshold. Including the corrections due to finite density that contributes to Pauli repulsion, the chemical potential

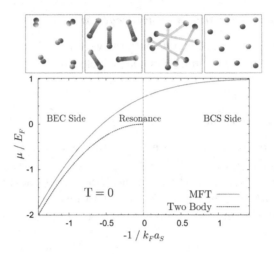

**Fig. 3.2.2.** The BCS-BEC crossover is schematically shown where the loosely bound fermion pairs in the BCS regime evolves to tightly bound bosonic molecules in the BEC regime on tuning the attraction. The chemical potential that signals the crossover is also shown, which in this process smoothly changes from the noninteracting value to that determined by the two-body problem.

and the excitation gap in this regime are given by

$$\mu \approx -\frac{1}{2a_s^2} + \pi \rho a_s \; ; \qquad \Delta_0{}^2 \approx \frac{4\pi\rho}{a_s}. \qquad (3.2.9)$$

The weak and strong attraction regimes are smoothly connected in the sense that one can go from one state to the other by adiabatically tuning the attraction, without encountering a phase transition. Hence, such a change of state is termed as a crossover — a BCS to BEC crossover induced by the attraction (see the schematic in Fig. 3.2.2). The chemical potential which smoothly switches from that of the noninteracting system to that determined by the two-body physics, upon increase of attraction signals the crossover (see Fig. 3.2.2). Throughout the crossover the pair wave function is a singlet, again owing to the nature of interaction.

### 2.D.3. *Collective excitations of the superfluid state*

A study of Gaussian fluctuations at zero temperature leads to two kinds of excitations — gapless sound mode and a gapped Anderson-Higgs mode. The phase stiffness is isotropic and is equal to $\rho/4$, independent of $a_s$. This is equivalent of saying that the superfluid density is same as the particle

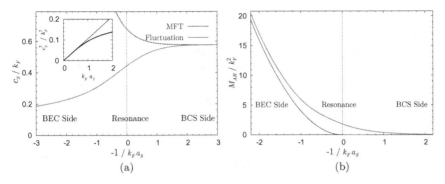

**Fig. 3.2.3.** Emergent properties across the BCS-BEC crossover in free vacuum. The vertical line corresponds to resonance. The speed of sound is shown in (a). The red curve is calculated using the Gaussian fluctuation theory described here. The blue curve is a result of mean field theory. The solid line in the inset is a result of Gaussian fluctuation calculation while the dashed straight line represents analytical approximation of Eq. (3.2.10). (b) The Anderson-Higgs mass obtained from Gaussian fluctuation calculation (red curve) is compared with the analytical approximation (black curve) of Eq. (3.2.10).

density. This result owes to the presence of Galilean invariance, in accordance to Leggett's result.[79,80]

In the BCS regime, the speed of sound mode is $c_s \approx k_F/\sqrt{3}$ and the Anderson-Higgs mass is exponentially small. The full evolution of both these quantities as a function of $a_s$ is shown in Fig. 3.2.3. $c_s$ can also be easily calculated within mean field theory (details not shown) and this result is also shown in the same figure. $c_s^2$ is inversely proportional to the compressibility of the system which is underestimated in the mean field theory due to its limitation to capture the low energy collective excitations. Hence, the mean field result is always higher than the one calculated through Gaussian fluctuations.

In the BEC regime where $\mu$ and $\Delta_0$ are given by Eq. (3.2.9), the sound-speed and the Anderson-Higgs mass are approximately given by

$$c_s^2 = \pi \rho a_s \quad \text{and} \quad M_{\text{AH}} = \frac{4}{a_s^2}. \qquad (3.2.10)$$

The agreement of these with the numerically calculated result is apparent in Fig. 3.2.3.

As discussed just now, the state obtained in this regime is a BEC of bosons which are nothing but strongly bound fermion pairs. The effective mass of these bosons is twice the fermion mass ($m_{\text{BEC}} = 2$) and the number density of bosons is half of the fermion density ($\rho_{\mathrm{BEC}} = \rho/2$). One can understand this state by a theory of bosons as follows.

Recall from the standard Bogoliubov theory of bosons[81] that a collection of bosons of mass $m_{\mathrm{B}}$ with number density $\rho_{\mathrm{B}}$ and a contact interaction described by a scattering length $a_{\mathrm{B}}$ has a superfluid ground state at zero temperature. The chemical potential of this system is $\mu_{\mathrm{B}} = \frac{4\pi a_{\mathrm{B}}}{m_{\mathrm{B}}}\rho_{\mathrm{B}}$ and the speed of sound is

$$c_{\mathrm{s}}^{\mathrm{B}} = \sqrt{\frac{\mu_{\mathrm{B}}}{m_{\mathrm{B}}}} = \sqrt{\frac{4\pi a_{\mathrm{B}}\rho_{\mathrm{B}}}{m_{\mathrm{B}}^2}}. \tag{3.2.11}$$

From Eq. (3.2.9), the chemical potential of the emergent bosons $\mu_{\mathrm{BEC}}$ (measured from the one set by the two-body binding energy) is $\mu_{\mathrm{BEC}} = 2\pi\rho a_{\mathrm{s}}$.[c] It is apparent that the speed of sound in this regime (Eq. (3.2.10)) is consistent with Eq. (3.2.11) from Bogoliubov theory, i.e., $c_{\mathrm{s}}^2 = \mu_{\mathrm{BEC}}/m_{\mathrm{BEC}} = 2\pi\rho a_{\mathrm{s}}/2 = \pi\rho a_{\mathrm{s}}$. This means that the state at small positive $a_{\mathrm{s}}$ is a BEC of bosons interacting with a contact interaction. Writing $c_{\mathrm{s}}^2 = 4\pi\, a_{\mathrm{BEC}}\, \rho_{\mathrm{BEC}}/(m_{\mathrm{BEC}})^2$ allows to calculate the boson-boson scattering length as

$$a_{\mathrm{BEC}} = 2a_{\mathrm{s}}. \tag{3.2.12}$$

Note here that the interaction among the emergent excitations (characterized by $a_{\mathrm{BEC}}$) depends on constituent interactions (characterized by $a_{\mathrm{s}}$).

### 2.D.4. Transition temperature

The superfluid transition temperature ($T_{\mathrm{c}}$) in the BCS regime is dictated by mean field theory, where the transition from a superfluid state to a normal state is governed by vanishing of the gap $\Delta_0$. On the other hand, in BEC regime, pair breaking effects are energetically expensive and the transition is determined by the phase fluctuations of the order parameter. Hence, $T_{\mathrm{c}}$ here can be estimated as the Bose condensation temperature of the tightly bound pairs treated as ideal (see Eq. (3.2.12)):

$$\frac{T_{\mathrm{c}}}{T_{\mathrm{F}}} = \left(\frac{16}{9\pi(\zeta(3/2))^2}\right)^{1/3} \frac{1}{m_{\mathrm{BEC}}}. \tag{3.2.13}$$

The $T_{\mathrm{c}}$ across the entire crossover calculated using the Nozieres-Schmitt-Rink (NSR) theory,[72,82,83] which is a Gaussian theory, is shown in Fig. 3.2.4

---

[c]The factor of 2 is to account for two fermions constituting each boson.

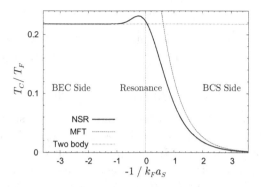

**Fig. 3.2.4.** Superfluid transition temperature across the BCS-BEC crossover in free vacuum. The black curve is calculated using NSR theory. The blue curve is a result of mean field theory. The red curve comes from the two-body analysis. The vertical line corresponds to resonance.

(black curve). $T_c$ calculated by the mean field theory is also shown (in blue). In the BCS regime, $T_c$ is well captured by the mean field theory. In this limit it can be shown that

$$T_c / \Delta_0^{(T=0)} \approx e^\gamma / \pi, \qquad (3.2.14)$$

where, $\Delta_0^{(T=0)}$ is the pairing amplitude at zero temperature and $\gamma$ is Euler's constant ($\approx 0.577$). The curve shown in red is the $T_c$ calculated using Eq. (3.2.13). It can be seen that this approach works quite well in the deep BEC regime.

What would happen to all this physics of interacting fermions in presence of RSOC of the kind discussed in Eq. (3.2.2)? What happens to the two-body problem? How does the many-body ground state evolve with RSOC? Are there novel excitations? How does the free vacuum phase diagram Fig. 3.2.4 extend in presence of RSOC? Is there any new physics? What implications do RSOC have in resolving important problems of condensed matter physics? Addressing these questions constitute the primary objective here.

## 3.3. The Two-body Problem

Having set the stage for discussions in the previous section, we proceed here with the two-body problem in RSOC. This forms the basis for the calculations and discussions of future sections.

### 3.A.  Noninteracting two-body problem

For any GFC, the single particle states are described by the quantum numbers of momentum $k$ and generalized helicity (called just as helicity from now on) $\alpha$ which takes on values $\pm$ :

$$|k\alpha\rangle = |k\rangle \otimes |\alpha\hat{k}_\lambda\rangle = C^\dagger_{k\alpha}|0\rangle \qquad (3.3.1)$$

where $|k\rangle$ is the usual plane-wave state, and $|\alpha\hat{k}_\lambda\rangle$ is the spin coherent state in the direction $\alpha\hat{k}_\lambda$, with $k_\lambda$ defined analogous to $p_\lambda$ of Eq. (3.2.2), $|0\rangle$ is the fermion vacuum and $C^\dagger_{k\alpha}$ is the operator that creates a fermion with momentum $k$ and helicity $\alpha$. The two helicity bands disperse as $\varepsilon_\alpha(k) = k^2/2 - \alpha|k_\lambda|$. For the spherical GFC the ground state will be degenerate and forms a 2 dimensional manifold. It is useful for future analysis to define the dispersion referred to the bottom of the +helicity band (with $\lambda_{\max} = \text{Maximum}(\lambda_x, \lambda_y, \lambda_z)$) as

$$\tilde{\varepsilon}_\alpha(k) = \varepsilon_\alpha(k) + \frac{\lambda_{\max}{}^2}{2}. \qquad (3.3.2)$$

For the noninteracting two-body problem, we observe that the total momentum and the relative momentum operators commute with the noninteracting Hamiltonian and moreover the interaction term also commutes with the total momentum operator, with COM momentum as the corresponding good quantum number. Hence, we construct states with specified COM momentum ($q = k_1 + k_2$), relative momentum ($k = (k_1 - k_2)/2$) and helicities as

$$|q\,k\,\alpha\,\beta\rangle \equiv C^\dagger_{(\frac{q}{2}+k)\alpha} C^\dagger_{(\frac{q}{2}-k)\beta}|0\rangle \qquad (3.3.3)$$

and eigen energies are given by $\varepsilon_{\alpha\beta}(q, k) = \varepsilon_\alpha\left(\frac{q}{2} + k\right) + \varepsilon_\beta\left(\frac{q}{2} - k\right)$. For a given $q$, the lowest energy is called the threshold energy. For spherical GFC, it is

$$\varepsilon_{th} = \begin{cases} -\frac{\lambda^2}{3} & \text{if } |q| < \frac{2\lambda}{\sqrt{3}}, \\ \frac{|q|^2}{4} - \frac{\lambda|q|}{\sqrt{3}} & \text{if } |q| \geq \frac{2\lambda}{\sqrt{3}}. \end{cases} \qquad (3.3.4)$$

The state in Eq. (3.3.3) is generically made out of both the singlet and triplet states. Since the interaction is in the singlet channel, it will be very useful to know how much the above state projects on to the singlet sector. This will be called the singlet amplitude and is defined as

$$A_{\alpha\beta}(q, k) = \langle q\,k\,s|q\,k\,\alpha\,\beta\rangle \qquad (3.3.5)$$

where, $|q\,k\,s\rangle = \frac{1}{\sqrt{2}}\left(C^\dagger_{\frac{q}{2}+k\,\uparrow} C^\dagger_{\frac{q}{2}-k\,\downarrow} - C^\dagger_{\frac{q}{2}+k\,\downarrow} C^\dagger_{\frac{q}{2}-k\,\uparrow}\right)|0\rangle$ is the singlet state with $q$ and $k$ as the COM and relative momenta. With this, the interaction (Eq. (3.2.5)) can be rewritten as

$$\mathcal{H}_v = \frac{v}{2V} \sum_{q} \sum_{k,k'} A_{\alpha\beta}(q,k) A^*_{\alpha'\beta'}(q,k') C^\dagger_{(\frac{q}{2}+k)\alpha} C^\dagger_{(\frac{q}{2}-k)\beta} C_{(\frac{q}{2}-k')\beta'} C_{(\frac{q}{2}+k')\alpha'}.$$

(3.3.6)

The singlet density of states (SDOS) is defined as

$$g_s(q,\omega) = \frac{1}{V} \sum_{k\,\alpha\,\beta} |A_{\alpha\beta}(q,k)|^2 \, \delta(\omega - \varepsilon_{\alpha\beta}(q,k)),$$

(3.3.7)

where $\delta$ is the Dirac delta function. When $q = 0$, for spherical GFC,

$$g_s(0,\omega) = \frac{\lambda^2 + 3\omega}{12\pi^2 \sqrt{\omega}}.$$

(3.3.8)

Here, $\omega$ is measured from the scattering threshold, below which SDOS is zero.

## 3.B. *Formulation of the interacting two-body problem*

The total momentum operator commutes with the Hamiltonian, rendering the system translationaly invariant. Therefore, the COM momentum $q$ is a good quantum number. Hence the ground state of the two particle problem with a specified $q$ can be expressed in terms of the eigen energy states of the noninteracting problem as

$$|GS^q\rangle = \sum_{k\,\alpha\,\beta} \gamma^q(k\,\alpha\,\beta) |q\,k\,\alpha\,\beta\rangle$$

(3.3.9)

with the ground state energy $E(q)$ given by the Schrodinger equation as $E(q) |GS^q\rangle = (\mathcal{H}_R + \mathcal{H}_v) |GS^q\rangle$. Therefore,

$$\sum_{k'\,\alpha'\,\beta'} \gamma^q(k'\,\alpha'\,\beta')\,(E(q) - \varepsilon_{\alpha'\beta'}(q,k')) |q\,k'\,\alpha'\,\beta'\rangle$$

$$= \sum_{k'\,\alpha'\,\beta'} \gamma^q(k'\,\alpha'\,\beta')\,\mathcal{H}_v |q\,k'\,\alpha'\,\beta'\rangle.$$

(3.3.10)

Operating $\langle q\,k\,\alpha\,\beta|$ from left and using orthonormality of the states $|q\,k\,\alpha\,\beta\rangle$,

$$\gamma^q(k\,\alpha\,\beta)\,(E(q) - \varepsilon_{\alpha\beta}(q,k)) = \sum_{k'\,\alpha'\,\beta'} \gamma^q(k'\,\alpha'\,\beta')\,\langle q\,k\,\alpha\,\beta|\,\mathcal{H}_v\,|q\,k'\,\alpha'\,\beta'\rangle.$$

Since $\mathcal{H}_v$ (see Eq. (3.2.5) and Eq. (3.3.6)) acts only in the singlet sector, we have $\gamma^q(k\,\alpha\,\beta) = \frac{v}{V} \frac{A^*_{\alpha\beta}(q,k)}{E(q)-\varepsilon_{\alpha\beta}(q,k)} \sum_{k'\,\alpha'\,\beta'} \gamma^q(k'\,\alpha'\,\beta') A_{\alpha'\beta'}(q,k')$.

Multiplying both sides by $A_{\alpha\beta}(q, k)$ and summing over $k$, $\alpha$ and $\beta$ one obtains the secular equation for the formation of a two-body bound state as[d] $\frac{1}{v} = \frac{1}{V} \sum_{k \, \alpha \, \beta} \frac{|A_{\alpha\beta}(q,k)|^2}{E(q) - \varepsilon_{\alpha\beta}(q,k)}$. Using regularization (Eq. (3.2.6)) and the idea of SDOS (Eq. (3.3.7)) this becomes,

$$
\frac{1}{4\pi a_s} = \frac{1}{V} \sum_{k} \left[ \left( \sum_{\alpha\beta} \frac{|A_{\alpha\beta}(q,k)|^2}{E(q) - \varepsilon_{\alpha\beta}(q,k)} \right) + \frac{1}{k^2} \right]
$$

$$
= \int_0^\infty d\omega \left( -\frac{g_s(q,\omega)}{\omega + E_b} + \frac{g_s^0(0,\omega)}{\omega} \right). \tag{3.3.11}
$$

Here $g_s^0(q, \omega)$ is the SDOS in free vacuum and $E_b$ is the binding energy defined by

$$
E(q) = \varepsilon_{th}(q) - E_b. \tag{3.3.12}
$$

In free vacuum the bound state is a singlet (Sec. 2.D.1). In contrast, it can be seen from Eq. (3.3.9) that in presence of RSOC, the normalized bound-state wave function is made up of both spatially symmetric singlet and spatially antisymmetric triplet pieces $|\psi_b\rangle = |\psi_s\rangle + |\psi_t\rangle$. We define a quantity called the triplet content which measures the weight of the wave function in the triplet sector as

$$
\eta_t = \langle \psi_t | \psi_t \rangle. \tag{3.3.13}
$$

As we shall see, this quantity plays an important role in many-body analysis also. The bound state wave function, when $q = 0$, is $|\psi_b\rangle \propto \sum_{k\alpha} \frac{\alpha}{2\tilde{\varepsilon}_\alpha(k) + E_b} C_{k\alpha}^\dagger C_{-k\alpha}^\dagger |0\rangle$, where $\tilde{\varepsilon}_\alpha(k)$ was defined in Eq. (3.3.2).

### 3.C.  *Results at zero center of mass momentum*

The formulation just presented reproduces the results of Sec. 2.D.1 pertaining to free vacuum. Recall that in free vacuum, a critical strength of attraction ($a_s > 0$) is needed to form a bound state of two fermions and the bound state is a singlet. The question of interest now is how RSOC affects the nature of the bound state of two fermions at $q = 0$, where we focus on spherical GFC.

Apart from translation and time reversal, this GFC has global (spatial + spin) rotational symmetries about all three axes generated by $J_i = L_i + \frac{1}{2}\tau_i$,

---

[d]The same secular equation can also be obtained by using a $T$-matrix analysis which follows very closely the analysis for the two-body problem without RSOC.[84]

where $L_i$ is the $i^{\text{th}}$ component of the orbital angular momentum. The secular equation (3.3.11) can be solved to obtain a closed form expression for the binding energy for any $a_s$:

$$E_b = \frac{1}{4} \left( \frac{1}{a_s} + \sqrt{\frac{1}{a_s{}^2} + \frac{4\lambda^2}{3}} \right)^2. \tag{3.3.14}$$

We find that **there is a bound state for *any* scattering length, negative or positive**, *i.e.*, $a_{sc} = 0^-$.

The bound state is a $J$-singlet ($J = 0$), composed of $L = 0$ spatial part with a spin singlet and $L = 1$ spatial part with a spin triplet (spin quantization axis along $\hat{r}$) and has the wave function

$$\psi_b(\boldsymbol{r}) \propto \frac{e^{-\sqrt{E_b}\,r}}{r} \left( \frac{\lambda}{\sqrt{3E_b}} \sin \frac{\lambda r}{\sqrt{3}} + \cos \frac{\lambda r}{\sqrt{3}} \right) | \uparrow\downarrow - \downarrow\uparrow\rangle$$

$$+ i \left( \left( \sqrt{E_b} + \frac{1}{r} \right) \sin \frac{\lambda r}{\sqrt{3}} - \frac{\lambda}{\sqrt{3}} \cos \frac{\lambda r}{\sqrt{3}} \right) \frac{e^{-\sqrt{E_b}\,r}}{\sqrt{E_b}\,r} | \uparrow\downarrow + \downarrow\uparrow\rangle_{\hat{r}}.$$

While $E_b$ determines the exponential decay of the spatial part, $\lambda$ determines the length scale of its spatial oscillations.

The different GFCs host varieties of wave functions. Some of the interesting ones where the wave function resembles those of $^3$He, also having spin nematic order consistent with the symmetries of the Hamiltonian are discussed in Ref. 46. It turns out that the strength of the critical attraction required to produce a two-body bound state is reduced even by the presence of a generic GFC and all oblate GFCs admit the special feature of

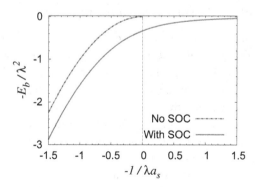

**Fig. 3.3.1.** The solid red line is the binding energy $E_b$ as a function of the scattering length $a_s$ and the RSOC strength $\lambda$ for spherical GFC (Eq. (3.3.14)). The dashed blue line is $E_b$ as a function of $a_s$ in free vacuum (Eq. (3.2.8)).

'bound state for any attraction'. Hence we can call RSOC as an "attraction amplifier". A detailed study of two-body scattering from a finite range box potential is carried out in Ref. 85.

The above phenomenon of promoted attraction induced by RSOC owes to the enhancement in SDOS $g_s(0,\omega)$ (Eq. (3.3.8)) due to infrared degeneracy it creates. While $g_s(0,\omega) \sim \sqrt{\omega}$ for $\omega \gg \lambda^2$, it is enhanced in the infrared ($\omega \ll \lambda^2$): $g_s(0,\omega) \sim 1/\sqrt{\omega}$. This infrared SDOS is similar to that in one spatial dimension, where the bound state forms for arbitrarily small attraction. This is also analogous to the Cooper instability[86] that lead to the celebrated BCS theory of superconductivity, where due to Pauli blocking, the lowest energy level accessible for the two particles is at the Fermi surface causing the enhancement of the infrared SDOS.

### 3.C.1. *The Rashbon state*

A very important aspect of the two-body problem at $q = 0$ is the **rashbon** state. It is the bound state of two fermions at resonant scattering length in presence of RSOC. All properties of this state are completely determined by the RSOC $\lambda$, and hence the name. For example, the binding energy of a rashbon is $E_b = \lambda^2 \mathcal{R}(\hat{\lambda})$, where $\mathcal{R}$ is a positive dimensionless function of $\hat{\lambda}$. The rashbon bound state also possesses a triplet content $\eta_t$ (Eq. (3.3.13)) that depends on the GFC. $\eta_t = 1/4$ for the spherical GFC. Certain other properties of rashbons will be studied in Sec. 3.D. The rashbon state plays a key role in the many body problem with RSOC, as discussed in the subsequent sections.

### 3.D. *Results at finite center of mass momentum*

Due to Galilean invariance, the two-body problem in the absence of RSOC at a nonzero COM momentum $q$ is straight forward. The threshold energy goes as $q^2/4$ and the binding energy remains unchanged. The question to address here pertains to the nature of two-body problem at a finite COM momentum in presence of RSOC, where there is no Galilean invariance.

The dispersion $E(q)$ calculated by solving Eq. (3.3.11) is in general anisotropic due to anisotropy in RSOC. However, here we restrict our analysis to spherical GFC, where due to its symmetry, $E(\boldsymbol{q}) = E(q)$, where $q = |\boldsymbol{q}|$. The dispersion obtained for various $a_s$ is shown in Fig. 3.3.2(a). A characteristic feature to be observed is: **For any attraction, however large, there exists a critical COM momentum, beyond which the bound state vanishes.**

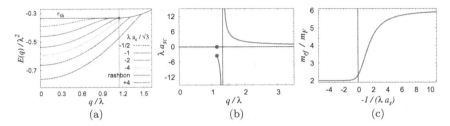

**Fig. 3.3.2.** (a) The dispersion for various scattering lengths in spherical GFC. (b) Critical scattering length ($a_{sc}$) as a function of COM momentum. (c) Mass of tightly bound fermion pairs in units of fermion mass ($m_F$). The vertical line indicates the resonant scattering length.

Consider a fixed momentum $q$. For all $q < q_0$, where $q_0 = 2\lambda/\sqrt{3}$, there is a bound state for *any attraction*. For $q > q_0$, a minimum attraction described by a critical scattering length $a_{sc}$ is necessary for the formation of a bound state. For $q = q_0^+$, $a_{sc} = -2\sqrt{3}/\lambda$. On increasing $q$, a stronger attraction is required to produce a bound state and when $q$ reaches $\sim 4\lambda/3$, a resonant attraction is necessary to produce a bound state. For $q \gtrsim 4\lambda/3$, a very strong attraction described by a small positive $a_s$ is necessary to produce a bound state. In fact, for $q \gg \lambda$, $1/a_s \simeq \sqrt{E_b + \lambda q/\sqrt{3}}$ and hence $E_b \simeq 1/a_s^2 - \lambda q/\sqrt{3}$, as long as $1/a_s > 1/a_{sc} \simeq \sqrt{\lambda q/\sqrt{3}}$. Thus, $a_{sc}$ scales as $1/\sqrt{\lambda q}$. The exact dependence of $a_{sc}$ on $q$, obtained numerically, is shown in Fig. 3.3.2(b).

The curvature of the dispersion $E(q)$ at $q = 0$ defines the effective low-energy inverse mass for the bound state: $m_{ef}^{-1} = \left. \frac{\partial^2 E(q)}{\partial q^2} \right|_{q=0}$. After some analysis:

$$m_{ef}^{-1} = \frac{2\sum_{k\alpha\beta} \frac{\partial^2 |A_{\alpha\beta}(q,k)|^2}{\partial q^2}\big|_{q=0}}{(E(0)-\varepsilon_{\alpha\beta}(0,k))^2} + \sum_{k\alpha} \frac{\partial^2 \varepsilon_{\alpha\alpha}(q,k)}{\partial q^2}\big|_{q=0}}{(E(0)-\varepsilon_{\alpha\alpha}(0,k))^2}}{\sum_{k\alpha} \frac{1}{(E(0)-\varepsilon_{\alpha\alpha}(0,k))^2}}.$$

In free vacuum, the first term in the numerator vanishes because of equal and opposite contributions from like and unlike helicities and recovers $m_{ef} = 2$. We are able to obtain an *analytical* expression for the mass of the bound state as

$$m_{ef} = \frac{6m_F}{7 + \frac{2\lambda^2}{E(0)} - 4\left(1 + \frac{\lambda^2}{3E(0)}\right)^{3/2}}, \tag{3.3.15}$$

shown in Fig. 3.3.2(c). Here $E(0)$ is analytically obtained through Eqs. (3.3.12), (3.3.4) and (3.3.14).

At a given $\lambda$, as expected, mass for a small positive $a_s$ is twice the fermion mass. Mass at resonance is the rashbon mass which is equal to $\frac{3}{7}(4 + \sqrt{2})m_F \approx 2.32m_F$. Interestingly, the value for mass in the small negative $a_s$ limit is (integer) 6.

The rich variety in the properties for generic and other special GFCs, including anisotropies in dispersion and mass are discussed in Ref. 48.

With reference to the physics of SDOS that explained the results at $q = 0$, the above results indicate that there is a sudden change in the infrared SDOS at $q = q_0$. The threshold energy for $q = q_0$ corresponds to the state where the relative momentum $k$ just vanishes. For $q < q_0$, the degenerate $k$ states produce a finite SDOS at the threshold. In particular, when $q = 0$, the SDOS diverges as $1/\sqrt{\omega}$ (Eq. (3.3.8)). Thus $a_{sc}$ vanishes for $q < q_0$. When $q$ crosses $q_0$, the SDOS at the threshold vanishes. This leads to a finite $a_{sc}$ accounting for the jump in the curve shown in Fig. 3.3.2(b).

The question then is why is there a depletion in the infrared SDOS at large $q$? When $q$ is very large, the vectors $\frac{q}{2} \pm k$ for those $k$s near the one that corresponds to the noninteracting ground state are nearly parallel. Hence the helicity states which are oriented along these vectors have a very low singlet component leading to the depletion of SDOS.

The essential physics of reduction in binding as a function of the COM momentum can be attributed to the **lack of Galilean invariance** in presence of RSOC. The system is Galilean invariant when $\lambda = 0$. In presence of RSOC, a Galilean transformation takes a finite COM momentum problem to a zero COM momentum problem with the spins coupled to a Zeeman magnetic field. But a Zeeman term is known to destroy pairing,[53] which is again due to the reduced infrared SDOS.

The results of this section suggest that pair breaking effects should be quite significant at the core of the vortices of superfluids in presence of RSOC. These results also play an important role in Sec. 3.7. The properties of rashbons calculated here will be later shown to be of profound significance in the many-body setting.

## 3.4. Rashba-spin-orbit Coupled Noninteracting Fermi Gas

The last section illustrated that RSOC can produce novel effects on a two fermion system. Given that the two-body problem itself is so rich, what happens to the many-body problem? Before entering into the interacting many-body problem, in this section we show that RSOC can have interesting effects even on a noninteracting gas of fermions. This section also forms a necessary background for further analysis.

## 4.A. *Fermi surface topology transition*

Let us consider a system of noninteracting fermions at $T = 0$ in spherical GFC. In free vacuum, $\lambda = 0$ and $\mu = E_F$.[e] The state consists of two Fermi spheres of radius $k_F$ in momentum space, one for each spin (Fig. 3.3.3(a)). In presence of RSOC, recall that, helicity is the good quantum number. At a given density $\rho$, the chemical potential of the noninteracting system $(\mu_{NI}(\lambda))$ decreases with increasing $\lambda$. We define $\lambda_T$ through $\mu_{NI}(\lambda_T) = \lambda_T{}^2/6$, which obtains $\lambda_T = \left(\sqrt{3}/2^{2/3}\right) k_F$. This is the $\lambda$ when $\mu_{NI}$ equals the bottom of the $-$helicity band.[f] When $\lambda < \lambda_T$, the noninteracting Fermi surfaces of the two helicities continue to be spheres, with the radius of the $+$helicity Fermi surface being larger than that of the $-$helicity Fermi surface (Fig. 3.3.3(b)). At $\lambda = \lambda_T$, the $-$helicity Fermi surface shrinks to a point

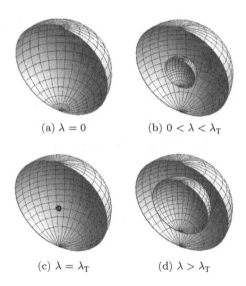

(a) $\lambda = 0$      (b) $0 < \lambda < \lambda_T$

(c) $\lambda = \lambda_T$      (d) $\lambda > \lambda_T$

**Fig. 3.3.3.** Fermi surface topology transition (FSTT) with increasing RSOC for spherical GFC. The figures here show only a sectioned half of the Fermi surfaces.

---

[e]Here and later, the chemical potential is measured from the bottom of the $+$helicity band.

[f]For a generic GFC, $\lambda_T$ is defined through $\mu_{NI}(\lambda_T) = \lambda_{max}{}^2/2$, see Eq. (3.3.2).

(Fig. 3.3.3(c)) and ceases to exist for any larger $\lambda$. For $\lambda > \lambda_T$, there is only +helicity Fermi sea and it is "a sphere with a hole", a region bounded by two concentric spherical Fermi surfaces (Fig. 3.3.3(d)).

Clearly, there is a **change in the topology** of the +helicity Fermi surface at $\lambda_T$. We call this the Fermi surface topology transition (FSTT), hence the subscript T in $\lambda_T$. For $\lambda < \lambda_T$ , there are low-energy excitations of both helicities, while for $\lambda > \lambda_T$ low-energy excitations are only of the +helicity.

For a generic GFC, the Fermi seas will be anisotropic with a shape determined by $\hat{\boldsymbol{\lambda}}$. FSTT associated with such a GFC is discussed in Ref. 47.

## 4.B.  *Chemical potential of the noninteracting system*

We next calculate $\mu_{\mathrm{NI}}(\lambda)$. With the introduction of the chemical potential $\mu$, the noninteracting (NI) Hamiltonian (Eq. (3.2.7) without the term $\mathcal{H}_v$) becomes

$$\mathcal{H}_{\mathrm{NI}} = \sum_{k\alpha} \xi_{k\alpha} C_{k\alpha}^{\dagger} C_{k\alpha} \tag{3.4.1}$$

where $\xi_{k\alpha} = \tilde{\varepsilon}_\alpha(\boldsymbol{k}) - \mu$. $\tilde{\varepsilon}_\alpha(\boldsymbol{k})$ was defined in Eq. (3.3.2) and $C_{k\alpha}^{\dagger}$ in Eq. (3.3.1). From this, the free energy turns out as

$$-PV = -T \sum_{k\,\alpha} \log\left(1 + e^{-\xi_{k\alpha}/T}\right), \tag{3.4.2}$$

where $P$ is the pressure. The chemical potential of the noninteracting system, $\mu_{\mathrm{NI}}(\lambda)$, is obtained by solving the number equation for $\mu$:

$$\rho = \frac{\partial P}{\partial \mu} = \frac{1}{V} \sum_{k\,\alpha} n_{\mathrm{F}}\left(\xi_{k\,\alpha}\right), \tag{3.4.3}$$

where, $n_{\mathrm{F}}(\xi) = 1/(\exp(\xi/T) + 1)$ is the Fermi distribution function.

### 4.B.1.  *Chemical potential at zero temperature*

At $T = 0$ Eq. (3.4.3) reduces for spherical GFC to $\rho = \sqrt{2\mu}(\lambda^2 + 2\mu)/ (3\pi^2)$, which can be inverted to:

$$\mu_{\mathrm{NI}}(\lambda) = \frac{1}{12}\left( 2^{2/3}\sqrt[3]{9\pi^2\sqrt{12\lambda^6\rho^2 + 729\pi^4\rho^4} + 2\lambda^6 + 243\pi^4\rho^2} - 4\lambda^2 \right.$$

$$\left. + \frac{2\sqrt[3]{2}\lambda^4}{\sqrt[3]{9\pi^2\sqrt{12\lambda^6\rho^2 + 729\pi^4\rho^4} + 2\lambda^6 + 243\pi^4\rho^2}} \right). \tag{3.4.4}$$

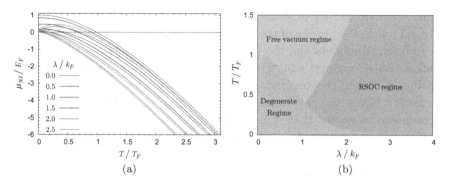

**Fig. 3.4.1.** (a) The chemical potential of the noninteracting system as a function of temperature for different RSOC strengths in spherical GFC. The solid lines are the exact results found by numerically solving Eq. (3.4.3). The dashed lines are the analytical approximations in the nondegenerate regime (see text). (b) Different regimes in RSOC — temperature space.

When $\lambda \ll \lambda_T$ and $\lambda \gg \lambda_T$, Eq. (3.4.4) can be approximated respectively as

$$\mu_{\mathrm{NI}}(\lambda) \approx E_F \left( 1 - \frac{1}{2^{\frac{1}{3}}} \left( \frac{\lambda}{\lambda_T} \right)^2 \right) \quad \text{and} \quad \mu_{\mathrm{NI}}(\lambda) \approx E_F \frac{2^{\frac{8}{3}}}{9} \left( \frac{\lambda_T}{\lambda} \right)^4.$$

### 4.B.2. *Effect of temperature on the chemical potential*

The exact noninteracting chemical potential as a function of temperature can be obtained by solving Eq. (3.4.3) numerically. The result for spherical GFC is shown in Fig. 3.4.1(a) as solid lines for various RSOC strengths.

When chemical potential is large and negative, the Fermi function can be approximated by the classical Boltzmann function to get an approximate expression in this regime as $\mu_{\mathrm{NI}} \simeq -T \log \left( \frac{\sqrt{T}(3T + \lambda^2)}{3\sqrt{2}\pi^{3/2}\rho} \right)$, shown as dashed lines in Fig. 3.4.1(a). Since, the approximate solution is obtained by approximating the Fermi function to the classical Boltzmann function, it is effectively like ignoring the quantum (or fermionic) nature of the particles. Hence the exact result being always higher than the approximate one owes to Pauli's exclusion principle.

The regime in which this approximation *cannot* be done is where the fermionic nature of the particles is important. This is depicted in Fig. 3.4.1(b) as the "degenerate regime". The boundary between the degenerate and the non-degenerate regime is determined by the noninteracting chemical potential at zero temperature (Eq. (3.4.4)). The non-degenerate regime further consists of two regimes, one in which the $1/\sqrt{\omega}$ part of the single particle density of states $(g_1(\omega) = 4g_s(\mathbf{0}, 2\omega)$, see Eq. (3.3.8)) plays crucial role in the physics of the system and the other in which the $\sqrt{\omega}$ part is sufficient to understand the physics. These two regimes are respectively called "RSOC regime" and "Free vacuum regime"; these are also depicted in Fig. 3.4.1(b). The temperature that separates these regimes is of the order $\lambda^2$. Given the constraints associated with cooling the system, it is reasonable to expect that it is in the RSOC regime where the effects of RSOC will be first visible in an experiment.

### 4.C. *Interesting effects in presence of a trap*

The presence of a trap is unavoidable in a cold atom experiment, which motivates us to include an isotropic-external-quadratic trapping potential in our discussions along with the homogeneous RSOC.

Local density approximation in this system yields that the cloud shrinks with increasing $\lambda$. It also predicts a spherical cloud even for anisotropic GFCs. However, a calculation of the cloud shape using exact eigen states shows that there is a characteristic anisotropy, inherited by the asymmetry of the GFC, in the cloud shape which increases with $\lambda$. This anisotropy can be explained using an approximation for the one particle energy levels obtained through an adiabatic formulation including the Pancharatnam-Berry phase effects. Thus, both size and shape of the cloud get affected by RSOC. These results should be useful in the design of cold atom experiments on fermions with RSOC.

Adiabatic formulation also reveals possibilities for the realization of interesting Hamiltonians by using RSOC in conjunction with a potential. We show that the use of spherical GFC with a harmonic trap produces a monopole field giving rise to a spherical-geometry-quantum-Hall like Hamiltonian in the momentum representation. A detailed discussion on such aspects is done in Ref. 49.

### 3.5. **Effect of RSOC on Interacting Fermions: Mean Field Analysis**

The FSTT taking place in the noninteracting system of fermions with RSOC suggests possibility of interesting physics in this system in presence of interactions. Here, to begin with, we explore this possibility mainly using a mean field theory.

## 5.A. Formulation

### 5.A.1. Mean field Hamiltonian

The full Hamiltonian Eq. (3.2.7) of the system can be rewritten as (see Eq. (3.4.1) and Eq. (3.2.5))

$$\mathcal{H} = \sum_{k\alpha} \xi_{k\alpha} C_{k\alpha}^{\dagger} C_{k\alpha} + \frac{\upsilon}{2V} \sum_q S^{\dagger}(q) S(q). \tag{3.5.1}$$

The interaction term being quartic in fermion operators, calls for an approximation for the analysis of the full Hamiltonian. Let us define $\delta S^{\dagger}(q) = S^{\dagger}(q) - \langle S^{\dagger}(q) \rangle$, where $\langle S^{\dagger}(q) \rangle$ is the expectation value of the operator $S^{\dagger}(q)$, and a similar definition for $\delta S(q)$. Mean field approximation amounts to substituting these into the above Hamiltonian and neglecting the term $\delta S^{\dagger}(q) \delta S(q)$. Along with this, we make the ansatz $\langle S^{\dagger}(q) \rangle = \langle S^{\dagger}(0) \rangle \delta_{q,0}$, where, $\delta_{q,0}$ is the Kronecker delta. Also note from Eq. (3.2.4) that $S^{\dagger}(0) = \frac{1}{\sqrt{2}} \sum_k \left( C_{k\uparrow}^{\dagger} C_{-k\downarrow}^{\dagger} - C_{k\downarrow}^{\dagger} C_{-k\uparrow}^{\dagger} \right) = \sqrt{2} \sum_k' \left( C_{k\uparrow}^{\dagger} C_{-k\downarrow}^{\dagger} - C_{k\downarrow}^{\dagger} C_{-k\uparrow}^{\dagger} \right) = \sqrt{2} \sum_{k\alpha}' \alpha C_{k\alpha}^{\dagger} C_{-k\alpha}^{\dagger}$, where prime over the summation symbol denotes sum over only half of the $k-$space and the invariance of the singlet state under spin-rotations has been exploited. With all these, the full Hamiltonian (Eq. (3.5.1)) after the mean field approximation becomes

$$\mathcal{H}_{\text{MF}} = \sum_{k\alpha} \xi_{k\alpha} C_{k\alpha}^{\dagger} C_{k\alpha} + \Delta_0 \sum_{k\alpha}' \alpha C_{k\alpha}^{\dagger} C_{-k\alpha}^{\dagger} + \Delta_0 \sum_{k\alpha}' \alpha C_{-k\alpha} C_{k\alpha} - \frac{V\Delta_0^{\,2}}{\upsilon}$$

where, $\Delta_0 = \frac{\upsilon \langle S^{\dagger}(0) \rangle}{\sqrt{2}V}$ is the excitation gap (also called pairing amplitude or order parameter) which, without loss of generality, is taken as real. Noting that $\xi_{k\alpha}$ has inversion symmetry, *i.e.*, $\xi_{-k\alpha} = \xi_{k\alpha}$,

$$\mathcal{H}_{\text{MF}} = \sum_{k\alpha}' \begin{pmatrix} C_{k\alpha}^{\dagger} & C_{-k\alpha} \end{pmatrix} \begin{bmatrix} \xi_{k\alpha} & \alpha\Delta_0 \\ \alpha\Delta_0 & -\xi_{k\alpha} \end{bmatrix} \begin{pmatrix} C_{k\alpha} \\ C_{-k\alpha}^{\dagger} \end{pmatrix} + \sum_{k\alpha}' \xi_{k\alpha} - \frac{V\Delta_0^2}{\upsilon}.$$

Bogoliubov transformation, $\gamma_{k\alpha 1} = u_{k\alpha} C_{k\alpha} - \alpha v_{k\alpha} C_{-k\alpha}^{\dagger}$, $\gamma_{k\alpha 2} = \alpha v_{k\alpha} C_{k\alpha} + u_{k\alpha} C_{-k\alpha}^{\dagger}$ with $u_{k\alpha}^2 = \frac{1}{2} \left( 1 + \frac{\xi_{k\alpha}}{E_{k\alpha}} \right)$, $v_{k\alpha}^2 = \frac{1}{2} \left( 1 - \frac{\xi_{k\alpha}}{E_{k\alpha}} \right)$ and $E_{k\alpha} = \sqrt{\xi_{k\alpha}^{\,2} + \Delta_0^{\,2}}$, leaves $\mathcal{H}_{\text{MF}} = \sum_{k\alpha}' E_{k\alpha} \left( \gamma_{k\alpha 1}^{\dagger} \gamma_{k\alpha 1} + \gamma_{k\alpha 2}^{\dagger} \gamma_{k\alpha 2} \right) + \sum_{k\alpha}' (\xi_{k\alpha} - E_{k\alpha}) - \frac{V\Delta_0^{\,2}}{\upsilon}.$ [g]

---

[g]For each helicity there are two branches of quasi-particle excitations labeled 1 and 2, making the total count four. However, these four branches are defined only in half of the momentum space, recovering the correct count of excitations.

### 5.A.2. *Gap equation and number equation*

With the above Hamiltonian, the free energy turns out as

$$-PV = \frac{1}{2}\sum_{k\alpha}(\xi_{k\alpha} - E_{k\alpha}) - \frac{V\Delta_0{}^2}{v} - T\sum_{k\alpha}\log\left(1 + e^{-E_{k\alpha}/T}\right).$$

Compare this with Eq. (3.4.2) for the noninteracting problem. On regularization (Eq. (3.2.6)) we obtain, $-P = \frac{1}{2V}\sum_{k\alpha}\left(\xi_{k\alpha} - E_{k\alpha} + \frac{\Delta_0{}^2}{k^2}\right) - \frac{\Delta_0{}^2}{4\pi a_s} - \frac{T}{V}\sum_{k\alpha}\log\left(1 + e^{-E_{k\alpha}/T}\right)$. Minimizing $P$ w.r.t. $\Delta_0$ obtains the gap equation

$$\frac{1}{4\pi a_s} = \frac{1}{2V}\sum_{k\alpha}\left(-\left(\frac{1 - 2\,n_F\,(E_{k\alpha})}{2E_{k\alpha}}\right) + \frac{1}{k^2}\right). \tag{3.5.2}$$

The number equation is obtained by $\rho = \frac{\partial P}{\partial \mu}$. This works out as

$$\rho = \frac{1}{V}\sum_{k\alpha}\frac{1}{2}\left(1 - \frac{\xi_{k\alpha}}{E_{k\alpha}}\left(1 - 2n_F\,(E_{k\alpha})\right)\right). \tag{3.5.3}$$

The simultaneous solution of Eq. (3.5.2) and Eq. (3.5.3) determines $\mu$ and $\Delta_0$ at a given $T$. At $T = 0$, the gap equation and the number equation reduce to

$$\frac{1}{4\pi a_s} = \frac{1}{2V}\sum_{k\alpha}\left(\frac{-1}{2E_{k\alpha}} + \frac{1}{k^2}\right) \quad \text{and} \quad \rho = \frac{1}{V}\sum_{k\alpha}\frac{1}{2}\left(1 - \frac{\xi_{k\alpha}}{E_{k\alpha}}\right). \tag{3.5.4}$$

### 5.A.3. *Ground state and its triplet content*

The ground state of the system is given by $|\Psi_G\rangle = \prod'_{k\alpha}(u_{k\alpha} + \alpha v_{k\alpha}c^\dagger_{k\alpha}c^\dagger_{-k\alpha})|0\rangle$, where $|0\rangle$ is the fermion vacuum. This can be (up to a normalization) rewritten as $|\Psi_G\rangle = e^{P^\dagger}|0\rangle$, where $P^\dagger$ is the pair creation operator given by $P^\dagger = \sum'_{k\alpha}\alpha\phi_{k\alpha}C^\dagger_{k\alpha}C^\dagger_{-k\alpha}$, where $\phi_{k\alpha} = v_{k\alpha}/u_{k\alpha}$. This can be reexpressed in terms of the singlet and triplet parts as

$$P^\dagger = \underbrace{\sum_k{}^{'}\phi_s(k)\left(C^\dagger_{k+}C^\dagger_{-k+} - C^\dagger_{k-}C^\dagger_{-k-}\right)}_{\text{singlet}} + \underbrace{\sum_k{}^{'}\phi_t(k)\left(C^\dagger_{k+}C^\dagger_{-k+} + C^\dagger_{k-}C^\dagger_{-k-}\right)}_{\text{triplet}}$$

$$\tag{3.5.5}$$

with $\phi_s(k) = \frac{1}{2}\left(\phi_{k+} + \phi_{k-}\right)$ and $\phi_t(k) = \frac{1}{2}\left(\phi_{k+} - \phi_{k-}\right)$. We define the triplet content $\eta_t$ for the many-body ground state as the weight of the triplet piece of the pair creation operator in Eq. (3.5.5). This may be compared with the definition of the triplet content in the two-body bound state (Eq. (3.3.13)).

## 5.B. *BCS-BEC crossover induced by RSOC*

Recall the results of free vacuum pertaining to the BCS-BEC crossover (Sec. 2.D.2) induced by tuning the attraction ($a_s$). The evolution of the ground state with RSOC at a *fixed* attraction is discussed here, focusing on spherical GFC and weak attraction regime ($k_F|a_s| \ll 1, a_s < 0$), where the effect of RSOC is most interesting.

When $k_F|a_s| \ll 1, a_s < 0$ and $\lambda = 0$, the usual BCS state of free vacuum described earlier is the ground state. For small $\lambda$ ($\lambda \ll \lambda_T$), the system is still described by the BCS theory. The gap equation in this regime yields an analytical approximation for the (exponentially small) gap as

$$\Delta_0 \simeq \frac{8\mu_{\mathrm{NI}}(\lambda)}{\exp\left(\frac{12\mu_{\mathrm{NI}}(\lambda)}{6\mu_{\mathrm{NI}}(\lambda)+\lambda^2}\right)} \exp\left(-\frac{3\pi\sqrt{\mu_{\mathrm{NI}}(\lambda)}}{\sqrt{2}|a_s|(6\mu_{\mathrm{NI}}(\lambda)+\lambda^2)}\right). \tag{3.5.6}$$

Evolutions of $\mu$ and $\Delta_0$ as a function of $\lambda$ are shown in Fig. 3.5.1(a) and 3.5.1(b). Figure 3.5.1(a) also shows the noninteracting chemical potential ($\mu_{\mathrm{NI}}(\lambda)$ described in Sec. 4.B.1), and the two-body energy $-E_b/2$. $\mu$ follows $\mu_{\mathrm{NI}}(\lambda)$ for $\lambda \ll \lambda_T$. **When $\lambda$ is comparable to $\lambda_T$, $\mu$ begins to deviate and around $\lambda \gtrsim \lambda_T$, it starts tending to $-E_b/2$, the value set by the two-body problem. This indicates a crossover from a BCS state to a BEC state** where the fermions form tight bosonic dimers and condense.

Further evidence to this crossover is obtained by evolution of $\eta_t$ shown in Fig. 3.5.1(c). The noninteracting ground state also has a triplet content. It increases monotonically with $\lambda$ and is equal to 1/2 for $\lambda \geq \lambda_T$. Attraction being in the *singlet* channel, the triplet content of the superfluid pair, as expected, is less than that of the noninteracting system, but has a similar qualitative behavior in the regime $\lambda \ll \lambda_T$. For $\lambda \gtrsim \lambda_T$, triplet content tends to that of the *two-body* bound-state wave function, demonstrating that **the pair wave function (Eq. (3.5.5)) tends to the two-body bound-state wave function.** This again signals the crossover to the BEC state — a **BCS-BEC crossover induced not by the attraction but by RSOC at a fixed attraction!**

The crossover regime overlaps with the regime of $\lambda$ where FSTT takes place. The qualitative nature of these results are similar even in the intermediate attraction regime. For a small positive $a_s$, the system is already a BEC even in free vacuum. However, the pair wave function is a singlet, *i.e.*, $\eta_t = 0$. By tuning $\lambda$, the system continues to be a BEC, but picks up a triplet component.

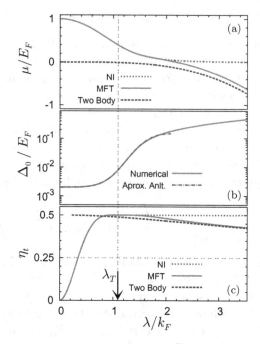

**Fig. 3.5.1.**   Evolution of the ground state with $\lambda$ at $k_F a_s = -1/4$ for the spherical GFC. (a) $\mu$ obtained from mean field theory (MFT) is compared with that of the noninteracting system (NI) and that set by the two-body binding energy $(-E_b/2)$. (b) Evolution of gap with $\lambda$. The analytical approximation (Eq. (3.5.6)) for $\lambda \ll \lambda_T$ is also shown. (c) Dependence of the triplet content on $\lambda$. This is compared with that of the noninteracting system and that of the two-body bound state.

### 5.C.  Bogoliubov spectrum

Now let us look at the dispersion of the Bogoliubov quasiparticles (Fig. 3.5.2) with particular focus on negative scattering lengths. We define the RSOC strength $\lambda_B$ (dependent on $a_s$) through $\mu(\lambda_B) = \mu_{NI}(\lambda_T)$.[h] At $\lambda = \lambda_B$, the nature of the dispersion will undergo a qualitative change.

When $\lambda = 0$, we obtain the standard BCS dispersion (Fig. 3.5.2(a)). When $\lambda < \lambda_B$, we see that there are low lying quasiparticle excitations of both helicities. In other words, the gap of excitation of quasiparticles for both helicities are equal (Fig. 3.5.2(b)). At $\lambda = \lambda_B$, the low lying $-$helicity excitation appears at $k = 0$ (Fig. 3.5.2(c)). For $\lambda > \lambda_B$, the negative helicity

---

[h]To be more precise, we are interested in those $a_s$ for which $\mu > 0$ in free vacuum. It is only then that this equation can have a root.

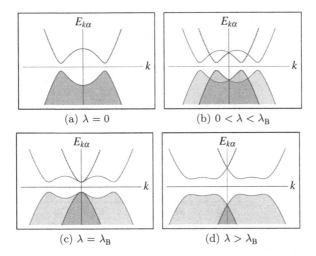

**Fig. 3.5.2.** Schematic evolution of the Bogoliubov quasiparticle dispersion with RSOC $\lambda$ in spherical GFC. Solid (red) lines correspond to +helicity quasiparticles, the dashed (blue) ones are for −helicity. The shading represents occupied states. The horizontal lines mark $\mu$.

quasiparticles have a higher gap and low lying quasiparticles are only of +helicity (Fig. 3.5.2(d)). $\lambda_B$ marks the transition point in the topology of the Bogoliubov dispersion and in the weak attraction regime, $\lambda_B \approx \lambda_T$.

## 5.D. *Rashbon Bose Einstein condensate (RBEC)*

Let us now study the BEC state in the $\lambda \to \infty$ limit. The parameter $1/\lambda a_s$, that dictates the physics of the two-body bound state (see Fig. 3.3.1 for example), tends to zero when $a_s \to \infty$ at a finite $\lambda$. We can thus expect the many body state obtained when $\lambda \gg$ maximum$(1/a_s, k_F)$ to be same as this state. Recall that the bound bosonic state of two fermions at resonance was termed as "rashbon" in Sec. 3.C.1. **Therefore, the properties of the BEC for $\lambda \to \infty$ are determined solely by $\lambda$ (the Rashba interaction), independent of $a_s$** (see Fig. 3.5.3). Hence we call this BEC as RBEC, standing for rashbon Bose Einstein condensate.

It is for this reason that certain properties of rashbons were enlisted in Sec. 3.3. Some more properties of RBEC will be discussed in Sec. 6.B. Detailed discussions and calculations including those for a generic GFC are done in Ref. 50.

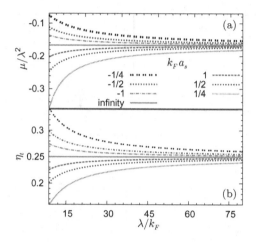

**Fig. 3.5.3.** Dependence of (a) chemical potential and (b) triplet content on $\lambda$ in spherical GFC in the RBEC regime. For all $a_s$, $\mu$ asymptotically attains the value set by the rashbon energy ($-\lambda^2/6$, see Eq. (3.3.14)) as $\lambda \to \infty$. $\eta_t$ also attains the rashbon value (1/4) independent of $a_s$.

## 5.E. *Estimation of transition temperature*

Recall from Sec. 2.D.4 that the transition temperature ($T_c$) of a weak superfluid is obtained by mean field theory through vanishing of the gap $\Delta_0$. From Eq. (3.5.2) and Eq. (3.5.3) this condition translates to simultaneously solving for $T$ and $\mu$ :

$$\frac{1}{4\pi a_s} = \frac{1}{2V} \sum_{k\,\alpha} \left( -\left( \frac{1 - 2\,n_F\,(\xi_{k\alpha})}{2\xi_{k\alpha}} \right) + \frac{1}{k^2} \right) \quad \text{and} \quad \rho = \frac{1}{V} \sum_{k\,\alpha} n_F\,(\xi_{k\alpha}).$$

$$(3.5.7)$$

In the regime where pair breaking effects are energetically costly, $T_c$ can be obtained as the BEC temperature of the bound bosonic pairs of fermions. This translates to

$$\frac{T_c}{T_F} = \left( \frac{16}{9\pi(\zeta(3/2))^2} \right)^{1/3} \frac{1}{m_{ef}},$$

$$(3.5.8)$$

where $m_{ef}$ is the effective mass of the bosonic pairs (see Sec. 3.D).

How these analyses apply to free vacuum and how much do they agree with the fluctuation calculation in their respective regimes was shown in Fig. 3.2.4. Now, we extend the calculations for the problem with RSOC, focusing on spherical GFC.[i] Refer to Fig. 3.5.4 which corresponds to a

---

[i]$T_c$ in extreme oblate GFC is calculated in Refs. 87 and 48.

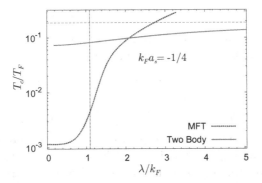

**Fig. 3.5.4.** Estimate of transition temperature in spherical GFC as a function of RSOC. Horizontal dashed line is the rashbon $T_c$. Vertical line corresponds to FSTT (Sec. 4.A).

fixed small negative $a_s$. For $\lambda \ll k_F$ the $T_c$ determined by the mean field theory obeys Eq. (3.2.14) and hence is exponentially small. However, for a large $\lambda$, in which regime we demonstrated that the system is a RBEC, $T_c$ is determined by Eq. (3.5.8). $m_{ef}$ calculated in Eq. (3.3.15) is used. What we find is quite remarkable. $T_c$ in this regime is of the order of Fermi temperature! In other words, **an exponentially small $T_c$ of a weak superfluid can be enhanced to the order of Fermi temperature by tuning RSOC without having to increase the attraction.** When $a_s$ is a small positive number, the $T_c$ for all values of $\lambda$ is given by Eq. (3.5.8).

Irrespective of $a_s$, when $\lambda \to \infty$, the $T_c$ tends to that of the rashbon which depends only on $\hat{\lambda}$ through the effective mass $m_{ef}$ (calculated for important GFCs in Ref. 48). It turns out that, among all the GFCs, the effective rashbon mass for the spherical GFC is largest and hence it has the smallest rashbon $T_c$ ($\approx 0.188 T_F$).

The conclusions made here will be verified by a more rigorous analysis including the fluctuation effects in Sec. 3.7.

## 3.6. Collective Excitations of the Superfluid

In the last section, we saw that the ground state admitted many novel physics. The natural question that arises now is regarding the excitations about this ground state, which will be addressed in this section using a Gaussian fluctuation theory with a functional integral framework.

## 6.A. *Formulation of fluctuations on the superfluid state*

The action corresponding to the full Hamiltonian (Eq. (3.5.1)) is $\mathcal{S}[\Psi] = \mathcal{S}_0 + \mathcal{S}_v$, where $\mathcal{S}_0$ is the action for the noninteracting system and is given by

$$\mathcal{S}_0 = \frac{1}{2} \sum_k \Psi^\star(k)(-G_0^{-1}(k, k'))\Psi(k'). \tag{3.6.1}$$

Here $\Psi^\star(k) = \left( c_+^*(k)\ c_+(-k)\ c_-^*(k)\ c_-(-k) \right)$ is the nambu vector composed of fermionic fields, $G_0$ is the noninteracting Green's function given by $G_0^{-1}(k, k') = \mathrm{Diag}(ik_n - \xi_+(\mathbf{k}), ik_n + \xi_+(\mathbf{k}), ik_n - \xi_-(\mathbf{k}), ik_n + \xi_-(\mathbf{k}))\delta_{k,k'}$, where (from now on) $k$ stands for the four vector $(ik_n, \mathbf{k})$, $ik_n$ being the Fermi Matsubara frequency.

$\mathcal{S}_v$ in the action describes the interaction among fermions as

$$\mathcal{S}_v = \frac{vT}{V} \sum_q S^\star(q)\, S(q), \tag{3.6.2}$$

where $S^\star(q) = \sum_{k,\alpha\beta} A_{\alpha\beta}(\mathbf{q}, \mathbf{k}) c_\alpha^* \left( \frac{q}{2} + k \right) c_\beta^* \left( \frac{q}{2} - k \right)$ is the singlet density in Fourier-Matsubara space (corresponding to Eq. (3.2.4)). (From now on) $q$ stands for the four vector $(iq_\ell, \mathbf{q})$ where $iq_\ell$ is the Bose Matsubara frequency, $\mathbf{q}$ is the COM momentum and $A_{\alpha\beta}(\mathbf{q}, \mathbf{k})$ was defined in Eq. (3.3.5). Performing Hubbard-Stratanovich transformation on $\mathcal{S}_v$ by introducing pair fields $\Delta(q)$, $\mathcal{S}[\Psi, \Delta] = \sum_k \Psi^\star(k)(-G^{-1}(k, k'))\Psi(k') - \frac{1}{v} \sum_q \Delta^\star(q)\, \Delta(q)$ where $G^{-1}(k, k') = G_0^{-1}(k, k') - \boldsymbol{\Delta}(k, k')$, and

$$\boldsymbol{\Delta}(k, k') = \begin{pmatrix} 0 & \Delta_{++}(k, k') & 0 & \Delta_{+-}(k, k') \\ \tilde{\Delta}_{++}(k, k') & 0 & \tilde{\Delta}_{+-}(k, k') & 0 \\ 0 & \Delta_{-+}(k, k') & 0 & \Delta_{--}(k, k') \\ \tilde{\Delta}_{-+}(k, k') & 0 & \tilde{\Delta}_{--}(k, k') & 0 \end{pmatrix} \tag{3.6.3}$$

with 
$$\Delta_{\alpha\beta}(k, k') = \sqrt{\frac{T}{V}} \sum_q \Delta(q) A_{\alpha\beta}\left( \mathbf{q}, \mathbf{k} - \frac{\mathbf{q}}{2} \right) \delta_{q, k-k'}$$

$$\tilde{\Delta}_{\alpha\beta}(k, k') = \sqrt{\frac{T}{V}} \sum_q \Delta^*(-q) A_{\beta\alpha}^*\left( -\mathbf{q}, \mathbf{k} - \frac{\mathbf{q}}{2} \right) \delta_{q, k-k'}. \tag{3.6.4}$$

The action is now quadratic in fermionic fields which can be integrated to yield

$$\mathcal{S}[\Delta] = -\ln\det[-G^{-1}] - \frac{1}{v} \sum_q \Delta^\star(q)\, \Delta(q).$$

The saddle point analysis of the action, which is equivalent to the mean field theory (Sec. 3.5) is achieved by looking for static and homogeneous solutions via the ansatz $\Delta^{\rm sp}(q) = \sqrt{2V/T}\Delta_0\,\delta_{q,0}$. With this ansatz for the saddle point, the Green's function $G(k,k')$ is

$$G(k,k') = \begin{pmatrix} G^p_+(k) & G^a_+(k) & 0 & 0 \\ -G^a_+(k) & G^h_+(k) & 0 & 0 \\ 0 & 0 & G^p_-(k) & G^a_-(k) \\ 0 & 0 & -G^a_-(k) & G^h_-(k) \end{pmatrix} \delta_{k,k'}\,,$$

where the matrix elements are

$$G^p_\alpha(k) = \frac{ik_n + \xi_{\boldsymbol{k}\,\alpha}}{(ik_n)^2 - E^2_{\boldsymbol{k}\,\alpha}}, G^h_\alpha(k) = \frac{ik_n - \xi_{\boldsymbol{k}\,\alpha}}{(ik_n)^2 - E^2_{\boldsymbol{k}\,\alpha}}, G^a_\alpha(k) = \frac{i\alpha\Delta_0}{(ik_n)^2 - E^2_{\boldsymbol{k}\,\alpha}}.$$

Recall that $E_{\boldsymbol{k}\,\alpha} = \sqrt{\xi_{\boldsymbol{k}\,\alpha}{}^2 + \Delta_0{}^2}$. The saddle point condition, after appropriate frequency sums and along with regularization (Eq. (3.2.6)) reproduces the mean field gap equation and the number equation derived in Sec. 5.A.2. The simultaneous solution of the gap equation and the number equation sets the values of $\Delta_0$ and $\mu$.

Collective excitations of the system are described by fluctuations about the saddle point state. We treat them at Gaussian level by introducing small fluctuations about the saddle point value of the pairing field, $\Delta(q) = \Delta^{\rm sp}(q) + \eta(q)$. After straightforward algebra, the action to quadratic order in $\eta$ becomes

$$S[\eta] = S^{\rm sp} + \frac{1}{2}\sum_q \begin{pmatrix} \eta^*(q) & \eta(-q) \end{pmatrix} \boldsymbol{\Pi}(q) \begin{pmatrix} \eta(q) \\ \eta^*(-q) \end{pmatrix},$$

where, $\quad \boldsymbol{\Pi}(q) = \begin{pmatrix} \Pi_{11}(q) & \Pi_{12}(q) \\ \Pi_{21}(q) & \Pi_{22}(q) \end{pmatrix}$

$\Pi_{11}(q) = \Pi_{22}(-q)$

$$= -\frac{1}{v} + \frac{T}{V}\sum_{k,\alpha\beta} |A_{\alpha\beta}(\boldsymbol{q},\boldsymbol{k})|^2 G^p_\alpha\left(iq_\ell + ik_n, \frac{\boldsymbol{q}}{2} + \boldsymbol{k}\right) G^h_\beta\left(ik_n, -\frac{\boldsymbol{q}}{2} + \boldsymbol{k}\right)$$

$\Pi_{12}(q) = \Pi_{21}(q) = \Pi_{12}(-q) = \Pi_{21}(-q)$

$$= -\frac{T}{V}\sum_{k,\alpha\beta} \alpha\beta|A_{\alpha\beta}(\boldsymbol{q},\boldsymbol{k})|^2 G^a_\alpha\left(iq_\ell + ik_n, \frac{\boldsymbol{q}}{2} + \boldsymbol{k}\right) G^a_\beta\left(ik_n, -\frac{\boldsymbol{q}}{2} + \boldsymbol{k}\right)$$

$$\tag{3.6.5}$$

Collective excitations of a superfluid can be described in terms of phase and amplitude oscillations. We, therefore, express $\eta$ in terms of two other fields: $\zeta$ (amplitude fluctuation) and $\phi$ (phase fluctuation), which are real in real space, as $\eta(q) = \Delta_0 \left( \zeta(q) + i\phi(q) \right)$, with $\zeta(-q) = \zeta^*(q)$ and $\phi(-q) = \phi^*(q)$. The action in terms of these two fields is

$$S[\zeta, \phi] = S^{\mathrm{sp}} + \frac{1}{2} \sum_q \left( \zeta^*(q) \ \phi^*(q) \right) \mathbf{\Gamma}(q) \begin{pmatrix} \zeta(q) \\ \phi(q) \end{pmatrix}$$

where, using Eq. (3.6.5), we find

$$\mathbf{\Gamma}(q) = \begin{pmatrix} \Gamma_{\zeta\zeta}(q) \ \Gamma_{\zeta\phi}(q) \\ \Gamma_{\phi\zeta}(q) \ \Gamma_{\phi\phi}(q) \end{pmatrix}$$

$$\Gamma_{\zeta\zeta}(q) = \Delta_0{}^2 \left( \Pi_{11}(q) + \Pi_{11}(-q) + 2\Pi_{12}(q) \right)$$

$$\Gamma_{\zeta\phi}(q) = i\Delta_0{}^2 \left( \Pi_{11}(q) - \Pi_{11}(-q) \right) = -\Gamma_{\phi\zeta}(q)$$

$$\Gamma_{\phi\phi}(q) = \Delta_0{}^2 \left( \Pi_{11}(q) + \Pi_{11}(-q) - 2\Pi_{12}(q) \right).$$

We preform the necessary frequency sums to obtain expressions for the $\Gamma$s. Here and henceforth in this section, we focus at zero temperature ($T = 0$). For small $q$ at $T = 0$, we have,

$$\Gamma_{\phi\phi}(iq_\ell, \mathbf{q}) = q_i K^{\mathrm{s}}_{ij} q_j - Z(iq_\ell)^2$$

$$\Gamma_{\zeta\phi}(iq_\ell, \mathbf{q}) = -iq_\ell X$$

$$\Gamma_{\zeta\zeta}(iq_\ell, \mathbf{q}) = U + q_i V_{ij} q_j - W(iq_\ell)^2$$

where the quantities $K^{\mathrm{s}}, Z, X, U, V, W$ depend on the saddle point values of $\Delta_0$ and $\mu$ and can be expressed as

$$Z = \frac{\Delta_0{}^2}{2V} \sum_{\mathbf{k}\,\alpha} \frac{1}{4E^3_{\mathbf{k}\,\alpha}} \qquad X = \frac{\Delta_0{}^2}{2V} \sum_{\mathbf{k}\,\alpha} \frac{\xi_{\mathbf{k}\,\alpha}}{2E^3_{\mathbf{k}\,\alpha}}$$

$$U = \frac{\Delta_0^4}{2V} \sum_{\mathbf{k}\,\alpha} \frac{1}{E^3_{\mathbf{k}\,\alpha}} \qquad W = Z - \frac{\Delta_0^4}{2V} \sum_{\mathbf{k}\,\alpha} \frac{1}{4E^5_{\mathbf{k}\,\alpha}}. \qquad (3.6.6)$$

$K^{\mathrm{s}}_{ij}$ is the phase stiffness given by[j]

$$K^{\mathrm{s}}_{ij} = \frac{\Delta_0{}^2}{2V} \sum_{\mathbf{k}\alpha} \frac{v^\alpha_i(\mathbf{k})v^\alpha_j(\mathbf{k})}{4E^3_{\mathbf{k}\,\alpha}} + \frac{2\Delta_0{}^2}{V} \sum_{\mathbf{k}} \frac{(\varepsilon_{\mathbf{k}+} - \varepsilon_{\mathbf{k}-})^2}{2E_{\mathbf{k}+}E_{\mathbf{k}-}(E_{\mathbf{k}+} + E_{\mathbf{k}-})} \tilde{S}_{ij}(\mathbf{k})$$

$$(3.6.7)$$

where $v^\alpha_i(\mathbf{k}) = \frac{\partial \varepsilon_\alpha(\mathbf{k})}{\partial k_i}$, and $\tilde{S}_{ij}(\mathbf{k})$ is a tensor that defines the singlet amplitude $A_{+-}(\mathbf{q}, \mathbf{k})$ for small $\mathbf{q}$ as $|A_{+-}(\mathbf{q}, \mathbf{k})|^2 = |A_{-+}(\mathbf{q}, \mathbf{k})|^2 \approx q_i \tilde{S}_{ij}(\mathbf{k}) q_j$.

The dispersion of the excitations is obtained by analytically continuing $iq_\ell \to \omega^+$ and solving $\det \mathbf{\Gamma}(\omega^+, \mathbf{q}) = 0$. We obtain two modes for a given

---

[j]The phase stiffness is directly related to the superfluid density $\rho_{\mathrm{s}}$.

$q = |q|\hat{q}$: a *gapless* sound mode and a *gapped* Anderson-Higgs (AH) mode. The speed of sound along direction $\hat{q}$ and the mass of the Anderson-Higgs mode $M_{AH}$ are obtained as

$$c_s^2(\hat{q}) = \frac{\hat{q}_i K_{ij}^s \hat{q}_j}{Z + \frac{X^2}{U}} \quad \text{and} \quad M_{AH}{}^2 = \frac{ZU + X^2}{ZW}. \tag{3.6.8}$$

Equations (3.6.7) and (3.6.8) are applicable to *any* $\lambda$ and $a_s$ at $T = 0$. Our formulation can readily be extended to finite temperatures and also reproduces the free vacuum results presented in Sec. 2.D.3. Recall that in free vacuum the BEC for small positive $a_s$ can be described by the Bogoliubov theory of interacting bosons, where the boson mass is twice the fermion mass and the effective boson-boson scattering length $a_{BEC}$ is proportional to $a_s$ (Eq. (3.2.12)). Now we discuss collective excitations of superfluids realized in presence of RSOC taking spherical GFC as an example.

### 6.B.  *Results in spherical GFC*

Here we show and discuss the evolution of $K^s$, $c_s$ and $M_{AH}$ with increasing $\lambda$ for different regimes of $a_s$.

In Fig. 3.6.1(a), we see that $K^s$ is $\rho/4$ when $\lambda = 0$ with a subsequent fall of order $\lambda^2/k_F{}^2$ for small $\lambda$. For small negative $k_F a_s$, the behaviour of $K^s$ is *nonmonotonic* with $\lambda$.[k] In the RBEC regime ($\lambda \gg$ maximum($k_F, 1/a_s$)), the

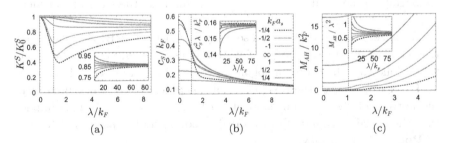

**Fig. 3.6.1.** Evolution of different quantities with increasing $\lambda$ for the spherical GFC and for various scattering lengths. (a). Phase stiffness $K^s$: $K_0^s = \rho/4$ is the phase stiffness in free vacuum. The inset shows that $K^s/K_0^s$ tends to $2/m^R$ for large $\lambda$ demonstrating the emergent Galilean invariance. (b). Sound speed $c_s$. (c). Mass of the Anderson-Higgs boson $M_{AH}$. The insets in (b) and (c) show that $c_s^2$ and $M_{AH}$ have the behaviours obtained in Eq. (3.6.10), independent of $a_s$ for large $\lambda$. The dashed vertical lines correspond to FSTT (Sec. 4.A).

---

[k]This is qualitatively same as the findings of Ref. 88 for extreme oblate GFC.

phase stiffness $K^s$ tends to $\rho^R/m^R$, where $\rho^R = \rho/2$ is the rashbon number density and $m^R$, the effective mass of rashbon, which from Eq. (3.3.15) is equal to $\frac{3}{7}(4+\sqrt{2})$ times the fermion mass. This means that, in the RBEC regime at low energies, the effective dispersion of rashbons (Sec. 3.D) is Galilean invariant, taking the form $\varepsilon_R(\boldsymbol{q}) = -E^R + \frac{|q|^2}{2m^R}$, where $E^R$ is the rashbon binding energy (equal to $\lambda^2/3$ for spherical GFC, see Eq. (3.3.14)). This is an instance of **emergent infrared symmetry** as this is consistent with Leggett's result[79,80] for a Galilean invariant system.

The sound speed (Fig. 3.6.1(b)) decreases monotonically with increasing $\lambda$ for all $a_s$ and the mass of the Anderson-Higgs boson is shown in Fig. 3.6.1(c). In the RBEC regime, the gap and the chemical potential are given to leading order by

$$\Delta_0^2 = \frac{2\pi}{\rho}\frac{\lambda}{\sqrt{3}} \quad \text{and} \quad \mu = -\frac{E^R}{2} + \pi\rho\frac{\sqrt{3}}{\lambda}. \tag{3.6.9}$$

Additional analysis provides

$$c_s^2 = \frac{2\pi\rho}{m^R}\left(\frac{\sqrt{3}}{\lambda}\right) \quad \text{and} \quad M_{AH} = \frac{2}{3}\lambda^2. \tag{3.6.10}$$

Analogous to the results of Sec. 5.D, in this limit, all these are independent of $a_s$.

A nonzero phase stiffness and sound propagation in RBEC suggest that rashbons are *interacting* bosons. We now explore the nature of this interaction following the route analogous to that described in Sec. 2.D.3.

From Eq. (3.6.9), the rashbon chemical potential $\mu^R$ (measured from the bottom of the rashbon band at $-E^R$) is $\mu^R = 2\pi\rho\frac{\sqrt{3}}{\lambda}$. This implies that the speed of sound (Eq. (3.6.10)) is **consistent with Eq. (3.2.11) of Bogoliubov theory**

$$c_s^2 = \frac{\mu^R}{m^R} = \frac{2\pi\rho}{m^R}\left(\frac{\sqrt{3}}{\lambda}\right) = \frac{4\pi}{(m^R)^2}\left(\frac{m^R\sqrt{3}}{\lambda}\right)\rho^R, \tag{3.6.11}$$

and the low energy properties of RBEC are similar to those of the usual Bogoliubov Bose fluid. This also demonstrates that RBEC is a condensate of rashbons with contact interaction described by the rashbon-rashbon scattering length[1]

$$a_R = \frac{m^R \sqrt{3}}{\lambda} = \frac{3\sqrt{3}(4 + \sqrt{2})}{7} \frac{1}{\lambda} \approx \frac{4}{\lambda}. \qquad (3.6.12)$$

Compare and contrast this with Eq. (3.2.12) of free vacuum. This result is remarkable because **the rashbon-rashbon interaction depends on a kinetic energy parameter** $\lambda$ **and is** *independent* **of the constituent-fermion-interaction parameter** $a_s$! To the best of our knowledge, it is unique of this kind.

Although this demonstration was in spherical GFC, the arguments made are applicable to other GFCs ($\hat{\lambda}$) as well. A quantitative analysis for a variety of GFCs along with a discussion on associated anisotropies is done in Ref. 50.

## 3.7. Fluctuation Theory of the Normal State: Gaussian and Beyond

In Sec. 5.E we estimated the superfluid $T_c$ using mean field theory (in the regime where $a_s < 0, |k_F a_s| \ll 1$ and $\lambda \ll k_F$) and two-body analysis (in the regime where $a_s > 0, k_F a_s \ll 1$ or $\lambda \gg k_F$), which we argued as valid in their respective regimes. An important outcome of that analysis was that the $T_c$ of weak superfluids can be enhanced to order of Fermi temperature by tuning RSOC. However, this claim is yet to be justified based on a calculation that includes fluctuations and is valid for all regimes of the parameters. Also, a study of the normal state of this system is most relevant as initial experiments with RSOC would realize a high temperature regime. Recall that, in free vacuum, a Gaussian fluctuation theory[72,82,83] yielded the superfluid $T_c$ (the black curve in Fig. 3.2.4). In this section we aim to obtain the full phase diagram including RSOC using a fluctuation theory. In the process we discuss the shortcomings of the Gaussian approximation while we also construct an approximate beyond-Gaussian theory.

### 7.A. *Formulating Gaussian fluctuations in the normal state*

The formulation here, upto Eq. (3.7.1), resembles to a great extent what is presented in Sec. 6.A before. Hence we only outline the formulation

---

[1]Further self consistent treatment of the theory[89,90] might be required for the accurate determination of the coefficient.

also borrowing the same/similar notations. We start again with the action $\mathcal{S}[\Psi] = \mathcal{S}_0 + \mathcal{S}_v + \mathcal{S}_F$, where $\mathcal{S}_0$ is the action for the noninteracting system (Eq. (3.6.1)) and $\mathcal{S}_v$ describes the fermion-fermion interaction (Eq. (3.6.2)). We introduce the forcing fields $F(q)$ in the action through $\mathcal{S}_F$ as $\mathcal{S}_F = \sqrt{T/V} \sum_q F(q)S^*(q) + F^*(q)S(q)$. Hubbard-Stratanovich transformation on $\mathcal{S}_v$ along with associated pairing fields $\eta(q)$ ($\Delta(q)$ were the pairing fields in the superfluid state) yields $\mathcal{S}[\Psi, \eta, F] = \sum_k \Psi^*(k)(-G^{-1}(k, k'))\Psi(k') - \frac{1}{v}\sum_q \eta^*(q)\,\eta(q)$, where $G^{-1}(k, k') = G_0^{-1}(k, k') - \gamma(k, k')$, $G_0$ is the non-interacting Greens function and $\gamma(k, k')$ is defined very similar to $\Delta(k, k')$ in Eqs. 3.6.3 and 3.6.4, with $\gamma(q) = \eta(q) + F(q)$. Integrating out the fermionic fields,

$$\mathcal{S}[\eta, F] = -\ln\det[-G^{-1}] - \frac{1}{v}\sum_q \eta^*(q)\,\eta(q). \qquad (3.7.1)$$

The formalism so far is exact. To study the physics of normal state, we expand this action about the saddle point where $\eta(q) = 0$ up to quartic order as

$$\mathcal{S} \approx -\ln\det[-G_0^{-1}] - \frac{1}{v}\sum_q \eta^*(q)\,\eta(q) + \sum_q \gamma^*(q)L(q)\gamma(q)$$

$$+ \sum_{q_1, q_2, q_3, q_4} \gamma^*(q_1)\gamma^*(q_2)K(q_1, q_2; q_3, q_4)\gamma(q_3)\gamma(q_4). \qquad (3.7.2)$$

The quantities $L$ and $K$ are derivatives of the action (Eq. (3.7.1)) to appropriate order in $\eta$. The arguments of $K$ satisfy momentum conservation.

Retaining only the first three terms in Eq. (3.7.2) produces the Gaussian fluctuation theory,[71,82,83] quadratic in $\eta$. Upon integration of the $\eta$ fields, we obtain the Gaussian action, $\mathcal{S}_G[F] = -\ln\det[-G_0^{-1}] + \sum_q \ln M(q) - \sum_q F^*(q)\chi(q)F(q)$, where $M(q) = L(q) - \frac{1}{v} = L(q) + \frac{1}{V}\sum_k \frac{1}{|k|^2} - \frac{1}{4\pi a_s}$, and $L(q) = \frac{1}{V}\sum_{k,\alpha,\beta}|A_{\alpha\beta}(q, k)|^2 \frac{1-n_F(\xi_{(q/2+k)\,\alpha})-n_F(\xi_{(q/2-k)\,\beta})}{iq_l-\xi_{(q/2+k)\,\alpha}-\xi_{(q/2-k)\,\beta}}$. The analysis also produces the pairing susceptibility $\chi(q) = L(q)\left(\frac{L(q)}{M(q)} - 1\right)$, whose divergence from the positive side up on the reduction of temperature indicates a pairing instability. We have verified numerically that the first such divergence of $\chi(0, q)$ occurs at $q = 0$, implying that the system is most susceptible to homogeneous pairing. $T_c$ is then obtained via the Thouless criterion[91]

$$-\frac{1}{4\pi a_s} - \frac{1}{4V}\sum_{k,\alpha}\left(\frac{1 - 2\,n_F(\xi_{k\,\alpha})}{\xi_{k\,\alpha}} - \frac{2}{|k|^2}\right) = 0.$$

This is identically the equation we encountered for $T_c$ even in the mean field theory (Eq. (3.5.7)). However the chemical potential that goes into the Thouless criterion (through $\xi_{k\,\alpha}$) makes all the difference. To get the chemical potential within Gaussian approximation we calculate the pressure $P = P_{\rm NI} + P_{\rm G}$ where, $P_{\rm NI}$ is the pressure corresponding to the noninteracting system (Eq. (3.4.2)) and $P_{\rm G}$ is the additional contribution arising from the Gaussian theory. With more analysis and analytically continuing $iq_l$ to real frequencies $(\omega^+)$, we obtain $P_{\rm G} = -\frac{1}{\pi V}\sum_q \int_{-\infty}^{\infty} \Delta\omega\, n_{\rm B}(\omega)\,\arg\left(M(\omega^+, q)\right)$, where $\arg(z)$ is the argument of the complex number $z$, $n_{\rm B}(x) = 1/(e^{x/T}-1)$ is the Bose function and $M(\omega^+, q)$ is the analytic continuation of $M(iq_l \to z, q)$ evaluated just above the real axis in the $z$-plane. The chemical potential (within Gaussian theory) is determined from the equation of state $\rho = \rho_t(T, \mu)$, where the number function $\rho_t(T, \mu)$ is given by

$$\rho_t(T,\mu) = \frac{\partial P}{\partial \mu} = \frac{1}{V}\sum_{k,\alpha} n_{\rm F}\left(\xi_{k\,\alpha}\right) - \frac{1}{\pi V}\sum_q \int_{-\infty}^{\infty} \Delta\omega\, n_{\rm B}(\omega)\,\frac{\partial\,\arg\left(M(\omega^+, q)\right)}{\partial \mu}.$$

$$(3.7.3)$$

## 7.B. *Executing the calculation*

In order to execute the calculations, the real and imaginary parts of $M$ form the essential components. The imaginary part is given by

$$\mathrm{Im}[M(\omega^+, q)] = -\frac{\pi}{V}\sum_{k,\alpha,\beta} |A_{\alpha\beta}(q, k)|^2$$

$$\times \left[1 - n_{\rm F}\left(\xi_{(q/2+k)\,\alpha}\right) - n_{\rm F}\left(\xi_{(q/2-k)\,\beta}\right)\right]\delta$$

$$\times \left[\omega - \xi_{(q/2+k)\,\alpha} - \xi_{(q/2-k)\,\beta}\right]. \qquad (3.7.4)$$

When the gas is dilute enough so that the $n_{\rm F}$ in the above equation can be neglected, $\mathrm{Im}[M]$ will essentially reduce to the two-particle SDOS (ignoring the shift of reference due to the chemical potential and an overall factor; see Eq. (3.3.7)). The presence of $n_{\rm F}$ signifies the effect on pairing arising from Pauli blocking. The real part of $M$ can be obtained by the Kramers-Krönig transformation as

$$\mathrm{Re}[M(\omega^+, q)] = -\frac{1}{\pi}\int_{-\infty}^{\infty} \Delta\zeta\frac{\mathrm{Im}[M(\zeta^+, q)]}{\omega - \zeta}. \qquad (3.7.5)$$

There exists a frequency $\omega_0(q)$ (referred here as scattering threshold) such that $\mathrm{Im}[M] = 0$ for all $\omega < \omega_0(q)$. This is given by

$$\omega_0(q) = -2\mu + \frac{(|q| - q_0)^2}{4}\,\Theta(|q| - q_0), \qquad (3.7.6)$$

where $\Theta$ is the step function. The quantity $M(z, q)$ may have an isolated zero below the scattering threshold at $z = \omega_{\rm b}(q)$ along the real axis; this signals the presence of a bosonic bound state of a fermion pair with COM momentum $q$. In analogy with Eq. (3.3.12), in this section we define the pole binding energy $E_b = \omega_0(q) - \omega_b(q)$, which should not be confused with the two-body binding energy although they are numerically same in the dilute limit.

It is useful to rewrite Eq. (3.7.3) by explicitly identifying the contributions to $\rho_t$

$$\rho_t(T, \mu) = \rho_F(T, \mu) + \rho_b(T, \mu) + \rho_c(T, \mu) \tag{3.7.7}$$

where $\rho_F(T, \mu) = \frac{1}{V} \sum_{k,\alpha} n_F(\xi_{k\,\alpha})$ is the fermion contribution, $\rho_b(T, \mu) = -\frac{1}{V} \sum_q n_B(\omega_b(q)) \frac{\partial \omega_b(q)}{\partial \mu}$ is the contribution from the bosonic poles, and $\rho_c(T, \mu) = -\frac{1}{\pi V} \sum_q \int_{\omega_0(q)}^{\infty} \Delta\omega \; n_B(\omega) \; \frac{\partial \arg(M(\omega^+, q))}{\partial \mu}$ is the contribution from the scattering continuum manifested as a branch cut of $M(z, q)$ along the real $z$-axis.

In what follows, where the results of the calculations are presented, we specialize to spherical GFC.[m] However, the qualitative conclusions are based on arguments that are applicable to all the oblate GFCs.

## 7.C. *Inadequacy of the Gaussian theory*

The Gaussian theory which is successful[76] in the description of the interacting Fermi gas in free vacuum, is not so in the presence of RSOC. A crucial question to address is whether the chemical potential (obtained by inverting Eq. (3.7.3)) is unique? In free vacuum, the solution is very much unique at least in the regime of interest. In other words, the number function is a monotonically increasing and a continuous function of the chemical potential.

Now, Fig. 3.7.1 shows the dependence of $\mu$ on $\lambda$ at a fixed low $T$ ($T < T_F$) and negative $a_s$ in presence of RSOC. The flabbergasting aspect of RSOC we encounter is that — the Gaussian theory has *two* solutions for $\mu$ for a given set of parameters. This is to say that for a given $a_s$, $\lambda$, $T$ and $\rho$ there can be more than one solution to $\mu$ that solves Eq. (3.7.3). For $\lambda \ll k_F$,

---

[m]It is worth mentioning that, unlike in free vacuum, an analytical expression for Im[$M$] is not available even in spherical GFC. On top of this, the multiple integrals in evaluation of $\rho_b$ and $\rho_c$ which are challenging for an efficient numerics even in free vacuum, are extremely involved in presence of RSOC due to various factors including principal valued integrals whose locations also have to be obtained numerically, sharp features and divergences in the integrand, among many. Performing these integrals is technically the hardest part of this work.

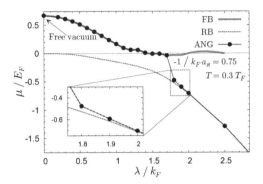

**Fig. 3.7.1.** Dependence of chemical potential ($\mu$) on RSOC strength: The Gaussian theory has two distinct solutions for $\mu$, called the free vacuum branch (FB) (higher $\mu$, solid line), and the rashbon branch (RB) (lower $\mu$, dashed line). The solid line with dots indicates $\mu$ in the approximate non-Gaussian (ANG) theory developed here. Non-Gaussian effects eliminate RB for $\lambda \lesssim k_F$.

one of the solutions, called the "free vacuum branch"(FB), is smoothly connected to that of the free vacuum found in previous works.[71,83] The other solution always has $\mu < 0$, even for $\lambda \ll k_F$. For large $\lambda$, chemical potential in this branch is determined by the rashbon dispersion, and hence called the rashbon branch (RB). Both the branches exist all the way upto $\lambda = 0$, except at $\lambda = 0$ where the RB solution does not exist. Also, the chemical potential along RB, which *always* has the lower free energy, approaches $0^-$ as $\lambda \to 0$. This suggests that the equilibrium state of the Gaussian theory with RSOC is not continuously connected to the free vacuum in the limit of $\lambda \to 0$!

Why are there multiple solutions? Before proceeding to understand the physics, we plot the number function as a function of $\mu$ in Fig. 3.7.2 for the set of parameters chosen in Fig. 3.7.1. Observe that in all the curves, when $\mu$ approaches $0^-$, the number function diverges. As soon as $\mu$ becomes positive, it comes back to a finite value and evolves smoothly thereafter.

At this instance one may wonder why is there a divergence in the number function at $\mu = 0^-$? It turns out that in Fig. 3.7.2, it is the pole term ($\rho_b(T, \mu)$ of Eq. (3.7.7)) which is causing the singularity. Here, $\rho_b$ diverges as $1/|\mu|$ for $\mu = 0^-$. Also, $\rho_b = 0$ (there is no pole for $M^{-1}$) when $\mu > 0$.

The next two natural questions are — why is there a divergence in $\rho_b$ when $\mu = 0^-$? and why does the pole suddenly go away when $\mu$ changes sign? To answer these, the pole term calls for a more careful attention. To

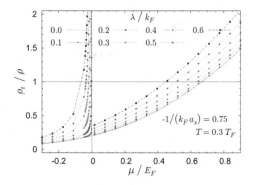

**Fig. 3.7.2.** The number function as a function of the chemical potential for different RSOC strengths $\lambda$ in spherical GFC at $-1/k_F a_s = 0.75$, $T = 0.3 T_F$.

this end we find these interesting results:

(1) In contrast to free vacuum, the pole formation is favored at smaller and not at larger values of $|\boldsymbol{q}|$.
(2) If $\mu < 0$ any $\boldsymbol{q}$ such that $0 \leq |\boldsymbol{q}| < q_0$ where $q_0 = 2\lambda/\sqrt{3}$ always admits a pole, *whatever* be $T$ and $a_s$.
(3) If $\mu > 0$ for all $\boldsymbol{q}$ such that $0 \leq |\boldsymbol{q}| < q_0$ the chances of having a pole is suppressed by RSOC.

The justifications for these points are as follows. The similarity between Im$[M]$ and SDOS was noted below Eq. (3.7.4). Arguing along the same lines, the similarity between Eq. (3.7.5) and the right hand side of the secular equation for two-body bound state (Eq. (3.3.11)) is also apparent. Hence the physics of the above results must be traceable from the two-body physics described in Sec. 3.D. When $\omega - \omega_0(\boldsymbol{q}) \to -\infty$, Re$[M]$ always diverges to $+\infty$ as $\sqrt{\omega_0(\boldsymbol{q}) - \omega}$, which is a free vacuum result. When $|\boldsymbol{q}| \gtrsim q_0$, owing to the depletion of Im$[M]$ in the infrared regime, the pole ceases to exist (Fig. 3.7.3(a)). On the other hand, when $|\boldsymbol{q}| < q_0$, Re$[M]$ is divergent not only at $\omega \to -\infty$ but also at $\omega = \omega_0(\boldsymbol{q})$, owing to the infrared divergence of Im[M]. If $\mu < 0$, this divergence at $\omega_0(\boldsymbol{q})$ is negative resulting always in a pole (Fig. 3.7.3(b)). Else if $\mu > 0$, this divergence is positive, thereby explaining the suppression of pole formation (Fig. 3.7.3(c)).[n]

---

[n]There is a possibility of double (or any even number of) poles as seen in Fig. 3.7.3(c). In the regime $|\boldsymbol{q}| < q_0$, $\omega_0(\boldsymbol{q}) < 0$ for $\mu > 0$ (see Eq. (3.7.6)). $\omega_b(\boldsymbol{q})$, being bounded above by $\omega_0(\boldsymbol{q})$, is then forced to be negative. But such poles are necessarily unphysical, because for a physically meaningful pole, $\omega_b(\boldsymbol{q}) \geq 0$ to ensure that Bose function that

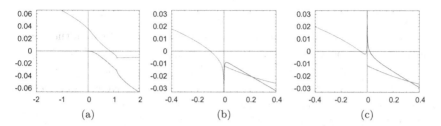

**Fig. 3.7.3.** The real (red curve) and the imaginary (blue curve) parts of $M(\omega, \boldsymbol{q})$ in units of $k_F$ as a function of $\omega - \omega_0(\boldsymbol{q})$ in units of $E_F$. All curves are at resonance, $\lambda = k_F$ and $T = 0.3\,T_F$. In (a) $|\boldsymbol{q}| = 1.5\,k_F$, $\mu = -0.01E_F$; (b) $|\boldsymbol{q}| = 0$, $\mu = -0.01E_F$; (c) $|\boldsymbol{q}| = 0$, $\mu = 0.01E_F$.

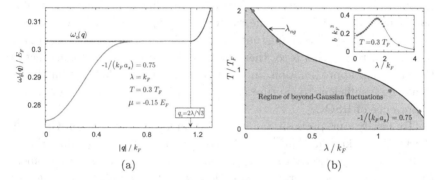

**Fig. 3.7.4.** (a). Energy dispersion of the bound boson (fermion-pair) in the Gaussian theory: The dashed black line is the scattering threshold $\omega_0(\boldsymbol{q})$ as a function of $|\boldsymbol{q}|$. (b). Regime where non-Gaussian effects are most wanted: At a given $T$ and $a_s$, non-Gaussian effects are crucial when $\lambda \lesssim \lambda_{\mathrm{ng}}(T, a_s)$. Inset shows the parameter $b$ that characterizes non-Gaussian effects at a fixed $T$.

The above analysis answers the second part of the question we are addressing which is regarding the sudden disappearance of the pole when $\mu$ turns positive. We now pursue on the first part of the question which is regarding the divergence of the pole term at $\mu = 0^-$.

First of all since $\mu$ is negative, from the analysis just presented, the existence of a pole is guaranteed in the regime $0 \le |\boldsymbol{q}| \le q_0$. It is evident from Fig. 3.7.4(a) that even for a largish negative $\mu$, in the regime $q_0/2 \lesssim |\boldsymbol{q}| \le q_0$, $\omega_\mathrm{b}(\boldsymbol{q}) \simeq \omega_0(\boldsymbol{q}) = -2\mu$. Hence, in this regime, $\omega_\mathrm{b}(\boldsymbol{q}) \simeq 2|\mu|$; $\frac{\partial \omega_\mathrm{b}(\boldsymbol{q})}{\partial \mu} \simeq -2$ ; $n_\mathrm{B}(\omega_\mathrm{b}(\boldsymbol{q})) \simeq \frac{1}{e^{2|\mu|/T} - 1}$. Also, in the limit $\mu \to 0^-$, $n_\mathrm{B}(\omega_\mathrm{b}(\boldsymbol{q})) \to T/2|\mu|$.

---

enters $\rho_\mathrm{b}$ is positive. We have verified numerically that such unphysical pole formation does not occur for the parameter regime discussed here.

Hence, as observed in Fig. 3.7.2, $\rho_b$ diverges as $\rho_b \simeq \frac{T}{2|\mu|}\frac{q_0^3}{3}$. Qualitatively, the bosons find a large number of finite $q$ states in each of which there is an enormous room provided by the Bose function to get occupied, forcing $\mu$ to be self-consistently negative resulting in RB not being smoothly connected to the free vacuum. Note also that FB does not have any contribution from $\rho_b$, since there is no bosonic bound state here for $\mu > 0$.

Let us now track the origin of obtaining the RB solution for $\lambda \ll k_F$. The existence of this branch when $\lambda \ll k_F$ entirely owes to a large set of finite $q$ states with vanishingly small pole binding energy. Imagine for a moment that such states were not there. Then the divergence in the pole term as seen in Fig. 3.7.2 will not be present. This then renders the solution for the chemical potential unique eliminating RB. But what would kill these weakly bound fully formed poles in the range $q_0/2 \lesssim |q| \leq q_0$? We propose that introducing non-Gaussian fluctuations would kill them and thereby cure the problem stated. If this is so, then the issue just described is an artifact of the Gaussian approximation. In other words, Gaussian fluctuation theory would be insufficient to capture all the essential aspects of the interaction in presence of RSOC.

To justify our proposition, we make an estimation of non-Gaussian effects. It is known that[83] the fourth order coefficient (called the $b$-coefficient, proportional to $K(q_1, q_2; q_3, q_4)$ in the limit of zero momentum) in the Landau free energy describes the repulsion among the bosons corresponding to the poles under investigation, which within the Gaussian approximation are noninteracting. It is given by $b = \frac{1}{4V}\sum_{k,\alpha}\left(\frac{1-2\,n_F(\xi_{k\,\alpha})}{\xi_{k\,\alpha}^3} - \frac{2\,n_F(\xi_{k\,\alpha})(1-n_F(\xi_{k\,\alpha}))}{T\,\xi_{k\,\alpha}^2}\right)$. The lowest order effect of $b$ would be to shift the energy of the bound state via a Hartree shift, *i.e.*, $\omega_b(q) \to \omega_b(q) + \kappa b^{1/3}\rho_b(T,\mu)$ where $\kappa = \frac{(2\pi)^{4/3}}{4-\sqrt{2}} \approx 4.484$. Hence, we can say that the weakly bound states survive the non-Gaussian fluctuations only if $E_b(q=0) \geq \kappa b^{1/3}\rho_b(T,\mu)$. Using this criterion we sketch the regime ($\lambda \leq \lambda_{ng}(T, a_s)$) in Fig. 3.7.4(b) where RB can be possibly eliminated by non-Gaussian effects. We find that the non-Gaussian effects, characterized by $b$ in the sense just described, are predominant in the regime where the RB branch should not have existed.

## 7.D. *Approximate non-Gaussian theory and the phase diagram*

A calculation involving beyond-Gaussian effects is highly involved and hence we now formulate an approximate non-Gaussian (ANG) theory that eliminates RB for $\lambda \lesssim \lambda_{ng}$, and provides a smooth evolution from the free vacuum state to the rashbon gas. Motivated by our earlier discussion, we allow only those bosonic states with $|\boldsymbol{q}| \leq q_b$ where $E_b(q_b) = \kappa b^{1/3}\rho_b^G$ and $\rho_b^G$ is the pole contribution calculated within the Gaussian approximation. We argue that these are the ones that are stable to non-Gaussian effects. We thus modify the equation of state (Eq. (3.7.3)) with $\rho_b^{ANG}(T,\mu) = -\frac{1}{V}\sum_{|\boldsymbol{q}|\leq q_b} n_B(\omega_b(\boldsymbol{q}))\frac{\partial \omega_b(\boldsymbol{q})}{\partial \mu}$.

Figure 3.7.1 shows the results of our ANG theory (see the curve marked ANG). As desired, it is successful in achieving a unique solution for $\mu$ at any $\lambda$ that connects smoothly to free vacuum for small $\lambda/\lambda_{ng}$ and to rashbon gas for large $\lambda/\lambda_{ng}$. The evolution should be smoother in a non-Gaussian theory treated without an approximation as above.

ANG theory is used to obtain the phase diagram shown in Fig. 3.7.5(a). Upon increase of $\lambda$, for small positive $a_s$ and resonant $a_s$ the $T_c$ evolves from that of the free vacuum to that of the rashbon-BEC. For small negative $a_s$ this evolution exhibits richer features as shown in detail in Fig. 3.7.5(b). They are: **large enhancement of $T_c$ upto order $T_F$** and non-monotonic dependence of $T_c$ on $\lambda$. For small $\lambda/k_F$, the $T_c$ follows the mean field value

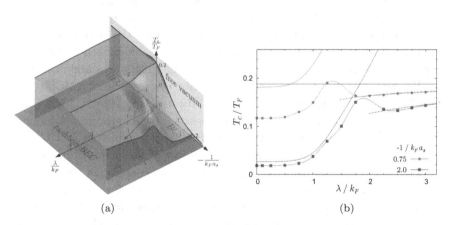

(a)                                    (b)

**Fig. 3.7.5.** (a). Phase diagram: Dependence of superfluid $T_c$ on $a_s$ and $\lambda$ (ANG theory). Crossover occurs in regime enclosed by the dashed lines. The dotted line indicates a candidate path that can be traced out in a cold atom experiment by varying the trap centre density at a fixed $a_s$ and $\lambda$. (b). Non-monotonic dependence of $T_c$ on $\lambda$ for weak attraction: Points indicate the $T_c$ calculated from our ANG theory (lines through the points are a guide to the eye). Dashed lines — $T_c$ from mean field theory. Dashed-dot lines — $T_c$ estimated from the condensation of the bound-fermion pairs (see Sec. 5.E). Thin horizontal line — condensation temperature of rashbons.

and for large $\lambda/k_F$, it is is determined by two-body physics, finally reaching that obtained from the rashbon mass, *independent* of $a_s$.

Figure 3.7.5(a) shows the crossover regime from a BCS like ground state to RBEC which also the regime where the superfluid with high $T_c$ occurs. It will be fantastic to use this remarkable feature of RSOC in material systems to make superconductors with high $T_c$.

## 3.8. Perspective

We saw via a two-body scattering problem that the two-body bound state exists in presence of SU(2) gauge fields that generate RSOC, however small, even in three spatial dimensions for arbitrary $s$-wave interaction strength. The nature of such a bound state in presence of $p$-wave and higher order interactions in systems with a larger gauge group, such as SU($N$) is worth investigating. We extended our study to finite COM momenta and showed that the two-body bound state ceases to exist even for arbitrarily strong attractions at high enough COM momenta. The consequences of this effect to topological defects in a rashbon condensate is also worth investigating.

When even a noninteracting system with the combined effect of a uniform RSOC and an external simple harmonic potential can produce effective Hamiltonians that are as novel as spherical geometry quantum Hall like Hamiltonians, imagination would be the only limit for the kind of Hamiltonians that can be engineered using in conjunction — inhomogeneous RSOC, different external potentials, interactions, Zeeman like terms and disorder.

We demonstrated that RSOC can induce a crossover from a BCS state to a BEC of rashbons even at a fixed small attraction. The rashbon condensate is a remarkable state. All its properties are determined solely by RSOC. Even the interaction between rashbons is determined by RSOC and is independent of the constituent fermion-fermion interaction. Interaction between emergent excitations being independent of constituent interactions is not so common in condensed matter physics or elsewhere.

We demonstrated the crucial role of beyond-Gaussian effects. We developed a simple theory that incorporates them in an approximate fashion. Using this theory we obtained the phase diagram of the system. A key result of our calculation is the demonstration of the enhancement of the exponentially small superfluid transition temperature with weak attraction to values comparable to Fermi temperature. This important point provides clues to producing superconductors with high transition temperatures. Further investigation along these lines would shed light on pseudogap features[92] which could be observed even at higher temperatures. The ANG theory provides motivation to treat beyond-Gaussian effects in a more detailed manner like the $G_0 - G$ or $G - G$ schemes.[93,94] The properties of a repulsive Fermi gas in the upper branch of the energy spectrum is also an interesting subject for research.

From an experimental point of view, taking the wavelength of the Raman lasers used for generating the spin-orbit coupling to be 500 nm, we estimate that a trap centre density of $2 \times 10^{12}$ cm$^{-3}$ would produce a rashbon condensate at low temperatures. Also, a wide Feshbach resonance whose width is much larger compared to $\lambda^2$ would be favourable to realize the rashbon.[95] Figure 3.4.1(b) and associated discussions would help experimentalists to identify most relevant regime to observe the first effects of RSOC in experiments. We urge the readers to refer to other chapters in this book for more information regarding experiments.

Another interesting line of research would be to use the findings here in real material systems. This could, for example, lead to the development of materials with high transition temperatures, even to room temperatures! Indeed this would be a significant direction ahead.

The general theoretical framework developed here provides a platform to address questions pertaining to interacting fermions in RSOC. Also, we sincerely hope that our results will motivate experimental research to find ways of realizing synthetic Rashba-spin-orbit coupling and thereby to uncover the fascinating physics of interacting fermions in its presence.

## Acknowledgements

The author thanks Vijay B. Shenoy for collaboration; Sudeep Kumar Ghosh, Adhip Agarwala and Sambuddha Sanyal for discussions; CSIR, India for financial support.

## References

1. P. A. Lee, N. Nagaosa, and X.-G. Wen, Doping a mott insulator: Physics of high-temperature superconductivity. *Rev. Mod. Phys.*, **78**, 17–85 (2006).
2. C. Nayak, S. H. Simon, A. Stern, M. Freedman, and S. Das Sarma, Non-abelian anyons and topological quantum computation. *Rev. Mod. Phys.*, **80**, 1083–1159 (2008).
3. M. Z. Hasan and C. L. Kane, Topological insulators. *Rev. Mod. Phys.*, **82**(4), 3045–3067 (2010).

4. R. Roy, $Z_2$ classification of quantum spin hall systems: An approach using time-reversal invariance. *Phys. Rev. B*, **79**, 195321 (2009).

5. R. Roy, Topological phases and the quantum spin hall effect in three dimensions. *Phys. Rev. B*, **79**, 195322 (2009).

6. A. Trabesinger, Quantum simulation. *Nat. Phys.*, **8**, 263–263 (2012).

7. J. I. Cirac and P. Zoller, Goals and opportunities in quantum simulation. *Nat. Phys.*, **8**, 264–266 (2012).

8. I. Bloch, J. Dalibard, and S. Nascimbene, Quantum simulations with ultracold quantum gases. *Nat. Phys.*, **8**, 267–276 (2012).

9. W. Ketterle and M. W. Zwierlein, Making, probing and understanding ultracold Fermi gases. *Nuovo Cimento Rivista Serie*, **31**, 247–422 (2008).

10. I. Bloch, J. Dalibard, and W. Zwerger, Many-body physics with ultracold gases. *Rev. Mod. Phys.*, **80**(3), 885–964 (2008).

11. S. Giorgini, L. P. Pitaevskii, and S. Stringari, Theory of ultracold atomic fermi gases. *Rev. Mod. Phys.*, **80**(4), 1215–1274 (2008).

12. T. Esslinger, Fermi-hubbard physics with atoms in an optical lattice. *Annual Review of Condensed Matter Physics*, **1**(1), 129–152 (2010).

13. M. H. Anderson, J. R. Ensher, M. R. Matthews, C. E. Wieman, and E. A. Cornell, Observation of bose-einstein condensation in a dilute atomic vapor. *Science*, **269**(5221), 198–201 (1995).

14. K. B. Davis, M. O. Mewes, M. R. Andrews, N. J. van Druten, D. S. Durfee, D. M. Kurn, and W. Ketterle, Bose-einstein condensation in a gas of sodium atoms. *Phys. Rev. Lett.*, **75**, 3969–3973 (1995).

15. N. Gemelke, X. Zhang, C.-L. Hung, and C. Chin, In situ observation of incompressible mott-insulating domains in ultracold atomic gases. *Nature*, **460**, 995–998 (2009).

16. W. S. Bakr, A. Peng, M. E. Tai, R. Ma, J. Simon, J. I. Gillen, S. Fölling, L. Pollet, and M. Greiner, Probing the superfluidtomott insulator transition at the single-atom level. *Science*, **329**(5991), 547–550 (2010).

17. C. A. Regal, M. Greiner, and D. S. Jin, Observation of resonance condensation of fermionic atom pairs. *Phys. Rev. Lett.*, **92**(4), 040403 (2004).

18. S. Shin, Yong-il, C. H., A. Schirotzek, and W. Ketterle, Phase diagram of a two-component fermi gas with resonant interactions. *Nature*, **451**, 689–693 (2006).

19. K. Maeda, G. Baym, and T. Hatsuda, Simulating dense qcd matter with ultracold atomic boson-fermion mixtures. *Phys. Rev. Lett.*, **103** (8), 085301 (2009).

20. Q. Baudouin, N. Mercadier, V. Guarrera, W. Guerin, and R. Kaiser, A cold-atom random laser. *Nat. Phys.*, **9**, 357–360 (2013).

21. D. C. McKay and B. DeMarco, Cooling in strongly correlated optical lattices: prospects and challenges. *Reports on Progress in Physics*, **74**(5), 054401 (2011).

22. T.-L. Ho and Q. Zhou, Squeezing out the entropy of fermions in optical lattices. *Proceedings of the National Academy of Sciences*, **106**(17), 6916–6920 (2009).

23. N. R. Cooper, Rapidly rotating atomic gases. *Advances in Physics*, **57**, 539–616 (2008).
24. J. Dalibard, F. Gerbier, G. Juzeliūnas, and P. Öhberg, *Colloquium*: Artificial gauge potentials for neutral atoms. *Rev. Mod. Phys.*, **83**, 1523–1543 (2011).
25. Y.-J. Lin, R. L. Compton, K. Jimenez-Garcia, J. V. Porto, and I. B. Spielman, Synthetic magnetic fields for ultracold neutral atoms. *Nature*, **462**(1), 628–632 (2009).
26. Y.-J. Lin, R. L. Compton, A. R. Perry, W. D. Phillips, J. V. Porto, and I. B. Spielman, Bose-einstein condensate in a uniform light-induced vector potential. *Phys. Rev. Lett.*, **102**(13), 130401 (2009).
27. Y.-J. Lin, K. Jimenez-Garcia, and I. B. Spielman, Spin-orbit-coupled bose-einstein condensates. *Nature*, **471**, 83–86 (2011).
28. R. A. Williams, L. J. LeBlanc, K. Jimenez-Garcia, M. C. Beeler, A. R. Perry, W. D. Phillips, and I. B. Spielman, Synthetic partial waves in ultracold atomic collisions. *Science*, **335**(6066), 314–317 (2012).
29. S. C. Ji, J. Y. Zhang, L. Zhang, Z. D. Du, W. Zheng, Y. J. Deng, H. Zhai, S. Chen, and J. W. Pan, Experimental determination of the finite-temperature phase diagram of a spin-orbit coupled bose gas. *Nat. Phys.*, **10**(4), 314–320 (2014).
30. P. Wang, Z.-Q. Yu, Z. Fu, J. Miao, L. Huang, S. Chai, H. Zhai, and J. Zhang, Spin-orbit coupled degenerate fermi gases. *Phys. Rev. Lett.*, **109**, 095301 (2012).
31. L. W. Cheuk, A. T. Sommer, Z. Hadzibabic, T. Yefsah, W. S. Bakr, and M. W. Zwierlein, Spin-injection spectroscopy of a spin-orbit coupled fermi gas. *Phys. Rev. Lett.*, **109**, 095302 (2012).
32. L. Huang, Z. Meng, P. Wang, P. Peng, S. L. Zhang, L. Chen, D. Li, Q. Zhou, and J. Zhang, Experimental realization of two-dimensional synthetic spin-orbit coupling in ultracold fermi gases. *Nat. Phys.*, **12**, 540–544 (2016).
33. Z. Meng, L. Huang, P. Peng, D. Li, L. Chen, Y. Xu, C. Zhang, P. Wang, and J. Zhang, Experimental observation of topological band gap opening in ultracold Fermi gases with two-dimensional spin-orbit coupling. ArXiv e-prints, **1511.08492** (2015).
34. D. Jaksch and P. Zoller, Creation of effective magnetic fields in optical lattices: the hofstadter butterfly for cold neutral atoms. *New J. Phys.*, **5**(1), 56 (2003).
35. K. Osterloh, M. Baig, L. Santos, P. Zoller, and M. Lewenstein, Cold atoms in non-abelian gauge potentials: From the hofstadter "moth" to lattice gauge theory. *Phys. Rev. Lett.*, **95**(1), 010403 (2005).
36. J. Ruseckas, G. Juzeliūnas, P. Öhberg, and M. Fleischhauer, Non-abelian gauge potentials for ultracold atoms with degenerate dark states. *Phys. Rev. Lett.*, **95**(1), 010404 (2005).
37. F. Gerbier and J. Dalibard, Gauge fields for ultracold atoms in optical superlattices. *New J. Phys.*, **12**(3), 033007 (2010).
38. D. L. Campbell, G. Juzeliūnas, and I. B. Spielman, Realistic rashba and dresselhaus spin-orbit coupling for neutral atoms. *Phys. Rev. A*, **84**, 025602 (2011).

39. Z.-F. Xu, L. You, and M. Ueda, Atomic spin-orbit coupling synthesized with magnetic-field-gradient pulses. *Phys. Rev. A*, **87**, 063634 (2013).

40. B. M. Anderson, I. B. Spielman, and G. Juzeliūnas, Magnetically generated spin-orbit coupling for ultracold atoms. *Phys. Rev. Lett.*, **111**, 125301 (2013).

41. T. D. Stanescu, B. Anderson, and V. Galitski, Spin-orbit coupled bose-einstein condensates. *Phys. Rev. A*, **78**, 023616 (2008).

42. T.-L. Ho and S. Zhang, Bose-einstein condensates with spin-orbit interaction. *Phys. Rev. Lett.*, **107**, 150403 (2011).

43. C. Wang, C. Gao, C.-M. Jian, and H. Zhai, Spin-orbit coupled spinor bose-einstein condensates. *Phys. Rev. Lett.*, **105**(16), 160403 (2010).

44. W. Cong-Jun, I. Mondragon-Shem, and Z. Xiang-Fa, Unconventional bose-instein condensations from spin-orbit coupling. *Chinese Physics Letters*, **28**(9), 097102 (2011).

45. R. Sachdeva and S. Ghosh, Density-wave–supersolid and mott-insulator–superfluid transitions in the presence of an artificial gauge field: A strong-coupling perturbation approach. *Phys. Rev. A*, **85**, 013642 (2012).

46. J. P. Vyasanakere and V. B. Shenoy, Bound states of two spin-$\frac{1}{2}$ fermions in a synthetic non-abelian gauge field. *Phys. Rev. B*, **83**(9), 094515 (2011).

47. J. P. Vyasanakere, S. Zhang, and V. B. Shenoy, Bcs-bec crossover induced by a synthetic non-abelian gauge field. *Phys. Rev. B*, **84** (1), 014512 (2011).

48. J. P. Vyasanakere and V. B. Shenoy, Rashbons: properties and their significance. *New J. Phys.*, **14**(4), 043041 (2012).

49. S. K. Ghosh, J. P. Vyasanakere, and V. B. Shenoy, Trapped fermions in a synthetic non-abelian gauge field. *Phys. Rev. A*, **84**, 053629 (2011).

50. J. P. Vyasanakere and V. B. Shenoy, Collective excitations, emergent galilean invariance, and boson-boson interactions across the bcs-bec crossover induced by a synthetic rashba spin-orbit coupling. *Phys. Rev. A*, **86**, 053617 (2012).

51. J. P. Vyasanakere and V. B. Shenoy, Fluctuation theory of rashba fermi gases: Gaussian and beyond. *Phys. Rev. B*, **92**, 121111 (2015).

52. H. Hu, L. Jiang, X.-J. Liu, and H. Pu, Probing anisotropic superfluidity in atomic fermi gases with rashba spin-orbit coupling. *Phys. Rev. Lett.*, **107**, 195304 (2011).

53. M. Gong, S. Tewari, and C. Zhang, Bcs-bec crossover and topological phase transition in 3d spin-orbit coupled degenerate fermi gases. *Phys. Rev. Lett.*, **107**, 195303 (2011).

54. M. Iskin and A. L. Subaş ı, Stability of spin-orbit coupled fermi gases with population imbalance. *Phys. Rev. Lett.*, **107**, 050402 (2011).

55. L. Han and C. A. R. Sá de Melo, Evolution from bcs to bec superfluidity in the presence of spin-orbit coupling. *Phys. Rev. A*, **85**, 011606 (2012).

56. S. Takei, C.-H. Lin, B. M. Anderson, and V. Galitski, Low-density molecular gas of tightly bound rashba-dresselhaus fermions. *Phys. Rev. A*, **85**, 023626 (2012).

57. N. Goldman, A. Kubasiak, A. Bermudez, P. Gaspard, M. Lewenstein, and M. A. Martin-Delgado, Non-abelian optical lattices: Anomalous quantum hall effect and dirac fermions. *Phys. Rev. Lett.*, **103**, 035301 (2009).

58. L. He and X.-G. Huang, Bcs-bec crossover in 2d fermi gases with rashba spin-orbit coupling. *Phys. Rev. Lett.*, **108**, 145302 (2012).

59. M. Gong, G. Chen, S. Jia, and C. Zhang, Searching for majorana fermions in 2d spin-orbit coupled fermi superfluids at finite temperature. *Phys. Rev. Lett.*, **109**, 105302 (2012).

60. J. P. A. Devreese, J. Tempere, and C. A. R. Sá de Melo, Quantum phase transitions and berezinskii-kosterlitz-thouless temperature in a two-dimensional spin-orbit-coupled fermi gas. *Phys. Rev. A*, **92**, 043618 (2015).

61. H. Zhai, Spin-orbit coupled quantum gases. *Int. J. Mod. Phys. B*, **26**, 1230001 (2012).

62. V. B. Shenoy and J. P. Vyasanakere, Fermions in synthetic non-abelian gauge potentials: rashbon condensates to novel hamiltonians. *Journal of Physics B: Atomic, Molecular and Optical Physics*, **46**(13), 134009 (2013).

63. A. V. Chaplik and L. I. Magarill, Bound states in a two-dimensional short range potential induced by the spin-orbit interaction. *Phys. Rev. Lett.*, **96**, 126402 (2006).

64. E. Cappelluti, C. Grimaldi, and F. Marsiglio, Topological change of the fermi surface in low-density rashba gases: Application to superconductivity. *Phys. Rev. Lett.*, **98**, 167002 (2007).

65. B. M. Anderson, G. Juzeliūnas, V. M. Galitski, and I. B. Spielman, Synthetic 3d spin-orbit coupling. *Phys. Rev. Lett.*, **108**, 235301 (2012).

66. C. J. Pethick and H. Smith. *Bose-Einstein Condensation in Dilute Gases*. (Cambridge University Press, 2004).

67. E. Braaten, M. Kusunoki, and D. Zhang, Scattering models for ultracold atoms. *Ann. Phys.*, **323**(7), 1770–1815 (2008).

68. P. W. Anderson, Random-phase approximation in the theory of superconductivity. *Phys. Rev.*, **112**, 1900–1916 (1958).

69. D. M. Eagles, Possible pairing without superconductivity at low carrier concentrations in bulk and thin-film superconducting semiconductors. *Phys. Rev.*, **186**(2), 456–463 (1969).

70. A. J. Leggett. Diatomic molecules and cooper pairs. In eds. A. Pekalski and R. Przystawa. *Modern Trends in the Theory of Condensed Matter*, pp. 13–27. Springer-Verlag, Berlin (1980).

71. C. A. R. Sá de Melo, M. Randeria, and J. R. Engelbrecht, Crossover from bcs to bose superconductivity: Transition temperature and time-dependent ginzburg-landau theory. *Phys. Rev. Lett.*, **71**(19), 3202–3205 (1993).

72. M. Randeria. Crossover from bcs theory to bose-einstein condenstion. In eds. A. Griffin, D. Snoke, and S. Stringari. *Bose-Einstein Condensation*, chapter 15, pp. 355–392. Cambridge University Press (1995).

73. J. R. Engelbrecht, M. Randeria, and C. A. R. Sá de Melo, Bcs to bose crossover: Broken-symmetry state. *Phys. Rev. B*, **55**, 15153–15156 (1997).

74. S. De Palo, C. Castellani, C. Di Castro, and B. K. Chakraverty, Effective action for superconductors and bcs-bose crossover. *Phys. Rev. B*, **60**, 564–573 (1999).

75. N. Dupuis, Berezinskii-kosterlitz-thouless transition and bcs-bose crossover in the two-dimensional attractive hubbard model. *Phys. Rev. B*, **70**, 134502 (2004).
76. W. Zwerger, Ed.. *The BCS-BEC Crossover and the Unitary Fermi Gas.* vol. 836. *Lecture Notes in Physics* (Springer-Verlag, Berlin, 2012).
77. M. J. H. Ku, A. T. Sommer, L. W. Cheuk, and M. W. Zwierlein, Revealing the superfluid lambda transition in the universal thermodynamics of a unitary fermi gas. *Science*, **335**(6068), 563–567 (2012).
78. M. Randeria and E. Taylor, Crossover from bardeen-cooper-schrieffer to bose-einstein condensation and the unitary fermi gas. *Annual Review of Condensed Matter Physics*, **5**(1), 209–232 (2014).
79. A. J. Leggett, On the superfluid fraction of an arbitrary many-body system at $t = 0$. *Journal of Statistical Physics*, **93**, 927–941 (1998).
80. A. J. Leggett. *Quantum Liquids: Bose Condensation and Cooper Pairing in Condensed-Matter Systems.* (Oxford University Press, 2006).
81. A. A. Abrikosov, L. P. Gor'kov, and I. Y. Dzyaloshinskii. *Quantum Field Theoretical Methods in Statistical Physics.* (Pergamon, 1965).
82. P. Noziéres and S. Schmitt-Rink, Bose condensation in an attractive fermion gas: From weak to stong coupling superconductivity. *J. Low Temp. Phys.*, **59**, 195–211 (1985).
83. M. Drechsler and W. Zwerger, Crossover from bcs-superconductivity to bose-condensation. *Annalen der Physik*, **504**(1), 15–23 (1992).
84. J. R. Taylor. *Scattering Theory.* (Dover Publications, New York, 2006).
85. X. Cui, Mixed-partial-wave scattering with spin-orbit coupling and validity of pseudopotentials. *Phys. Rev. A*, **85**, 022705 (2012).
86. L. N. Cooper, Bound electron pairs in a degenerate fermi gas. *Phys. Rev.*, **104**, 1189–1190 (1956).
87. Z.-Q. Yu and H. Zhai, Spin-orbit coupled fermi gases across a feshbach resonance. *Phys. Rev. Lett.*, **107**, 195305 (2011).
88. K. Zhou and Z. Zhang, Opposite effect of spin-orbit coupling on condensation and superfluidity. *Phys. Rev. Lett.*, **108**, 025301 (2012).
89. H. Hu, X.-J. Liu, and P. D. Drummond, Equation of state of a superfluid fermi gas in the bcs-bec crossover. *Europhys. Lett.*, **74** (4), 574–580 (2006).
90. R. B. Diener, R. Sensarma, and M. Randeria, Quantum fluctuations in the superfluid state of the bcs-bec crossover. *Phys. Rev. A*, **77**(2), 023626 (2008).
91. D. J. Thouless, Perturbation theory in statistical mechanics and the theory of superconductivity. *Annals of Physics*, **10**(4), 553–588 (1960).
92. J. P. Gaebler, J. T. Stewart, T. E. Drake, D. S. Jin, A. Perali, P. Pieri, and G. C. Strinati, Observation of pseudogap behaviour in a strongly interacting fermi gas. *Nat. Phys.*, **6**(8), 569–573 (2010).
93. Q. Chen, J. Stajic, S. Tan, and K. Levin, Bcsbec crossover: From high temperature superconductors to ultracold superfluids. *Physics Reports*, **412**(1), 1–88 (2005).

94. G. C. Strinati. *Pairing Fluctuations Approach to the BCS-BEC Crossover*, In ed. W. Zwerger. *The BCS-BEC Crossover and the Unitary Fermi Gas*, vol. 836. *Lecture Notes in Physics*, chapter 4, pp. 99–126. Springer-Verlag, Berlin (2012).

95. V. B. Shenoy, Feshbach resonance in a synthetic non-abelian gauge field. *Phys. Rev. A*, **89**, 043618 (2014).

# Chapter 4

# Pairing Superfluidity in Spin-Orbit Coupled Ultracold Fermi Gases

Xingze Qiu and Wei Yi*

*Key Laboratory of Quantum Information CAS,*
*University of Science and Technology of China,*
*Hefei, Anhui, 230026, China*

In this chapter, we review recent progresses on the study of pairing superfluidity in ultracold Fermi gases with synthetic spin-orbit coupling. Recent studies have shown that different forms of spin-orbit coupling in various spatial dimensions can lead to a wealth of novel pairing orders. A common theme of these variations is the emergence of new pairing mechanisms which are direct results of spin-orbit-coupling-modified single-particle dispersion spectra. As different configurations can give rise to single-particle dispersion spectra with drastic differences in symmetry, spin dependence and low-energy density of states, novel Fermi-surface topologies and pairing instabilities arise, which lead to interesting pairing orders.

## 4.1. Introduction

The recent experimental realization of synthetic gauge field in ultracold atomics gases has greatly extended the horizon of quantum simulation in these systems.[1–16] A particularly important case is the implementation of synthetic spin-orbit coupling (SOC), a non-Abelian gauge field, in these systems, where the internal degrees of freedom of the atoms are coupled to the atomic center-of-mass motional degrees of freedom. In condensed-matter materials, SOC plays a key role in many interesting phenomena, such as the quantum spin Hall effects, topological insulators, and topological superconductors.[17–19] Although the form of the synthetic SOC currently realized

---

*Correspondence addressed to: wyiz@ustc.edu.cn.

in cold atoms differs crucially from those in condensed-matter systems, there exist various theoretical proposals on realizing synthetic SOC which can induce topologically nontrivial phases.[20–27] Thus, the hope of simulating the various topological phases, the topological superfluid state in particular, in the highly controllable environment of an ultracold atomic gas stimulated intensive theoretical studies on spin-orbit coupled Fermi gases.[28–43] Furthermore, recent studies suggest that other exotic superfluid phases and novel phenomena can be engineered with carefully designed configurations.[44–57] As such, SOC has a great potential of becoming a powerful tool of quantum control in ultracold atomic gases.

In this Chapter, we focus on the zero-temperature pairing physics in a spin-orbit coupled ultracold Fermi gas. We will discuss the exotic superfluid phases in systems with different spatial dimensions and with different forms of SOC. A fundamentally important effect of SOC is the modification of the single-particle dispersion spectra.[58–61] We will start from there and show how this effect leads to interesting pairing phases such as the topological superfluid state, the various gapless superfluid states, the SOC-induced Fulde-Ferrell (FF) state, and the topological FF state. We will also touch upon the topic of exotic few-body states in spin-orbit coupled Fermi systems whose stability also benefits from the SOC-modified single-particle dispersion.

## 4.2. Synthetic Gauge Field and Synthetic Spin-orbit Coupling

We start by briefly introducing the implementation of synthetic SOCs. Similar to the generation of other types of artificial gauge potentials in atomic systems, the implementation of synthetic SOC is based on the adiabatic theorem and the associated geometrical phase.[62] By engineering the coupling between atomic internal states with the external laser or magnetic fields, the atoms experience adiabatic potentials when moving through space. The resulting geometrical phase emerging in the effective Hamiltonian then gives rise to different kinds of synthetic gauge potentials.

To see this, we follow the formalism in Ref. 63 and consider the general quantum mechanical problem of a single atom coupled to a laser field, with the Hamiltonian

$$H = \frac{\boldsymbol{P}^2}{2m}\hat{1} + U, \qquad (4.2.1)$$

where $m$ is the atomic mass, $U$ describes the coupling between the atom and the spatially varying laser field, and $\hat{1}$ is the identity operator spanning the subspace of the atomic internal degrees of freedom. Without loss of generality, we assume the atom-laser coupling involves $N$ internal states of the atom. It follows that $U$ has $N$ eigen states $|\chi_n(\mathbf{r})\rangle$ and eigenvalues $\epsilon_n(\mathbf{r})$ ($n = 1, \ldots, N$), which are essentially the dressed states and the adiabatic potentials, respectively. We also assume that the first $q$ atomic dressed states are degenerate (or quasi-degenerate), and that these levels are well-separated from the remaining $N - q$ states. Invoking the adiabatic theorem to study the dynamics of the atomic wave function within the degenerate ground-state subspace leads to the Schrödinger's equation

$$i\hbar\frac{\partial}{\partial t}\tilde{\Psi} = \left[\frac{(-i\hbar\nabla - \boldsymbol{A})^2}{2m} + V + W\right]\tilde{\Psi}, \qquad (4.2.2)$$

where $\tilde{\Psi} = (\psi_1, \ldots, \psi_q)^T$ is the projected wave function in the degenerate subspace, and $\{\boldsymbol{A}, V, W\}$ are $q \times q$ matrices with their elements defined by

$$\begin{aligned}
\boldsymbol{A}_{mn} &= i\hbar\langle\chi_m|\nabla\chi_n\rangle, \\
V_{mn} &= \langle\chi_m|U|\chi_n\rangle = \epsilon_n\delta_{mn}, \\
W_{mn} &= \frac{1}{2m}\Sigma_{l=q+1}^{N}\boldsymbol{A}_{ml} \cdot \boldsymbol{A}_{ln}.
\end{aligned} \qquad (4.2.3)$$

Here, $\boldsymbol{A}$ and $W$ are identified as the vector and the scalar gauge potential, respectively. Importantly, a unitary transformation $S(\mathbf{r})$ can be applied to the degenerate ground-state subspace, under which the wave function and the gauge potentials transform as

$$\begin{aligned}
\tilde{\Psi} &\to S(\mathbf{r})\tilde{\Psi}, \\
\boldsymbol{A} &\to S(\mathbf{r})\boldsymbol{A}S^\dagger(\mathbf{r}) - i\hbar(\nabla S(\mathbf{r}))S^\dagger(\mathbf{r}), \\
V &\to S(\mathbf{r})VS^\dagger(\mathbf{r}), \\
W &\to S(\mathbf{r})WS^\dagger(\mathbf{r}).
\end{aligned} \qquad (4.2.4)$$

If the ground state is non-degenerate ($q = 1$), $\boldsymbol{A}$ is a vector, and the resulting gauge potential is Abelian. Such Abelian synthetic gauge fields can give rise to synthetic electro-magnetic fields. On the other hand, if the ground state is degenerate, $\boldsymbol{A}$ is a matrix, and the resulting gauge potential is in general non-Abelian. Synthetic SOCs are implemented by generating non-Abelian gauge potentials.

Experimentally, adiabatic potentials are typically generated by coupling the internal states of an atom with lasers, and it was Spielman's group at NIST that had first realized a uniform vector gauge potential in a BEC of $^{87}$Rb atoms.[1] The scheme utilizes Raman lasers to couple hyperfine states in the ground-state manifold of $^{87}$Rb, such that when an atom jumps from one internal state to another via the Raman process, its center-of-mass momentum also changes. Applying a similar scheme, vortices were later observed in a BEC of $^{87}$Rb with synthetic magnetic field, followed by the implementation of synthetic electric field.[2,3] In 2010, via a slightly modified Raman scheme, Spielman's group was able to generate synthetic SOC, a non-Abelian version of the synthetic gauge field, in a BEC of $^{87}$Rb atoms.[4,64,65] This was soon followed by the realization of synthetic SOC in ultracold Fermi gases by the ShanXi group and the MIT group in 2012.[5,6] While the SOCs in these early experiments have been one dimensional, two-dimensional SOCs have also been realized very recently both in free space and in a lattice configuration. The guiding principle of all of these experiments, however, reply on the generation of the geometric phase.

As a concrete example, we discuss below the realization of the one dimensional synthetic SOC. We consider the system in one dimension with no external potentials, where two atomic states in the ground-state hyperfine manifold are coupled via a typical Λ-scheme Raman process (see Fig. 4.2.1). The first leg of the Raman coupling is a control laser with a Rabi frequency $\Omega_c$, which is spatially independent. The second leg of the Raman coupling is a probe laser with the Rabi frequency $\Omega_p(x) = \Omega_p e^{ikx}$.

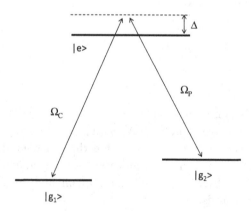

**Fig. 4.2.1.** Level scheme for the Λ-type Raman process. $|g_1\rangle$ and $|g_2\rangle$ are hyperfine states within the ground state hyperfine manifold, while $|e\rangle$ is the electronically excited state.

The Hamiltonian of system can be written in an appropriate rotating frame

$$H = \left(\frac{P^2}{2m}\right)\hat{1} + U = \left(\frac{P^2}{2m}\right)\hat{1} + \hbar \begin{pmatrix} 0 & 0 & \frac{\Omega_c}{2} \\ 0 & 0 & \frac{\Omega_p}{2}e^{ikx} \\ \frac{\Omega_c}{2} & \frac{\Omega_p}{2}e^{-ikx} & \Delta \end{pmatrix}. \qquad (4.2.5)$$

Diagonalizing $U$, we get the dressed state: $|\psi_1\rangle = (-e^{-ikx}\cos\alpha, \sin\alpha, 0)^T$, $|\psi_2\rangle = (e^{-ikx}\sin\alpha\cos(\theta/2), \cos\alpha\cos(\theta/2), -e^{-ikx}\sin(\theta/2))^T$, and $|\psi_3\rangle = (e^{-ikx}\sin\alpha\sin(\theta/2), \cos\alpha\sin(\theta/2), e^{-ikx}\cos(\theta/2))^T$.

Here $\theta = \arctan(\sqrt{\Omega_c^2 + \Omega_p^2}/\Delta)$, $\alpha = \arctan(\Omega_c/\Omega_p)$. While $|\psi_1\rangle$ corresponds to the dark state of the $\Lambda$-scheme, $|\psi_2\rangle$ and $|\psi_3\rangle$ are the bright state. Their corresponding adiabatic potentials are $E_1 = 0$, $E_2 = \hbar(\Delta - \sqrt{\Delta^2 + \Omega_c^2 + \Omega_p^2})/2$, and $E_3 = \hbar(\Delta + \sqrt{\Delta^2 + \Omega_c^2 + \Omega_p^2})/2$. In the large-detuning limit $|\Delta| \gg \Omega_c, \Omega_p$, $|\psi_1\rangle$ and $|\psi_2\rangle$ span a quasi-degenerate subspace $|\Psi\rangle = \psi^1|\psi_1\rangle + \psi^2|\psi_2\rangle$. Therefore, following Eq. (4.2.2) and Eq. (4.2.3), we get

$$i\hbar\frac{\partial}{\partial t}\begin{pmatrix}\psi^1 \\ \psi^2\end{pmatrix} = \left(\frac{(-i\hbar\nabla - A)^2}{2m} + V + W\right)\begin{pmatrix}\psi^1 \\ \psi^2\end{pmatrix}, \qquad (4.2.6)$$

with

$$A = \frac{\hbar k}{2}\begin{pmatrix} 2\cos^2\alpha & -\sin 2\alpha\cos(\theta/2) \\ -\sin 2\alpha\cos(\theta/2) & 2\sin^2\alpha + \cos^2\alpha(1 - \cos\theta)\end{pmatrix},$$

$$V = \begin{pmatrix} E_1 & 0 \\ 0 & E_2 \end{pmatrix},$$

$$W = \frac{\hbar^2 k^2}{2m}\begin{pmatrix} \sin^2\alpha\cos^2\alpha\sin^2(\theta/2) & \sin\alpha\cos^3\alpha\sin\theta\sqrt{1 - \cos\theta}/ \\ & (2\sqrt{2}) \\ \sin\alpha\cos^3\alpha\sin\theta\sqrt{1 - \cos\theta}/ & \cos^4\alpha\sin^2\theta/4 \\ (2\sqrt{2}) & \end{pmatrix}.$$

$$(4.2.7)$$

While a similar scheme was proposed in Ref. 7, it is essentially equivalent to the experimentally realized one-dimensional SOC in the large detuning limit, where the synthetic gauge field above can be related to the experimental SOC via a gauge transformation. In the following, we adopt the more popular two-level picture of the experimentally realized SOC and demonstrate this point explicitly.

Starting from the Hamiltonian Eq. (4.2.5) and adiabatically eliminating the excited state $|e\rangle$, we have

$$H = \left(\frac{\boldsymbol{P}^2}{2m}\right)\hat{1} + U' = \left(\frac{\boldsymbol{P}^2}{2m}\right)\hat{1} + \frac{\hbar}{4\Delta}\begin{pmatrix} \Omega_c^2 & \Omega_c\Omega_p e^{-ikx} \\ \Omega_c\Omega_p e^{ikx} & \Omega_p^2 \end{pmatrix}. \quad (4.2.8)$$

One may then apply Eq. (4.2.2) and Eq. (4.2.3) to find the corresponding $\boldsymbol{A}$. We choose three different gauges by defining three different bases. The first set of basis $\Psi_0 := \{|\psi_0^1\rangle, |\psi_0^2\rangle\} = \{(1,0)^T, (0,1)^T\}$, which is the natural spin basis. The second $\Psi_1 := \{|\psi_1^1\rangle, |\psi_1^2\rangle\} = \{(e^{-ikx/2}, 0)^T, (0, e^{ikx/2})^T\}$, which is the basis chosen in most of the literature concerning one-dimensional (1D) SOC. The third $\Psi_2 := \{|\psi_2^1\rangle, |\psi_2^2\rangle\} = \{(-e^{-ikx}\cos\alpha, \sin\alpha)^T, (e^{-ikx}\sin\alpha, \cos\alpha)^T\}$, which is essentially the dark-state eigen vector of the $\Lambda$-scheme $|\psi_1\rangle$ in the large-detuning limit. Here, $\alpha$ follows the same definition after Eq. 4.2.5.

With the system wave function given by $|\Psi\rangle = \sum_{j=1,2} \psi_i^j |\psi_i^j\rangle (i = 0, 1, 2)$, from the Schrödinger's equation, we have

$$i\hbar\frac{\partial}{\partial t}\begin{pmatrix} \psi_i^1 \\ \psi_i^2 \end{pmatrix} = \left[\frac{(-i\hbar\nabla - \boldsymbol{A}_i)^2}{2m} + V_i\right]\begin{pmatrix} \psi_i^1 \\ \psi_i^2 \end{pmatrix}. \quad (4.2.9)$$

We can now evaluate the gauge potential $\boldsymbol{A}_i$ and the adiabatic potential $V_i$ under the corresponding basis. Note that $W = 0$ under all three bases, as we are only concerned with dynamics within the degenerate subspace. Under $\Psi_0$,

$$\boldsymbol{A}_0 = \begin{pmatrix} 0 & 0 \\ 0 & 0 \end{pmatrix},$$

$$V_0 = S_0 U S_0^\dagger = -\frac{\hbar}{4\Delta}\begin{pmatrix} \Omega_c^2 & \Omega_c\Omega_p e^{-ikx} \\ \Omega_c\Omega_p e^{ikx} & \Omega_p^2 \end{pmatrix}. \quad (4.2.10)$$

Under $\Psi_1$,

$$\boldsymbol{A}_1 = S_1\boldsymbol{A}_0 S_1^\dagger - i\hbar(\nabla S_1)S_1^\dagger = \frac{\hbar k}{2}\begin{pmatrix} 1 & 0 \\ 0 & -1 \end{pmatrix},$$

$$V_1 = S_1 U S_1^\dagger = -\frac{\hbar}{4\Delta}\begin{pmatrix} \Omega_c^2 & \Omega_c\Omega_p \\ \Omega_c\Omega_p & \Omega_p^2 \end{pmatrix}. \quad (4.2.11)$$

Under $\Psi_2$,

$$\boldsymbol{A}_2 = S_2\boldsymbol{A}_0 S_2^\dagger - i\hbar(\nabla S_2)S_2^\dagger = \frac{\hbar k}{2}\begin{pmatrix} 2\cos^2\alpha & -\sin 2\alpha \\ -\sin 2\alpha & 2\sin^2\alpha \end{pmatrix},$$

$$V_2 = S_2 U S_2^\dagger = -\frac{\hbar}{4\Delta}\begin{pmatrix} 0 & 0 \\ 0 & \Omega_c^2 + \Omega_p^2 \end{pmatrix}. \quad (4.2.12)$$

Here $S_i = \left( \psi_i^1 \; \psi_i^2 \right)^\dagger$, and it is straightforward to check that the gauge potentials under different bases can be transformed into one another according to Eq. (4.2.4). Here, the gauge potential $\boldsymbol{A}_1$ corresponds to the experimentally realized 1D SOC, with the effective Hamiltonian given by $V_1$. And it is straightforward to demonstrate that in the large-detuning limit, $\boldsymbol{A}$ in Eq. 4.2.7 reduces to $\boldsymbol{A}_2$, $W$ in Eq. 4.2.7 vanishes, and the adiabatic potential $V$ in Eq. 4.2.7 becomes $V_2$.

## 4.3. Single-particle Dispersion Under SOC

A fundamental effect of SOC in ultracold Fermi gases is the modification of the single-particle dispersion. Under SOC, both the symmetry and the low-energy density-of-states are different, leading to unconventional BEC in Bose systems, and exotic pairing states in Fermi systems with attractive interactions. In this section, we will focus on the single-particle dispersion under typical forms of SOC, and discuss their potential impact on many-body systems.

### 3.A. *Rashba SOC*

The single-particle Hamiltonian under a Rashba SOC can be written as

$$H = \sum_{\mathbf{k},\sigma} \epsilon_{\mathbf{k}} a_{\mathbf{k},\sigma}^\dagger a_{\mathbf{k},\sigma} + \sum_{\mathbf{k}} \left[ \alpha \left( k_x - i k_y \right) a_{\mathbf{k},\uparrow}^\dagger a_{\mathbf{k},\downarrow} + H.C. \right], \qquad (4.3.1)$$

where $\epsilon_{\mathbf{k}} = \hbar^2 k^2 / 2m$, $a_{\mathbf{k},\sigma}^\dagger$ $(a_{\mathbf{k},\sigma})$ is the creation (annihilation) operator for the pseudo-spin $\sigma = \{\uparrow,\downarrow\}$, $\alpha$ is the SOC strength, and $H.C.$ stands for Hermitian conjugate. The pseudo-spin here is related to the adiabatic potential, and its exact relation with the hyperfine states is scheme dependent. The Hamiltonian can be diagonalized by introducing the annihilation operators in the so-called helicity basis

$$a_{\mathbf{k},+} = \frac{1}{\sqrt{2}} \left( e^{i\varphi_{\mathbf{k}}} a_{\mathbf{k},\uparrow} + a_{\mathbf{k},\downarrow} \right), \qquad (4.3.2)$$

$$a_{\mathbf{k},-} = \frac{1}{\sqrt{2}} \left( e^{i\varphi_{\mathbf{k}}} a_{\mathbf{k},\uparrow} - a_{\mathbf{k},\downarrow} \right), \qquad (4.3.3)$$

with $\varphi_{\mathbf{k}} = \arg(k_x + i k_y)$. The diagonalized Hamiltonian

$$H = \sum_{\mathbf{k},\lambda=\pm} \xi_\lambda a_{\mathbf{k},\lambda}^\dagger, \qquad (4.3.4)$$

where $\xi_\pm = \epsilon_{\mathbf{k}} \pm \alpha k$ are the resulting single-particle eigen energies.

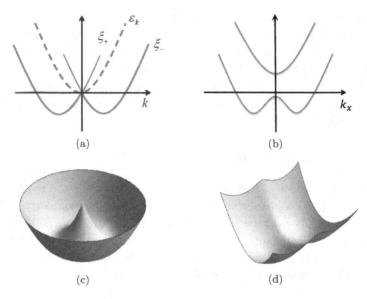

**Fig. 4.3.1.** Illustration of the single-particle dispersion spectra under SOC. (a) The dashed curve represents the free-particle spectra ($\epsilon_k$); the solid curves are the spectra for the helicity branches ($\xi_\pm$). (b) The spectra of the helicity branches under SOC and an out-of-plane Zeeman field ($h\sigma_z$). (c) The single-particle spectrum of the lower helicity branch under the Rashba SOC, three-dimensional view. (d) The single-particle spectrum of the lower helicity branch under the 1D SOC, three-dimensional view.

The single-particle spectra under Rashba SOC is illustrated in Fig. 4.3.1, where it is clear that SOC breaks the inversion symmetry and splits the spin spectra into two helicity branches. Due to the symmetry of the Rashba SOC, points of the lowest energy in the lower helicity branch form a ring in momentum space, and the ground state is infinitely degenerate in this case [see Fig. 4.3.1(c)]. Correspondingly, the density-of-states for a Rashba spin-orbit coupled spectra in three dimensions is a constant at the lowest energy. This peculiar single-particle spectra and density of states naturally lead to interesting many-body phenomena.[58–61]

For noninteracting bosons, as the ground-state degeneracy is infinite, BEC is no longer possible at zero temperature even in three dimensions. However, a weak interaction can induce a spontaneous symmetry breaking, and the bosons will condense to either one point or two opposite points on the degenerate ring in momentum space, depending on the interaction parameters. This leads to the so-called plane-wave phase and the stripe phase in a uniform interacting Bose gas with Rashba SOC.[58,59]

For a noninteracting Fermi gas at zero temperature, atoms occupy all the low-energy states up to the Fermi energy. As the atom number density increases, the topology of the Fermi surface changes and the system undergoes a Lifshitz transition. Furthermore, it has been shown that for atoms with attractive $s$-wave interaction between the spin species and with Rashba SOC, the two-body bound state energy is enhanced due to the increased density of states at low energies.[60,61] In fact, a two-body bound state exists in three dimensions even in the weak-coupling limit, similar to the case of a two-dimensional problem without SOC, suggesting an effective reduction of dimensions.

### 3.B. *One-dimensional SOC*

The single-particle Hamiltonian under a typical 1D SOC can be written as

$$H = \sum_{\mathbf{k},\sigma} \epsilon_{\mathbf{k}} a^{\dagger}_{\mathbf{k},\sigma} a_{\mathbf{k},\sigma} + \sum_{\mathbf{k}} h \left( a^{\dagger}_{\mathbf{k},\uparrow} a_{\mathbf{k},\downarrow} + H.C. \right)$$
$$+ \sum_{\mathbf{k}} \alpha k_x \left( a^{\dagger}_{\mathbf{k}\uparrow} a_{\mathbf{k}\uparrow} - a^{\dagger}_{\mathbf{k}\downarrow} a_{\mathbf{k}\downarrow} \right), \qquad (4.3.5)$$

where $a^{\dagger}_{\mathbf{k},\sigma}$ ($a_{\mathbf{k},\sigma}$) is the creation (annihilation) operator for different hyperfine states. The SOC parameter $\alpha$ is related to the momentum transfer of the Raman process, and the effective Zeeman field $h$ is proportional to the effective Rabi-frequency. The helicity basis that diagonalizes the Hamiltonian becomes

$$a_{\mathbf{k},\pm} = \pm\beta^{\pm}_{\mathbf{k}} a_{\mathbf{k},\uparrow} + \beta^{\mp}_{\mathbf{k}} a_{\mathbf{k},\downarrow}, \qquad (4.3.6)$$

with $\beta^{\pm}_{\mathbf{k}} = \left[ \sqrt{h^2 + \alpha^2 k_x^2} \pm \alpha k_x \right]^{1/2} / \sqrt{2} [h^2 + \alpha^2 k_x^2]^{1/4}$. The single-particle dispersion of the helicity branches $\xi_{\pm} = \epsilon_{\mathbf{k}} \pm \sqrt{h^2 + \alpha^2 k_x^2}$. When $h < m\alpha^2/\hbar^2$, the degenerate ground state manifold consists of two points in momentum space, in contrast to a ring under the Rashba SOC [see Fig. 4.3.1(d)]. The spin components of the helicity branches are momentum dependent under the 1D SOC, which is also different from that of the Rashba SOC. Although these differences can lead to drastically different many-body effects, as we will show later, cold atomic gases under the 1D SOC still preserve many of the interesting properties of those under the Rashba SOC. For example, in a BEC under the 1D SOC, the stripe phase and the plane-wave phase can still be identified in the finite-temperature phase diagram.[12] While in a noninteracting degenerate Fermi gas, topological changes of the Fermi surface have been observed experimentally (see Fig. 4.3.2).

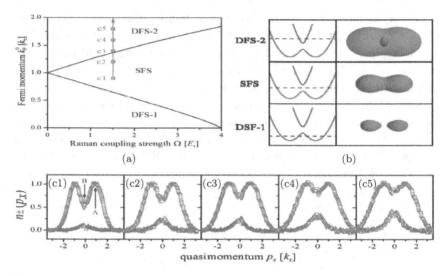

**Fig. 4.3.2.** Experimental observation of the Lifshitz transition for a noninteracting degenerate Fermi gas with SOC.[5] Reprinted with permission from the American Physical Society.

## 3.C. *Other forms of SOC*

Besides the Rashba and the 1D SOC mentioned above, the isotropic SOC of the form $\mathbf{k} \cdot \boldsymbol{\sigma}$, or the three-dimensional SOC, has also been discussed recently by various authors.[26,27,40,42,50,51,60,66] For bosons, spin textures of the ground state under an isotropic SOC may acquire interesting topology. For fermions, it is expected that the system under appropriate parameters can behave like three-dimensional topological insulators or Weyl semimetals. Additionally, the application of an effective Zeeman field can deform the single-particle dispersion spectra along the direction of the field. As we will see later, this would lead to novel pairing states under attractive inter-particle interactions. Other more exotic forms of synthetic SOC exist and have been discussed in the literature.[67–69] Perhaps most importantly for the study of topological matter, two-dimensional SOCs have recently been experimentally realized both in free space and in lattice configurations. In Ref. 15, by considering a tripod scheme and using the dark states as pseudo-spins, the effective Hamiltonian near some special point in momentum space can be written in the form of a Rashba SOC. By manipulating the laser parameters, it is later shown that the gap between the two helicity branches can be controlled, which opens up the possibility of realizing topologically non-trivial pairing state once attractive interaction and pairing can be introduced. On the other hand, in Ref. 16, by considering Raman-assisted hopping on a two-dimensional lattice, a two-dimensional SOC, which is not of the Rashba type, can be realized. The ground-state of the non-interacting system is a chiral topological insulator, for which the band topology has already been confirmed experimentally. We note that a common feature of these SOCs is the absence of counterparts in naturally occurring condensed-matter systems, which makes the quantum simulation of systems with exotic forms of SOC in ultracold atomic gases more appealing.

## 4.4. Pairing Physics Under SOC

With the understanding of single-particle dispersions under SOC, we now examine the pairing physics in an attractively interacting Fermi gas with SOC. We start with the conventional BCS pairing mechanism in the weak-coupling limit, which should help us to appreciate the fundamentally new pairing mechanisms in a spin-orbit coupled Fermi gas.

### 4.A. *Interbranch pairing and intrabranch pairing*

It is well known that a two-component Fermi sea without spin imbalance becomes unstable in the presence of a small attractive interaction. The resulting Cooper instability leads to a BCS ground state, which is the basis for understanding conventional superconductivity in most metals at low enough temperatures. In ultracold Fermi gases, pairing superfluidity has been experimentally observed thanks to the Feshbach resonance technique.[70,71] The effective BCS mean-field Hamiltonian of such a system can be written as

$$
H - \mu N = \sum_{\mathbf{k}} \begin{bmatrix} a_{\mathbf{k}\uparrow}^{\dagger} & a_{-\mathbf{k}\downarrow} \end{bmatrix} \begin{bmatrix} \epsilon_{\mathbf{k}} - \mu & \Delta \\ \Delta^* & -(\epsilon_{\mathbf{k}} - \mu) \end{bmatrix} \begin{bmatrix} a_{\mathbf{k}\uparrow} \\ a_{-\mathbf{k}\downarrow}^{\dagger} \end{bmatrix}
$$
$$
+ \sum_{\mathbf{k}} (\epsilon_{\mathbf{k}} - \mu) - \frac{|\Delta|^2}{U}, \tag{4.4.1}
$$

where $U$ is the bare interaction rate, $\Delta = U \langle a_{-\mathbf{k}\downarrow} a_{\mathbf{k}\uparrow} \rangle$ is the pairing order parameter, and $\mu$ is the chemical potential, which, in the weak-coupling

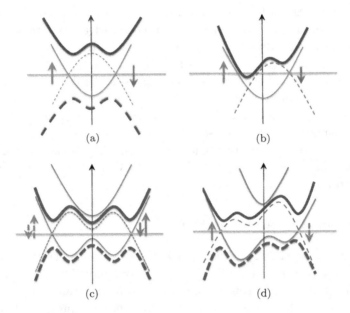

**Fig. 4.4.1.** Schematics of the pairing mechanisms. (a) Illustration of the BCS pairing mechanism. (b) Illustration of the conventional FF pairing mechanism. (c) Pairing mechanism under SOC and an out-of-plane Zeeman field. (d) Pairing mechanism of the SOC-induced FF state with both out-of-plane and in-plane Zeeman fields. In all figures, the thin red (blue) curve represents the particle (hole) dispersion of the noninteracting system, the thick solid (dashed) curve represents the quasiparticle (quasihole) dispersion under the pairing interaction. The horizontal green line is the Fermi energy. See Ref. 52.

limit, is essentially the Fermi energy of the system. Under the Hamiltonian (4.4.1), the BCS pairing mechanism in the weak-coupling limit can be schematically illustrated as Fig. 4.4.1(a). The dispersion of the spin-up particles crosses that of the spin-down holes at the Fermi surface, where the pairing mean field $\Delta$ couples the two branches, leading to an avoided crossing and the quaiparticle (quasihole) dispersions. From the positions of the avoided crossings, we see that the pairing state in this case has zero center-of-mass momentum. In the presence of a spin imbalance, or an effective Zeeman field, the Fermi surface mismatch between the two spin species can lead to competition between the BCS pairing mechanism and the Fulde-Ferrell-Larkin-Ovchinnikov pairing mechanism.[72] In the latter case, the dispersion of the minority spin species is shifted so that pairing can take place near the Fermi surface [see Fig. 4.4.1(b)], thus leading to a finite center-of-mass momentum pairing state. In both of these cases, an important feature is that pairing involves both the particle and the hole branches of different spin species. We address this kind of pairing mechanism as interbranch pairing. Note in Fig. 4.4.1(b) and in the rest of the review, we consider the simple case where pairing occurs with a single-valued center-of-mass momentum, the so-called Fulde-Ferrell (FF) state.

Under SOC, with different single-particle dispersions, a new pairing mechanism emerges, which becomes more transparent and also physically more interesting in the presence of effective Zeeman fields. Extending the BCS mean-field theory, we may write down the effective mean-field Hamiltonian of a two-component Fermi gas under the Rashba SOC and an effective out-of-plane Zeeman field

$$
H_{\text{eff}} = \frac{1}{2} \sum_{\mathbf{k}} \begin{bmatrix} \xi_{\mathbf{k}} + h & 0 & \alpha k e^{i\varphi_{\mathbf{k}}} & \Delta \\ 0 & -(\xi_{\mathbf{k}} + h) & -\Delta^* & \alpha k e^{-i\varphi_{\mathbf{k}}} \\ \alpha k e^{-i\varphi_{\mathbf{k}}} & -\Delta & \xi_{\mathbf{k}} - h & 0 \\ \Delta^* & \alpha k e^{i\varphi_{\mathbf{k}}} & 0 & -(\xi_{\mathbf{k}} - h) \end{bmatrix} + \sum_{\mathbf{k}} (\epsilon_{\mathbf{k}} - \mu) - \frac{|\Delta|^2}{U},
$$

$$(4.4.2)$$

where $h$ is the effective Zeeman field along the $z$ direction. The Hamiltonian above is written in the pseudo-spin basis $\left\{ a_{\mathbf{k}\uparrow}, a^{\dagger}_{-\mathbf{k}\uparrow}, a_{\mathbf{k}\downarrow}, a^{\dagger}_{-\mathbf{k}\downarrow} \right\}^T$. Under SOC, the mean-field ground state of the system has both $s$- and $p$-wave pairing components, as both $\sum_{\mathbf{k}} \langle a_{-\mathbf{k}\downarrow} a_{\mathbf{k}\uparrow} \rangle$ and $\sum_{\mathbf{k}} \langle a_{-\mathbf{k}\sigma} a_{\mathbf{k}\sigma} \rangle$ $(\sigma = \uparrow, \downarrow)$ are finite.[31,61] Starting from the Hamiltonian above, we now examine the pairing mechanism in the weak-coupling limit.

In the absence of any Zeeman fields, as illustrated in Fig. 4.3.1(a), the two helicity branches cross at the origin in momentum space. Under an effective out-of-plane Zeeman field, i.e., when $h$ is finite, the helicity branches are coupled and a gap opens up at the origin [see Fig. 4.4.1(c)]. When the chemical potential lies in this gap, as is the case in Fig. 4.4.1(c), pairing can occur within the lowest helicity branch, leading to the opening of pairing gaps at the crossings of the particle and hole dispersions of the lower branch. This is possible since SOC mixes different spin species so that $s$-wave pairing between spin-up and spin-down atoms can happen within the same helicity branch. We call this pairing mechanism intrabranch pairing. As we will demonstrate in the next section, for a Rashba SOC, this is the pairing scenario that results in the exotic topological superfluid state. When the chemical potential lies above the gap, or when the interaction becomes stronger, interbranch pairing becomes important and competes with intrabranch pairing. From the location of the avoided crossings, we see that pairing states in this case have zero center-of-mass momentum.

When an in-plane Zeeman field is added, the single-particle dispersion spectra become deformed. As shown in Fig. 4.4.1(d), in this case, intra-branch pairing naturally leads to pairing states with finite center-of-mass momenta. Apparently, these SOC-induced FF pairing states are the result of interplay between SOC and Fermi surface asymmetry, and are different in mechanism from the conventional FF (cFF) states discussed previously.

With these qualitative analysis in the weak-coupling limit, we have already seen the various possibilities of exotic pairing states under SOC. In the following, let us see some more concrete examples.

### 4.B. Topological superfluid state and gapless superfluid state

Perhaps the most important exotic pairing state in a spin-orbit coupled Fermi system is the topological superfluid state in two dimensions, orig-inally investigated in the context of semiconductor/superconductor het-erstructures with Rashba SOC, $s$-wave pairing order and an out-of-plane Zeeman field.[17-19] As has been shown in Refs. 28, 29, SOC breaks the inversion symmetry and mixes up different spin components in the helicity branches. The introduction of the external Zeeman field breaks the time-reversal symmetry and opens up a gap between different helicity branches. In the weak-coupling limit, when the Fermi surface of the system lies in this gap, the ground state of the system becomes a topological superfluid state with Majorana zero modes on the boundaries or at the cores of vortex excitations. Importantly, these non-Abelian Majorana zero modes are pro-tected by a bulk gap and are therefore useful for fault-tolerant topological quantum computation.[73]

The stabilization of the topological superfluid state can be understood as the direct result of intrabranch pairing, which leads to a mixture of $s$- and $p$-wave symmetries in the pairing order parameter.[31,61] In parame-ter regions where the $p$-wave pairing dominates, the $p_x \pm i p_y$ symmetry of the pairing order parameter inherited from the Rashba SOC allows one to map the system onto a $p$-wave superfluid with the corresponding symmetry.[29]

In the context of ultracold atomic gases, a natural question is whether the topological superfluid state and the associated Majorana zero modes can be prepared and probed. Although the experimental realizaiton of the Rashba SOC has not been achieved yet, there have been various theoretical proposals on its implementation in cold atoms, which gives rise to even more theoretical characterizations of the topological superfluid phase and the related phase transitions in the context of ultracold Fermi gases. In two dimensions, it is found that a topological superfluid phase which supports Majorana zero modes can be stabilized.[28,29] In a uniform system, these different phases will give rise to phase-separated states with various first-order boundaries (see Fig. 4.4.2). In a trapping potential, which is more relevant to the experimental conditions, these different phases naturally separate in space, leading to shell structures similar to those in a polarized Fermi gas.[32]

Another intriguing finding on the mean-field phase diagram Fig. 4.4.2 is the absence of a normal phase, even in the limit of large magnetic field $\alpha k_F/h \to 0$. Instead of being zero in the case without SOC, here, the order parameter decreases exponentially with increasing $h$, but remains finite within the entire SF and TSF regime. This somewhat counterintuitive result is an artifact of the mean-field approach, which completely neglects quantum and thermal fluctuations. A more careful analysis incorporating particle-hole fluctuations in the large-Zeeman-field limit shows that there

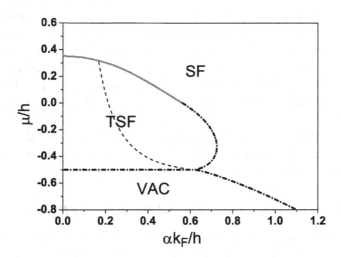

**Fig. 4.4.2.** Typical phase diagram in the $\alpha$-$\mu$ plane for a two-dimensional Fermi gas under Rashba SOC and an out-of-plane Zeeman field $h$. The first-order phase transition is shown in red solid curve while the second-order phase transitions are shown in dash-dotted black curves. The thin dashed curve in the topological superfluid (TSF) region marks the $\Delta/h = 10^{-3}$ threshold. The effective Zeeman field $h$ is taken to be the energy unit, while the unit of momentum $k_h$ is defined through $\hbar^2 k_h^2/(2m) = h$. Here, VAC represents the vacuum. See Ref. 32.

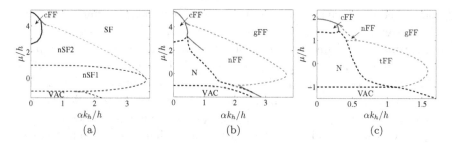

**Fig. 4.4.3.** Typical phase diagrams for a two-dimensional Fermi gas under: (a) 1D SOC without two-photon detuning (no in-plane Zeeman field); (b) 1D SOC with finite two-photon detuning (finite in-plane Zeeman field); (c) Rashba SOC with both the out-of-plane and in-plane Zeeman field. The solid curves are first-order boundaries, the dashed curves represent continuous phase boundaries. Here, VAC represents the vacuum, and N represents normal state. See Refs. 38 and 52.

exits a so-called polaron-molecule transition, suggesting that the true ground state is a normal phase, at least in the weak-coupling regime.[47]

When the Rashba SOC is replaced by the less symmetric 1D SOC, the topological superfluid phase would be replaced by various nodal superfluid phases (nSF1 and nSF2), with either two or four discrete nodal points along the direction with no SOC. The mean-field zero-temperature phase diagram in two dimensions is shown in Fig. 4.4.3(a). Note that the cFF phase is due to the interbranch pairing mechanism and not the SOC-induced FF states that we have discussed previously. In three dimensions, neither the Rashba SOC nor the 1D SOC would lead to a topological superfluid, as the pairing superfluid phases therein are either fully gapped trivial superfluid (SF) state or nodal superfluid phase. Interestingly, for gapless superfluid states in a three dimensional Fermi gas with the 1D SOC, the nodal points typically form closed surfaces in momentum space.[49]

### 4.C. *SOC-induced FF pairing and topological FF state*

An important consequence of the intrabranch pairing is the dramatically enhanced FF pairing states under SOC and Fermi surface asymmetry. As illustrated in Fig. 4.4.1(c), under an additional in-plane Zeeman field, the Fermi surface is deformed, such that it no longer has inversion symmetry along the axis of the transverse field. In the weakly interacting limit, it is clear that a simple BCS pairing state with zero center-of-mass momentum becomes energetically unfavorable. This opens up the possibility of an exotic FFLO-like pairing state with finite center-of-mass momentum.[52] From the general argument above, we may further infer that such a pairing state is a natural result of the co-existence of SOC and Fermi surface asymmetry, and should generally exist in such systems, regardless of the exact type of SOC in the system. Indeed, these exotic pairing states have been reported to exist in various systems of different dimensions and with different forms of SOC.[36–42,49–51,74,75]

To illustrate this, we first consider an experimentally relevant system, where the 1D SOC is imposed on a two-dimensional Fermi gas with effective axial and transverse Zeeman fields. Similar to the Rashba case, a mean-field description of the pairing states here can be seen as a natural extension of the standard BCS theory

$$
H_{\text{eff}} = \frac{1}{2} \sum_{\mathbf{k}} \begin{pmatrix} \lambda_{\mathbf{k}}^+ & 0 & h & \Delta_{\mathbf{Q}} \\ 0 & -\lambda_{\mathbf{Q}-\mathbf{k}}^+ & -\Delta_{\mathbf{Q}}^* & -h \\ h & -\Delta_{\mathbf{Q}} & \lambda_{\mathbf{k}}^- & 0 \\ \Delta_{\mathbf{Q}}^* & -h & 0 & -\lambda_{\mathbf{Q}-\mathbf{k}}^- \end{pmatrix}
$$
$$
+ \sum_{\mathbf{k}} \xi_{|\mathbf{Q}-\mathbf{k}|} - \frac{|\Delta_{\mathbf{Q}}|^2}{U}, \qquad (4.4.3)
$$

where $\lambda_{\mathbf{k}}^\pm = \xi_{\mathbf{k}} \pm \alpha k_x \mp h_x$, the order parameter $\Delta_{\mathbf{Q}} = U \sum_{\mathbf{k}} \langle a_{\mathbf{Q}-\mathbf{k}\downarrow} a_{\mathbf{k}\uparrow} \rangle$. The Hamiltonian (4.4.3) has been written under the hyperfine-spin basis $\left\{ a_{\mathbf{k}\uparrow}, a_{\mathbf{Q}-\mathbf{k}\uparrow}^\dagger, a_{\mathbf{k}\downarrow}, a_{\mathbf{Q}-\mathbf{k}\downarrow}^\dagger \right\}^T$. Again, the SOC parameter $\alpha$ is related to the momentum transfer of the Raman process in the 1D-SOC scheme, and the effective Zeeman field $h$ and $h_x$ are proportional to the effecitve Rabi-frequency and the two-photon detuning of the Raman lasers, respectively. The zero-temperature thermodynamic potential can be obtained by diago-nalizing the effective Hamiltonian

$$
\Omega = \sum_{\mathbf{k}} \xi_{|\mathbf{Q}-\mathbf{k}|} + \sum_{\mathbf{k},\nu} \theta(-E_{\mathbf{k},\nu}^\eta) E_{\mathbf{k},\nu}^\eta - \frac{|\Delta_{\mathbf{Q}}|^2}{U}, \qquad (4.4.4)
$$

where the quasiparticle (quasihole) dispersion $E_{\mathbf{k},\nu}^\eta$ ($\nu = 1, 2, \eta = \pm$) are the eigenvalues of the matrix in Hamiltonian (4.4.3), and $\theta(x)$ is the Heaviside step function.

From the previous general analysis in the weak-coupling limit, the BCS pairing states with zero center-of-mass momentum would become unsta-ble against an FF pairing state under the Fermi surface asymmetry. This implies an instability of the BCS state with finite $h_x$. This point can be demonstrated by performing a small $\mathbf{Q}$ expansion around the local

minimum in the thermodynamic potential landscape that corresponds to
the BCS pairing state

$$\Omega(\Delta, Q_x) = \Omega_0(\Delta) + \Omega_1(\Delta)Q_x + \Omega_2(\Delta)Q_x^2 + \mathcal{O}(Q_x^3), \qquad (4.4.5)$$

where we have assumed $\mathbf{Q} = (Q_x, 0)$. It is then straightforward to demonstrate numerically that for $h_x = 0$, we have $\Omega_1 = 0$, $\Omega_2 > 0$; while for
$h_x \neq 0$, we have $\Omega_1 \neq 0$, which has an opposite sign to that of $h_x$. This
is a direct evidence that the BCS pairing state in the presence of Fermi
surface asymmetry and SOC becomes unstable against an FF state, with
the center-of-mass momentum $\mathbf{Q}$ opposite to the direction of the transverse
field $h_x$. A qualitative understanding of these FF states is that the combination of SOC and Fermi surface asymmetry shifts the local minima that
corresponds to BCS pairing states ($Q = 0$) onto the finite-$\mathbf{Q}$ plane. This
is further reflected by the observation that the magnitude of the center-of-
mass momentum $Q$ decreases as Fermi surface asymmetry becomes smaller,
i.e., with increasing chemical potential $\mu$ or SOC strength $\alpha$.[52,53]

We show the mean-field phase diagram in Fig. 4.4.3(b). Recalling
Fig. 4.4.2, it is clear that the topological superfluid state under the Rashba
SOC is now replaced by a nodal FF (nFF) state, while the trivial superfluid state is replaced by a gapped FF (gFF) state. These SOC-induced FF
states originate from intrabranch pairing, and are different in mechanism
from the conventional FF state.

Among these SOC-induced FF states, perhaps the most interesting case
is the topological FF (tFF) state, where the pairing state can have nonzero
center-of-mass momentum and topologically nontrivial properties simultaneously.[36–39] This exotic pairing phase can be understood as derived from a
typical topological superfluid phase in two dimensions, where Rashba SOC,
$s$-wave pairing order and an out-of-plane Zeeman field coexist. With the
addition of another effective Zeeman field in the transverse direction, the
Fermi surface becomes asymmetric, and according to the preceding analysis, the ground state of the system necessarily acquires a nonzero center-
of-mass momentum. More importantly, the ground state would inherit all
topological properties from the topological superfluid state, provided that
the deformation of the Fermi surface should not be drastic enough to close
the bulk gap. The phase diagram of a two-dimensional Fermi gas under
Rashba SOC and cross Zeeman fields is shown in Fig. 4.4.3(c). Apparently,
the stability region of the tFF state roughly corresponds to the topological
superfluid state in Fig. 4.4.2.

## 4.5. Engineering Novel States

An advantage of ultracold atomic gases is their highly tunable parameters. While SOC provides the possibility of engineering single-particle dispersion, in principle, one can also change the interaction strength via the well-established Feshbach resonance technique. In fact, when combining SOC with other available tools in ultracold atomic gases, we can engineer highly nontrivial states.

### 5.A. *Engineering FF state*

As an example, we consider a three-component Fermi gas, where one fermion species ('impurity') is tuned close to a wide Feshbach resonance with one of the spin-species in a two-component Fermi gas under the 1D SOC.[54] The pairing mechanism here is illustrated in Fig. 4.5.1. Under the 1D SOC, the spin components in both helicity branches are momentum dependent. The pairing states naturally acquire a nonzero center-of-mass momentum, which is dependent on the position of the Fermi surface as well as the interaction strength. For instance, in the weak-coupling limit, when the Fermi surface lies in the lower branch, the center-of-mass momentum is negative, since the pairing is lower-branch dominated; while if the Fermi

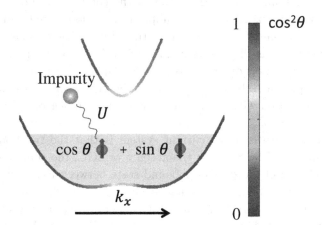

**Fig. 4.5.1.** Schematics of the three-component Fermi system in Sec. 5.A. The impurity atoms interact only with the spin-up atoms in a spin-orbit coupled Fermi gas. The fractions of spin components in both helicity branches are momentum dependent, as characterized by $\cos\theta = -\beta_{\mathbf{k}}^-$ and $\sin\theta = \beta_{\mathbf{k}}^+$, where $\beta_{\mathbf{k}}^\pm$ is defined in Sec. 4.3. See Ref. 54.

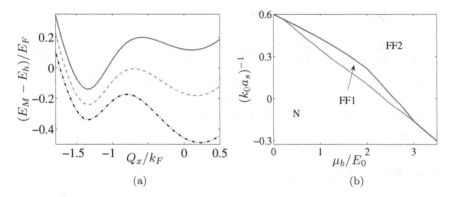

**Fig. 4.5.2.** (a) Competition between molecules of different center-of-mass momenta as the actual Fermi energy changes. Molecular energies $E_M$ relative to the Fermi energy $E_h$ as functions of the center-of-mass momentum $Q_x$, with $h/E_F = 1.2$ (solid), $h/E_F = 1$ (dashed), and $h/E_F = 0.8$ (dash-dotted). The unit of energy is the Fermi energy $E_F$ of a two-component, noninteracting Fermi gas in the absence of SOC and with the same total number density. (b) Typical phase diagram in the $(\mu_b/E_0, (k_0 a_s)^{-1})$ plane in a Li-K-K system. The red (blue) curve shows the transition boundary between the normal state (N) and FF states with different center-of-mass momentum (FF1 and FF2). See Ref. 54.

surface lies in the upper branch, the center-of-mass momentum can be positive, as the ground state is the result of the competition between lower-branch-dominated pairing and the upper-branch-dominated pairing. This can lead to the competition between pairing states with different center-of-mass momentum, as illustrated in Fig. 4.5.2. Apparently, the FF pairing states in this system is different from the SOC-induced FF states discussed previously. Qualitatively, we are replacing the Fermi surface asymmetry in the previous section with the asymmetry in the spin degrees of freedom of the helicity basis.

A notable feature of the current system is that the two-body bound state is suppressed by SOC. In the absence of SOC, the spin-down atoms are decoupled, and the two-body bound state between the 'impurity' and the spin-up atom emerges at resonance with $a_s^{-1} = 0$, where $a_s$ is the scattering length between the spin-up atom and the 'impurity' atom. Under SOC, this two-body bound state threshold is pushed toward the BEC limit with $a_s^{-1} > 0$ (see Fig. 4.5.3). Roughly, as SOC establishes 'correlation' between the spin-up and the spin-down atoms, it is more difficult for the 'impurity' and the spin-up atoms to combine and form a bound state. As we will see later, a more rigorous description of this peculiar 'correlation'

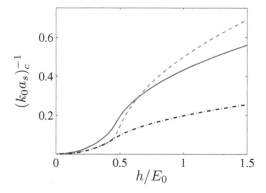

**Fig. 4.5.3.** Two-body bound state threshold for varying SOC parameter $h$ and a fixed $\alpha k_0/E_0 = 1$. The mass ratio $\eta = m_a/m_b$, where $m_b$ is the mass of the impurity atom, $m_a$ is the mass of the atoms under SOC. Here, $\eta = 1$ (solid), $\eta = 6/40$ (dashed), and $\eta = 40/6$ (dash-dotted). The unit of energy $E_0 = 2m_a\alpha^2/\hbar^2$, and the unit of momentum $k_0 = 2m_a\alpha/\hbar^2$. See Ref. 54.

is the symmetry in the SOC-modified single-particle dispersion. This suppression of two-body bound state naturally leads to the question of the stability of three-body bound states, i.e., whether a three-body bound state can be more stable than the two-body bound state in certain parameter region?

### 5.B. *Exotic trimer states under SOC*

A short answer to the question above is: not with the 1D SOC.[54] However, it is possible to stabilize three-body bound states, or trimers, if we consider other forms of SOC, especially those with a high symmetry. Indeed, if we replace the 1D SOC with the Rashba SOC in the previously discussed three-component Fermi system, trimers can be stabilized even in the absence of any two-body bound states.[56] The stabilization of this so-called Borromean state can be understood from the special symmetry of the Rashba-modified single-particle dispersion spectra (see Fig. 4.5.4). As illustrated in Fig. 4.3.1(c), under the Rashba SOC, the degenerate subspace of the single-particle ground state forms a ring in momentum space. With such a spectral symmetry, the scattering within the lowest-energy subspace is blocked due to the total momentum conservation (see Fig. 4.5.4), which effectively suppresses the formation of a two-body bound state. In contrast, the three-body scattering is not blocked. Under the 1D SOC, the single-particle dispersion spectra have less symmetry. Hence, although the

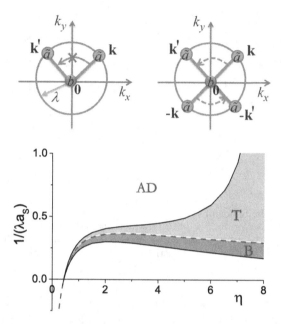

**Fig. 4.5.4.** (Upper panel) Illustration of the Borromean binding mechanism in the the three-component Fermi system in Sec. 5.B. In this $\tilde{a} - \tilde{a} - b$ system, $\tilde{a}$ atoms are subject to the Rashba SOC, and $b$ serves as the 'impurity'. (Upper left) The two-body scattering within the lowest-energy subspace of the $\tilde{a} - b$ system cannot proceed due to the conservation of total momentum. (Upper right) The three-body scattering within the lowest energy subspace of the $\tilde{a} - \tilde{a} - b$ system is allowed (green dashed arrows). (Lower panel) Phase diagram for the system in Sec. 5.B on the $(1/(\lambda a_s), \eta)$ plane, where $\lambda = m_a \alpha / \hbar^2$. The lower and upper solid curves respectively represent the threshold of Borromean ('B') state and the boundary at which the ordinary trimer ('T') merges into the atom-dimer continuum ('AD'). The blue dashed curve is the dimer threshold, which is also the boundary between 'B' and 'T' for $\eta \geqslant 0.39$. See Ref. 56.

two-body bound states are also suppressed in those systems, they remain to be energetically more favorable than the trimers.

Borromean states have been found before in different physical systems. However, previously studied Borromean states are nonuniversal, with properties dependent on the short-range details of the two-body interaction potential.[76-78] Here, the Borromean states are universal, with properties only dependent on the two-body $s$-wave scattering length and the SOC coupling strength. Furthermore, as the symmetry-facilitated trimer formation is robust in the presence of a Fermi surface,[57] it is reasonable to expect interesting many-body states with exotic few-body correlations near the trimer threshold. The universal Borromean state in the spin-orbit coupled

Fermi system is thus highly nontrivial. Note that universal trimers have also been identified in a related system, where an atom under isotropic SOC interacts with two other atoms.[55] Both of these cases suggest that exotic few-body states may be stabilized by introducing SOC as another dimension of control.

## 4.6. Summary

In this Chapter, we have focused on the general pairing mechanism in a spin-orbit coupled Fermi gas. As SOC changes the single-particle dispersion spectra, pairing superfluidity in the system takes on many new and exotic forms: topological superfluid, SOC-induced FF state, topological superfluid so on so forth. Besides these exotic pairing states, SOC can also lead to novel universal trimer states, which features exotic few-body correlations upon which even more interesting many-body phases may emerge. In all these cases, by modifying the single-particle dispersion, SOC proves to be a powerful tool of quantum control, which, when combined with the outstanding tunability of ultracold atomic gases, is playing an increasingly important role in fulfilling the potential of quantum simulation with ultracold atomic gases.

## References

1. Y.-J. Lin, R. L. Compton, A. R. Perry, W. D. Phillips, J. V. Porto, and I. B. Spielman, *Phys. Rev. Lett.*, **102**, 130401 (2009).
2. Y.-J. Lin, R. L. Compton, K. Jiménez-García, W. D. Phillips, J. V. Porto, I. B. Spielman. *Nature (London)*, **462**, 628 (2009).
3. Y.-J. Lin, R. L. Compton, K. Jiménez-García, W. D. Phillips, J. V. Porto, and I. B. Spielman. *Nat. Phys.*, **7**, 531 (2011).
4. Y.-J. Lin, K. Jiménez-García, and I. B. Spielman. *Nature (London)*, **471**, 83 (2011).
5. P. Wang, Z.-Q. Yu, Z. Fu, J. Miao, L. Huang, S. Chai, H. Zhai, and J. Zhang. *Phys. Rev. Lett.*, **109**, 095301 (2012).
6. L. W. Cheuk, A. T. Sommer, Z. Hadzibabic, T. Yefsah, W. S. Bakr, and M. W. Zwierlein. *Phys. Rev. Lett.*, **109**, 095302 (2012).
7. X.-J. Liu, M. F. Borunda, X. Liu, and J. Sinova. *Phys. Rev. Lett.*, **102**, 046402 (2009).
8. J.-Y. Zhang, S.-C. Ji, Z. Chen, L. Zhang, Z.-D. Du, B. Yan, G.-S. Pan, B. Zhao, Y. Deng, H. Zhai, S. Chen, and J.-W. Pan. *Phys. Rev. Lett.*, **109**, 115301 (2012).
9. C. Qu, C. Hamner, M. Gong, C. Zhang, and P. Engels. *Phys. Rev. A*, **88**, 021604(R) (2013).

10. R. A. Williams, M. C. Beeler, L. J. LeBlanc, K. Jimenez-Garcia, and I. B. Spielman. *Phys. Rev. Lett.*, **111**, 095301 (2013).
11. Z. Fu, L. Huang, Z. Meng, P. Wang, L. Zhang, S. Zhu, H. Zhai, P. Zhang, and J. Zhang. *Nat. Phys.*, **10**, 110 (2014).
12. S.-C. Ji, J.-Y. Zhang, L. Zhang, Z.-D. Du, W. Zheng, Y.-J. Deng, H. Zhai, S. Chen, and J.-W. Pan. *Nat. Phys.*, **10**, 314 (2014).
13. A. J. Olson, S.-J. Wang, R. J. Niffenegger, C.-H. Li, C. H. Greene, and Y. P. Chen. *Phys. Rev. A*, **90**, 013616 (2014).
14. Z. Meng, L. Huang, P. Peng, D. Li, L. Chen, Y. Xu, C. Zhang, P. Wang, and J. Zhang. *Phys. Rev. Lett.*, **117**, 235304 (2016).
15. L. Huang, Z. Meng, P. Wang, P. Peng, S.-L. Zhang, L. Chen, D. Li, Q. Zhou, and J. Zhang. *Nat. Phys.*, **12**, 540 (2016).
16. Z. Wu, L. Zhang, W. Sun, X.-T. Xu, B.-Z. Wang, S.-C. Ji, Y. Deng, S. Chen, X.-J. Liu, and J.-W. Pan. *Science*, **354**, 83 (2016).
17. M. Z. Hasan and C. L. Kane. *Rev. Mod. Phys.*, **82**, 3045 (2010).
18. X.-L. Qi and S.-C. Zhang. *Rev. Mod. Phys.*, **83**, 1057 (2011).
19. J. Alicea. *Rep. Prog. Phys.*, **75**, 076501 (2012).
20. D. L. Campbell, G. Juzeliūnas, and I. B. Spielman. *Phys. Rev. A*, **84**, 025602 (2011).
21. J. D. Sau, R. Sensarma, S. Powell, I. B. Spielman, and S. Das Sarma. *Phys. Rev. B*, **83**, 140510(R) (2011).
22. Z. F. Xu and L. You. *Phys. Rev. A*, **85**, 043605 (2012).
23. X.-J. Liu, K. T. Law, and T. K. Ng. *Phys. Rev. Lett.*, **112**, 086401 (2014).
24. B. M. Anderson, I. B. Spielman, and G. Juzeliūnas. *Phys. Rev. Lett.*, **111**, 125301 (2013).
25. Z.-F. Xu, L. You, and M. Ueda. *Phys. Rev. A*, **87**, 063634 (2013)
26. Y. Li, X. Zhou, and C. Wu. *Phys. Rev. B*, **85**, 125122 (2012).
27. B. M. Anderson, G. Juzeliūnas, V. M. Galitski and I. B. Spielman. *Phys. Rev. Lett.*, **108**, 235301 (2012).
28. C. Zhang, S. Tewari, R. M. Lutchyn, and S. Das Sarma. *Phys. Rev. Lett.*, **101**, 160401 (2008).
29. M. Sato, Y. Takahashi, and S. Fujimoto. *Phys. Rev. Lett.*, **103**, 020401 (2009).
30. M. Gong, S. Tewari, and C. Zhang. *Phys. Rev. Lett.*, **107**, 195303 (2011).
31. L. Dell'Anna, G. Mazzarella, and L. Salasnich. *Phys. Rev. A*, **84**, 033633 (2011).
32. J. Zhou, W. Zhang, W. Yi. *Phys. Rev. A*, **84**, 063603 (2011).
33. M. Gong, G. Chen, S. Jia, and C. Zhang. *Phys. Rev. Lett.*, **109**, 105302 (2012).
34. R. Liao, Y. Yi-Xiang, and W.-M. Liu. *Phys. Rev. Lett.*, **108**, 080406 (2012).
35. L. He and X.-G. Huang. *Phys. Rev. Lett.*, **108**, 145302 (2012).
36. C. Chen. *Phys. Rev. Lett.*, **111**, 235302 (2013).
37. C. Qu, Z. Zheng, M. Gong, Y. Xu, L. Mao, X. Zou, G. Guo, and C. Zhang. *Nat. Commun.*, **4**, 2710 (2013).
38. W. Zhang and W. Yi. *Nat. Commun.*, **4**, 2711 (2013).
39. X.-J. Liu and H. Hu. *Phys. Rev. A*, **88**, 023622(R) (2013).
40. Y. Xu, R.-L. Chu and C. Zhang. *Phys. Rev. Lett.*, **112**, 136402 (2014).

41. Y. Cao, S.-H. Zou, X.-J. Liu, G.-L. Long and H. Hu. *Phys. Rev. Lett.*, **113**, 115302 (2014).
42. H. Hu, L. Dong, Y. Cao, H. Pu, and X.-J. Liu. *Phys. Rev. A*, **90**, 033624 (2014).
43. J.-S. Pan, X.-J. Liu, W. Zhang, W. Yi, and G.-C. Guo. *Phys. Rev. Lett.*, **115**, 045303 (2015).
44. M. Iskin and A. L. Subasi. *Phys. Rev. Lett.*, **107**, 050402 (2011).
45. W. Yi and G.-C. Guo. *Phys. Rev. A*, **84**, 031608(R) (2011).
46. L. Han and C. A. R. Sá de Melo. *Phys. Rev. A*, **85** 011606(R) (2012).
47. W. Yi and W. Zhang. *Phys. Rev. Lett.*, **109**, 140402 (2012).
48. X. Yang and S. Wan. *Phys. Rev. A*, **85**, 023633 (2012).
49. M. Iskin and A. L. Subasi. *Phys. Rev. A*, **87**, 063627 (2013).
50. L. Dong, L. Jiang, and H. Pu. *New J. Phys.*, **15**, 075014 (2013).
51. X.-F. Zhou, G.-C. Guo, W. Zhang, and W. Yi. *Phys. Rev. A*, **87**, 063606 (2013).
52. F. Wu, G.-C. Guo, W. Zhang, and W. Yi. *Phys. Rev. Lett.*, **110**, 110401 (2013).
53. F. Wu, G.-C. Guo, W. Zhang, and W. Yi. *Phys. Rev. A*, **88**, 043614 (2013).
54. L. Zhou, X. Cui, and W. Yi. *Phys. Rev. Lett.*, **112**, 195301 (2014).
55. Z.-Y. Shi, X. Cui, and H. Zhai. *Phys. Rev. Lett.*, **112**, 013201 (2014).
56. X. Cui and W. Yi. *Phys. Rev. X*, **4**, 031026 (2014).
57. X. Qiu, X. Cui, and W. Yi. *Phys. Rev. A*, **94**, 051604 (2016).
58. C. Wang, C. Gao, C. Jian, and H. Zhai. *Phys. Rev. Lett.*, **105**, 160403 (2010).
59. C. Wu, I. Mondragon-Shem, and X.-F. Zhou. *Chin. Phys. Lett.*, **28** 097102 (2011).
60. J. P. Vyasanakere, S. Zhang, and V. B. Shenoy. *Phys. Rev. B*, **84**, 014512 (2011).
61. Z.-Q. Yu and H. Zhai. *Phys. Rev. Lett.*, **107**, 195305 (2011).
62. J. Dalibard, F. Gerbier, G. Juzeliūnas, and P. Öhberg. *Rev. Mod. Phys.*, **83**, 1523 (2011).
63. J. Ruseckas, G. Juzeliūnas, P. Öhberg, and M. Fleischhauer, PRL **95**, 010404 (2005).
64. X.-J. Liu, M. F. Borunda, X. Liu, and J. Sinova. *Phys. Rev. Lett.*, **102**, 046402 (2009).
65. I. B. Spielman. *Phys. Rev. A*, **79**, 063613 (2009).
66. Y. Li, X. Zhou, and C. Wu. *Phys. Rev. A*, **93**, 033628 (2016).
67. N. Goldman, F. Gerbier, and M. Lewenstein. *J. Phys. B*, **46**, 134010 (2013).
68. N. Goldman, G. Juzeliunas, P. Öhberg, and I. B. Spielman. *Rep. Prog. Phys.*, **77**, 126401 (2014).
69. J. Chen, H. Hu, and Gao Xianlong. *Phys. Rev. A*, **90**, 023619 (2014).
70. W. Ketterle and M. W. Zwierlein, *Making, probing and understanding ultracold Fermi gases*, Ultracold Fermi Gases, Proceedings of the International School of Physics "Enrico Fermi", Course CLXIV, Varenna, 20–30 June 2006, edited by M. Inguscio, W. Ketterle, and C. Salomon (IOS Press, Amsterdam) (2008).

71. C. Chin, R. Grimm, P. Julienne, and E. Tiesinga. *Rev. Mod. Phys.*, **82**, 1225 (2010).
72. P. Fulde and R. A. Ferrell. *Phys. Rev.*, **135**, A550 (1964); A. I. Larkin and Y. N. Ovchinnikov. *Sov. Phys. JETP*, **20**, 762 (1965).
73. C. Nayak, S. H. Simon, A. Stern, M. Freedman, and S. Das Sarma. *Rev. Mod. Phys.*, **80**, 1083 (2008).
74. K. Michaeli, A. C. Potter, and P. A. Lee. *Phys. Rev. Lett.*, **108**, 117003 (2012).
75. Z. Zheng, M. Gong, X. Zou, C. Zhang, and G.-C. Guo. *Phys. Rev. A*, **87**, 031602(R) (2013).
76. V. Efimov, *Yad. Fiz.* **12**, 1080 (1970); *Sov. J. Nucl. Phys.*, **12**, 589 (1971).
77. M. V. Zhukov, B. V. Danilin, D. V. Fedorov, J. M. Bang, I. S. Thompson, and J. S. Vaagen. *Phys. Rep.*, **231**, 151 (1993).
78. E. Braaten and H.-W. Hammer. *Phys. Rep.*, **428**, 259 (2006).

# Chapter 5

# Superfluid Properties of a Spin-orbit Coupled Fermi Gas

Juan Yao[1], Shanshan Ding[2], Zhenhua Yu[3], and Shizhong Zhang[2]

[1]*Institute for Advanced Study, Tsinghua University,
Beijing 100084, China*
[2]*Department of Physics and Center of Theoretical and Computational
Physics, The University of Hong Kong, Hong Kong, China*
[3]*School of Physics and Astronomy, Sun Yat-Sen University,
Zhuhai 519082, China*

In this chapter, we discuss the superfluid properties of a spin-orbit coupled Fermi gas. After reviewing the single particle aspects of the Rashba
coupled spin-orbit Fermi gas, we investigate the ground state properties along the Bose-Einstein condensate to Bardeen-Cooper-Schrieffer
crossover. The collective excitations and anisotropic sound modes are
calculated. We calculate the normal and superfluid densities and investigate the structure of quantum vortex and its associated excitations.

## 5.1. Introduction

Superfluidity is one of the most fascinating phenomena in many-body
physics where quantum mechanical effects, usually governing microscopic
physics, manifests itself on a truly macroscopic scale. Examples of superfluidity fall into two broad categories: bosons and fermions. In the former case,
the Bose statistics of the particles leads to the Bose-Einstein condensation
at low temperatures and interaction effects further stabilize the superfluid
response, giving rise to an non-zero superfluid critical velocity.[1–3] In the latter case, Fermi statistics forbids condensation into a single fermion state,
but weak attraction between fermions can result in the condensation of pairs
of fermions. The simplest example is an $s$-wave superconductor where electrons with opposite spins form Cooper pairs which condense below a critical
temperature.[4] In this case, Cooper pairs do not have internal degrees of freedom and the resultant superfluid behaves very much like that of bosons,

for example, as in Helium four.[3] An example of richer phenomena is the $p$-wave superfluidity of Helium three, where the spins of the two Helium three atoms form a triplet; the internal degrees of freedom leads to multitude of textures of the order parameters and as a result, much more complicated superfluid behaviour. We note, however, that in both cases, the spin and orbital degrees of freedom are decoupled at the single particle level, and the much weaker dipolar interaction which does couple spin and orbit in Helium three is relevant only to determine the condensate structure among the nearly degenerate multitude of the superfluid states.

An entirely different situation arises in the context of spin-orbit coupled quantum gases. In this case, strong spin-orbit couplings of the Rashba or Dresselhaus type are engineered by using light-atom interactions.[9-23] The single particle states involved are then naturally given in the so-called helicity basis in which the orientation of the spin and the direction of the momentum of each particle are intimately locked. One thus expects that the superfluid response of the system which is related to the movement of mass density should be intimately connected with its spin (magnetic) properties. Indeed, this is found to be the case in bosons in the presence of spin-orbit coupling.[24] In particular, it is found that when the system goes from the so-called plane wave state to the zero-momentum state, the superfluid density vanishes at the transition point while the condensate density is still significant. Furthermore, exact relations are also found between the super-fluid density and the magnetic susceptibilities. For fermions with spin-orbit coupling, even though no analogous phase transition (from plane-wave to zero momentum) exists, the effects of spin-orbit coupling on the superfluid is nonetheless important.[25-31]

In this chapter, we investigate two effects associated with superfluidity, namely its response to a linear moving and to a rotational boundary conditions. The former is related to the concepts of normal and superfluid densities and the later is related to the appearance of quantized vortices.

## 5.2. Single Particle Hamiltonian

For definiteness, we consider a two-component Fermi gas subject to a Rashba-type spin-orbit coupling. The single-particle Hamiltonian is given by

$$\hat{H}_{\text{SOC}} = \frac{\hat{k}^2}{2m}\mathbf{I} + \lambda(\boldsymbol{\sigma} \times \hat{\mathbf{k}})_z - h\sigma_z = \frac{\hat{k}^2}{2m}\mathbf{I} + \lambda(\sigma_x \hat{k}_y - \sigma_y \hat{k}_x) - h\sigma_z.$$

$$(5.2.1)$$

Here $m$ is the mass of the fermion atom. $\lambda$ denotes the strength of the Rashba spin-orbit coupling and $h$ is the external Zeeman field. $\mathbf{I}$ stands for the two-by-two identity matrix and $\sigma_i$ $(i = x, y, z)$ are the Pauli matrices. We take $\hbar = 1$ throughout. Expressing the above Hamiltonian in the momentum space,

$$\hat{H}_{\text{SOC}} = \frac{\hat{k}^2}{2m}\mathbf{I} + \mathbf{h_k} \cdot \boldsymbol{\sigma}, \tag{5.2.2}$$

we can identify the effective Zeeman fields

$$\mathbf{h_k} = (\lambda k_y, -\lambda k_x, -h). \tag{5.2.3}$$

The eigenstates of the single particle Hamiltonian correspond to either spin aligned with the effective field $\mathbf{h_k}$ or against it. The eigen-energies are given by

$$E_{\pm} = \frac{k^2}{2m} \pm |\mathbf{h_k}| = \frac{k^2}{2m} \pm \sqrt{\lambda^2 k_{\perp}^2 + h^2}, \tag{5.2.4}$$

in which $k_{\perp}^2 = k_x^2 + k_y^2$ and $k^2 = k_{\perp}^2 + k_z^2$.

## 2.A. *Ground state degeneracy*

When $h = 0$, the dispersion relation Eq. (5.2.4) is reduced to $E_{\pm} = \frac{k^2}{2m} \pm \lambda k_{\perp}$. For the lower branch $E_-$, the lowest energy occurs at $k_{\perp} = m\lambda$. In this case, the system possesses a degeneracy of single particle ground states with the minima forming a circle with radius $k_{\perp} = m\lambda$. For non-zero $h$, there are two cases depending on the strength of Rashba spin-orbit coupling. For $h \geq m\lambda^2$, a unique ground state exists at $k_{\perp} = 0$ as show in Fig. 5.2.1(a). For $h < m\lambda^2$, similar to the case with $h = 0$, the ground states form a degenerate circle with a reduced radius $k_{\perp} = \sqrt{m^2\lambda^4 - h^2}/\lambda$, as shown in Fig. 5.2.1(b).

## 2.B. *Density of states*

For the two branches of the single particle states, the corresponding density of states is given by

$$\mathcal{N}_{\pm}(E) = \frac{1}{V}\sum_{\mathbf{k}} \delta(E - E_{\pm}), \tag{5.2.5}$$

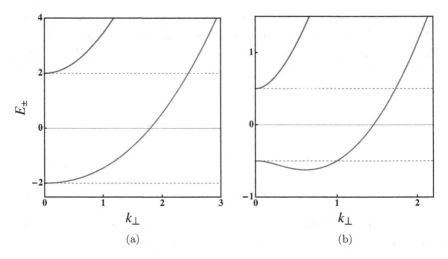

**Fig. 5.2.1.** Dispersion relation Eq. (5.2.4) for $E_\pm$ in terms of $k_\perp$ with $k_z = 0$ for various $\lambda$. (a) Unique minimum with $m\lambda^2 = 0.5h$. (b) Multi-minima with $m\lambda^2 = 2h$. Dashed lines donote $E_\pm = \pm h$.

with

$$\mathcal{N}_+(E) = \frac{m}{4\pi^2}\left\{2\sqrt{2m(E-h)} + 2m\lambda \arctan\left[\frac{m\lambda^2 + h}{\sqrt{2m\lambda^2(E-h)}}\right]\right.$$

$$\left. - \pi m\lambda \right\}\theta(E-h), \qquad\qquad (5.2.6a)$$

$$\mathcal{N}_-(E) = \frac{m}{4\pi^2}\left\{2\sqrt{2m(E+h)} + 2m\lambda \arctan\left[\frac{m\lambda^2 - h}{\sqrt{2m\lambda^2(E+h)}}\right]\right.$$

$$\left. + \pi m\lambda - \frac{m^2\lambda}{2\pi}\right\}\theta(E+h) + \frac{m^2\lambda}{2\pi}, \qquad (5.2.6b)$$

in which $\theta(E)$ is the Heaviside step function. The upper branch contributes to the density of states only when $E > h$.

For $m\lambda^2 \leq h$ with a unique ground state [Fig. 5.2.1(a)], the density of states as a function of energy $E$ is plotted in Fig. 5.2.2. When $E < h$, only the lower branch $E_-$ contributes to the density of states. Starting from $E = h$ [the dashed line in Fig. 5.2.2], both $E_+$ and $E_-$ make contributions to the total density of states. In this case, the behaviour of the density of states

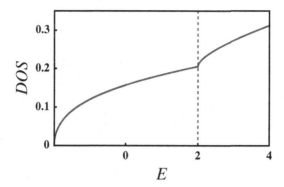

**Fig. 5.2.2.** Density of states (DOS) as a function of energy $E$ for $m\lambda^2 = h/2$. The dashed line indicates $E = h$.

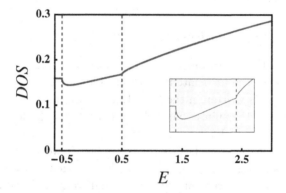

**Fig. 5.2.3.** Density of states (DOS) as a function of energy $E$ for $m\lambda^2 = 2h$. The dashed lines indicate $E = \pm h$. Inset: a close-up look of DOS for $-h < E < h$.

is similar to a three-dimensional free Fermi gases. When $m\lambda^2 > h$, due to the large strength of spin-orbit coupling, the density of states at low energy is much more complicated as plotted in Fig. 5.2.3. At low energy $E < -h$, the density of states is a constant. For $E > h$, qualitative behaviour of the density of states is similar to the case with $m\lambda^2 \leq h$. Between $-h$ and $h$ (the two dashed lines in Fig 5.2.3), the density of states is non-monotonic. This interval can be adjusted by changing $h$ and vanishes when $h = 0$. In this case, the density of states is similar to that of two-dimensional free Fermi gases and the spin-orbit coupling effectively reduces the dimensionality from three to two.

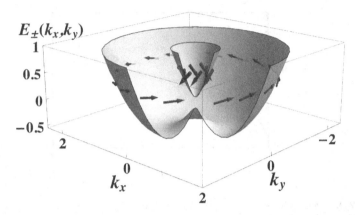

**Fig. 5.2.4.**  Single particle spectra of the upper branch $E_+$ and the lower branch $E_-$ as a function of $k_x$ and $k_y$ for $k_z = 0$. The arrows at each branch denotes the spin direction at a constant energy $E = 0.5$.

## 2.C.  *Spin-momentum locking*

Setting $k_z = 0$; the full dispersion relation is shown in Fig. 5.2.4, where the spin directions of the associated states are indicated by the arrows. Once the Hamiltonian is written in the form Eq. (5.2.2), it is easy to see that for the upper branch the spin is parallel to $\mathbf{h_k}$ while for the lower branch the spin is anti-parallel to $\mathbf{h_k}$.

When $h = 0$, the spin lies in the $xy$-plane. If we write the spin part of the corresponding state as a superposition of eigenstates of $\sigma_z$, the spin up and spin down component will have equal weight. When $h \neq 0$, the $z$ component of the spin is non-zero, as shown in Fig. 5.2.4. For the upper branch, the spin is pointing slightly downwards. Each pair of helicity states $\mathbf{k}$ and $-\mathbf{k}$ have generally different fractions of spin up and spin down due to the non-zero external magnetic field $h$.

## 5.3.  Contact Interaction

For atoms employed in ultracold atomic experiments, the typical range of inter-atomic interactions $r_0$ is around $50 \sim 100\text{Å}$, much smaller than the inter-particle spacing which is almost always greater than $1000\text{Å}$. In quantum degenerate regime where the de Broglie wavelength is much larger than the inter-particle spacing, typical scattering momentum $k$ satisfies $kr_0 \ll 1$ and low energy collisions occur. Though inter-atomic interactions are complicated in general, simplification arises due to this low energy character of the collisions. An effective contact potential can be used to describes the interaction between the long-wavelength and low-frequency degrees of freedom while the shorter wavelengths can be taken into account through a renormalization procedure.[32]

We thus consider the $s$-wave interaction modeled by

$$\hat{V} = g \int d\mathbf{r} \psi_\uparrow^\dagger(\mathbf{r})\psi_\downarrow^\dagger(\mathbf{r})\psi_\downarrow(\mathbf{r})\psi_\uparrow(\mathbf{r}), \tag{5.3.1}$$

where $g$ is the coupling constant which can be renormalized to the physical parameter, the $s$-wave scattering length $a_s$, using

$$\frac{1}{g} = \frac{m}{4\pi a_s} - \frac{1}{V}\sum_{\mathbf{k}} \frac{m}{k^2}. \tag{5.3.2}$$

In the momentum space, Eq. (5.3.1) can be written as

$$\hat{V} = \frac{g}{V}\sum_{\mathbf{k},\mathbf{l},\mathbf{q}} C_{\mathbf{k}+\mathbf{q}/2\uparrow}^\dagger C_{-\mathbf{k}+\mathbf{q}/2\downarrow}^\dagger C_{-\mathbf{l}+\mathbf{q}/2\downarrow} C_{\mathbf{l}+\mathbf{q}/2\uparrow}, \tag{5.3.3}$$

where $C_{\mathbf{k}\sigma} = \int d\mathbf{r}\psi_\sigma(\mathbf{r})e^{-i\mathbf{k}\cdot\mathbf{r}}/\sqrt{V}$. Detailed analysis of the usage of zero-range model in a spin-orbit coupled quantum gas can be found in the following references.[33–41]

## 5.4. BEC-BCS Crossover of a Spin-orbit Coupled Two-component Fermi Gas

In this section, we consider pairing of a spin-orbit coupled Fermi gas with attractive interaction between opposite spins. Introducing the mean field pairing order parameter $\Delta = \frac{g}{V}\sum_{\mathbf{k}} \langle C_{-\mathbf{k}\downarrow}C_{\mathbf{k}\uparrow}\rangle$, we can re-write the interaction Hamiltonian as

$$\hat{V}_{\mathrm{MF}} = \Delta\sum_{\mathbf{k}}[C_{\mathbf{k}\uparrow}^\dagger C_{-\mathbf{k}\downarrow}^\dagger + C_{-\mathbf{k}\downarrow}C_{\mathbf{k}\uparrow}] - \frac{V}{g}\Delta^2. \tag{5.4.1}$$

The full mean field Hamiltonian is thus given by

$$\hat{H}_{\mathrm{MF}} = \hat{K} + \hat{V}_{\mathrm{MF}} \tag{5.4.2}$$

with

$$\begin{aligned}
\hat{K} &= \hat{H}_{\mathrm{SOC}} - \mu\hat{N} \\
&= \sum_{\mathbf{k}} \left[ (\xi_{\mathbf{k}} - h)C_{\mathbf{k}\uparrow}^\dagger C_{\mathbf{k}\uparrow} + \lambda k_\perp e^{i\varphi(\mathbf{k})}C_{\mathbf{k}\uparrow}^\dagger C_{\mathbf{k}\downarrow} \right. \\
&\quad \left. + \lambda_R k_\perp e^{-i\varphi(\mathbf{k})}C_{\mathbf{k}\downarrow}^\dagger C_{\mathbf{k}\uparrow} + (\xi_{\mathbf{k}} + h)C_{\mathbf{k}\downarrow}^\dagger C_{\mathbf{k}\downarrow} \right],
\end{aligned} \tag{5.4.3}$$

$\hat{H}_{SOC}$ is given by Eq. (5.2.1) and $\varphi(\mathbf{k}) = \text{Arg}[k_y + ik_x]$ and $\xi_k = \frac{k^2}{2m} - \mu$ is the free fermion energy with respect to the chemical potential $\mu$.

## 4.A. Helicity basis

Spin-orbit coupling can induce effective $p$-wave interactions in a system with $s$-wave interactions. For our system, this effect can be clearly seen by using the helicity basis that are the eigenstates of $\hat{H}_{SOC}$. The non-interacting part of the full Hamiltonian Eq. (5.4.3) is diagonal in the helicity basis

$$\hat{K} = \sum_{\mathbf{k}} \left[ \zeta_{\mathbf{k}+} b_{\mathbf{k}+}^\dagger b_{\mathbf{k}+} + \zeta_{\mathbf{k}-} b_{\mathbf{k}-}^\dagger b_{\mathbf{k}-} \right]. \tag{5.4.4}$$

The helicity basis operators $\gamma_{\mathbf{k}\pm}$ are related to the operators $C_{\mathbf{k}\sigma}$ ($\sigma = \uparrow, \downarrow$) via the transformation

$$\begin{bmatrix} b_{\mathbf{k}+} \\ b_{\mathbf{k}-} \end{bmatrix} = \hat{U}^\dagger \begin{bmatrix} C_{\mathbf{k}\uparrow} \\ C_{\mathbf{k}\downarrow} \end{bmatrix}, \tag{5.4.5}$$

where $\hat{U}$ is constructed from the eigenvectors of $\hat{K}$ with

$$\hat{U} \equiv \begin{bmatrix} u_{\mathbf{k}+} & u_{\mathbf{k}-} \\ v_{\mathbf{k}+} & v_{\mathbf{k}-} \end{bmatrix}, \tag{5.4.6}$$

as

$$\hat{K} \begin{bmatrix} u_{\mathbf{k}\pm} \\ v_{\mathbf{k}\pm} \end{bmatrix} = \zeta_{\mathbf{k}\pm} \begin{bmatrix} u_{\mathbf{k}\pm} \\ v_{\mathbf{k}\pm} \end{bmatrix}. \tag{5.4.7}$$

It is straightforward to find $\zeta_{\mathbf{k}\pm} = E_\pm - \mu$ with $E_\pm$ given in Eq. (5.2.4) and the eigenvectors are

$$\begin{bmatrix} u_{\mathbf{k}+} \\ v_{\mathbf{k}+} \end{bmatrix} = \begin{bmatrix} \sin\dfrac{\theta(\mathbf{k})}{2} e^{i\varphi(\mathbf{k})} \\ \cos\dfrac{\theta(\mathbf{k})}{2} \end{bmatrix}, \tag{5.4.8a}$$

$$\begin{bmatrix} u_{\mathbf{k}-} \\ v_{\mathbf{k}-} \end{bmatrix} = \begin{bmatrix} \cos\dfrac{\theta(\mathbf{k})}{2} \\ -\sin\dfrac{\theta(\mathbf{k})}{2} e^{-i\varphi(\mathbf{k})} \end{bmatrix}, \tag{5.4.8b}$$

with

$$\sin\theta(\mathbf{k}) = \frac{\lambda k_\perp}{\sqrt{h^2 + \lambda^2 k_\perp^2}}. \tag{5.4.9}$$

Likewise, the mean field interaction $\hat{V}_{MF}$ in the helicity basis becomes

$$
\hat{V}_{MF} = \Delta \sum_{\mathbf{k}} \left[ \frac{\sin\theta(\mathbf{k})}{2} e^{i\varphi(\mathbf{k})} b_{-\mathbf{k}+} b_{\mathbf{k}+} + \frac{\sin\theta(\mathbf{k})}{2} e^{-i\varphi(\mathbf{k})} b_{-\mathbf{k}-} b_{\mathbf{k}-} \right.
$$
$$
\left. + \cos\theta(\mathbf{k}) b_{-\mathbf{k}+} b_{\mathbf{k}-} + \text{h.c.} \right] - \frac{V}{g} \Delta^2.
$$

(5.4.10)

In the limiting case when $h = 0$, $\theta(\mathbf{k}) = \pi/2$, $\hat{V}_{MF}$ is reduced to

$$
\hat{V}_{MF} = \Delta \sum_{\mathbf{k}} \left[ \frac{1}{2} e^{i\varphi(\mathbf{k})} b_{-\mathbf{k}+} b_{\mathbf{k}+} + \frac{1}{2} e^{-i\varphi(\mathbf{k})} b_{-\mathbf{k}-} b_{\mathbf{k}-} \right] - \frac{V}{g} \Delta^2,
$$

(5.4.11)

where pairing only occurs between the same helicity basis. In this case, the system possess an effective $p_x + i p_y$ symmetry. In the opposite limit with $h$ goes to infinity, $\theta = 0$, $\hat{V}_{MF}$ now is reduced to

$$
\hat{V}_{MF} = \Delta \sum_{\mathbf{k}} [b_{-\mathbf{k}+} b_{\mathbf{k}-} + \text{h.c.}] - \frac{V}{g} \Delta^2,
$$

(5.4.12)

where only $b_{-\mathbf{k}+} b_{\mathbf{k}-}$ remains; this is the familiar form of the $s$-wave pairing. For finite $h$, the system possesses a mixture of $s$-wave and $p$-wave pairings.

### 4.B. *Ground state properties*

To understand the effects of the spin-orbit coupling on the BCS-BEC crossover in the ground state, we use the Bogoliubov-de Gennes (BdG) formalism to solve the gap function and chemical potential. The full mean-field Hamiltonian of Eq. (5.4.2) can be written in Nambu formalism, using $\Psi^\dagger(\mathbf{k}) = [C_{\mathbf{k}\uparrow}^\dagger, C_{\mathbf{k}\downarrow}^\dagger, C_{-\mathbf{k}\downarrow}, C_{-\mathbf{k}\uparrow}]$,

$$
\hat{H}_{MF} = \frac{1}{2} \sum_{\mathbf{k}} \Psi^\dagger(\mathbf{k}) \mathcal{M} \Psi(\mathbf{k}) + \frac{1}{2} \sum_{\mathbf{k}} K_{\uparrow\uparrow} + \frac{1}{2} \sum_{\mathbf{k}} K_{\downarrow\downarrow} - \frac{g}{V} \Delta^2
$$

(5.4.13)

where

$$
\mathcal{M} = \begin{bmatrix} K_{\uparrow\uparrow} & K_{\uparrow\downarrow} & \Delta & 0 \\ K_{\downarrow\uparrow} & K_{\downarrow\downarrow} & 0 & -\Delta \\ \Delta & 0 & -K_{\downarrow\downarrow} & K_{\uparrow\downarrow} \\ 0 & -\Delta & K_{\downarrow\uparrow} & -K_{\uparrow\uparrow} \end{bmatrix}
$$

(5.4.14)

and $K_{\sigma\sigma'}$ with $\sigma$ $(\sigma') = \uparrow, \downarrow$ is the corresponding matrix element of $\hat{K}$ given by Eq. (5.4.3). One thing worth noting is that $K_{\uparrow\downarrow}(-\mathbf{k}) = -K_{\uparrow\downarrow}(\mathbf{k})$. In a compact form

$$\mathcal{M} = \begin{bmatrix} \hat{K} & \Delta\sigma_z \\ \Delta\sigma_z & \hat{K}' \end{bmatrix}, \tag{5.4.15}$$

where $\hat{K}' = -(\sigma_y \hat{K} \sigma_y)^*$ with $K^*_{\uparrow\downarrow} = K_{\downarrow\uparrow}$. Suppose that $\mathcal{M}$ has an eigenvalue of $\mathcal{M}$ with an eigenstate $(\mathbf{u}, \mathbf{v})^T$, where $\mathbf{u} = (u_\uparrow, u_\downarrow)$ and $\mathbf{v} = (v_\downarrow, v_\uparrow)$, namely,

$$\hat{K}\mathbf{u} + \Delta\sigma_z\mathbf{v} = E\mathbf{u}, \tag{5.4.16a}$$

$$\Delta\sigma_z\mathbf{u} + \hat{K}'\mathbf{v} = E\mathbf{v}. \tag{5.4.16b}$$

It is straightforward to check that

$$\mathcal{M}\begin{bmatrix} \sigma_y\mathbf{v}^* \\ \sigma_y\mathbf{u}^* \end{bmatrix} = -E\begin{bmatrix} \sigma_y\mathbf{v}^* \\ \sigma_y\mathbf{u}^* \end{bmatrix}. \tag{5.4.17}$$

Namely, the spectrum of Eq. (5.4.13) always comes in pairs with energy $E$ and $-E$.

Now we define the quasi-particle operators as

$$\gamma^\dagger_{\mathbf{k}\sigma} = \sum_{\sigma'} C^\dagger_{\mathbf{k}\sigma'} u_{\sigma\sigma'} + C_{-\mathbf{k}\sigma'} v_{\sigma\sigma'}. \tag{5.4.18}$$

The diagonalization of Eq. (5.4.15) would give a set of two positive and two negative eigenvalues and their corresponding eigenstates. We can write the mean field Hamiltonian in terms of the fermionic quasi-particle operators $\gamma$ and $\gamma^\dagger$,

$$\hat{H}_{\text{MF}} = \sum_{\mathbf{k},\sigma} \gamma^\dagger_{\mathbf{k}\sigma} \gamma_{\mathbf{k}\sigma} E_\sigma(\mathbf{k}) + \frac{1}{2}\sum_{\mathbf{k},\sigma}[K_{\sigma\sigma} - E_\sigma(\mathbf{k})] - \frac{g}{V}\Delta^2, \tag{5.4.19}$$

where $E_\sigma(\mathbf{k})$ are the positive eigenvalues of $\mathcal{M}$. The first term of the diagonal Hamiltonian above describes the excitations and the remaining terms give the ground state energy at zero temperature.

Solving Eq. (5.4.16), we find that the excitation energies are given by

$$E_\pm(\mathbf{k}) = \sqrt{\xi_\mathbf{k}^2 + \lambda^2 p_\perp^2 + h^2 + \Delta^2 \pm 2\sqrt{(h^2 + \lambda^2 p_\perp^2)\xi_\mathbf{k}^2 + h^2\Delta^2}} \tag{5.4.20}$$

$$\equiv \sqrt{\xi_\mathbf{k}^2 + \lambda^2 p_\perp^2 + h^2 + \Delta^2 \pm 2E_0^2},$$

where $E_0 = \left[(h^2 + \lambda^2 p_\perp^2)\xi_\mathbf{k}^2 + h^2\Delta^2\right]^{1/4}$. At finite temperatures $T = 1/k_B\beta$, with $k_B$ being the Boltzmann constant, the expectation value $\langle \gamma^\dagger_{\mathbf{k}\sigma}\gamma_{\mathbf{k}\sigma}\rangle =$

$f(E_\sigma(\mathbf{k})) = (\exp \beta E_\sigma(\mathbf{k}) + 1)^{-1}$. The gap $\Delta$ and chemical potential $\mu$ for the ground state should satisfy the following conditions,

$$\frac{\partial \left( \left\langle \hat{H}_{\mathrm{MF}} \right\rangle - TS \right)}{\partial \Delta} = 0, \tag{5.4.21a}$$

$$-\frac{\partial \left( \left\langle \hat{H}_{\mathrm{MF}} \right\rangle - TS \right)}{\partial \mu} = N, \tag{5.4.21b}$$

where $S$ is the entropy

$$S = -k_B \sum_{E_\sigma(\mathbf{k})} \left\{ [1 - f(E_\sigma(\mathbf{k}))] \ln[1 - f(E_\sigma(\mathbf{k}))] + f(E_\sigma(\mathbf{k})) \ln[f(E_\sigma(\mathbf{k}))] \right\}. \tag{5.4.22}$$

The conditions Eq. (5.4.21b) lead to the following gap and number equations,

$$\frac{Vm}{4\pi a_s} = -\sum_{E_\sigma(\mathbf{k})} \left( 1 + \sigma \frac{h^2}{E_0^2} \right) \frac{\tanh \frac{\beta E_\sigma}{2}}{4E_\sigma} + \sum_k \frac{1}{2\epsilon_k}, \tag{5.4.23a}$$

$$N = -\frac{1}{2} \sum_{E_\sigma(\mathbf{k})} \left( 1 + \sigma \frac{h^2 + \lambda^2 p_\perp^2}{E_0^2} \right) \xi_k \frac{\tanh \frac{\beta E_\sigma}{2}}{E_\sigma(k)} + \sum_k 1, \tag{5.4.23b}$$

in which $g$ is replaced by $a_s$ according to Eq. (5.3.2). Through the BCS-BEC crossover, the evolution of the gap parameter and the chemical potential are shown in Fig. 5.4.1 for various strength of spin-orbit coupling with $h = 0$. Non-zero Zeeman fields incline to suppress pairing. However, due to the spin-orbit coupling, pairing can be still maintained as has been shown in Ref. 42.

## 4.C.  BCS-BEC crossover with spin-orbit coupling: collective excitations

To anticipate what we would need later for the calculation of the superfluid density, and also to make the discussions complete, we calculate the pairing fluctuation contributions to the thermodynamic potential. We express the pairing field as the summation of its mean field value and its fluctuation as

$$\Delta(x) = \Delta_0 + \delta\Delta(x), \tag{5.4.24a}$$

$$\Delta(x)^* = \Delta_0 + \delta\Delta(x)^*. \tag{5.4.24b}$$

In the path integral formalism (see appendix), the terms in the action $S$ of first order $\delta\Delta$ is automatically zero due to the saddle point approximation.

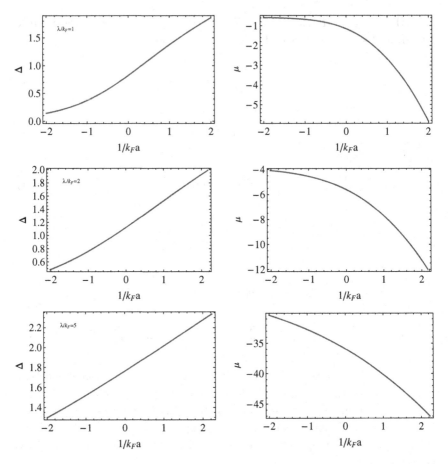

**Fig. 5.4.1.** The gap parameter $\Delta$ and the chemical potential $\mu$ as functions of $1/k_F a_s$ through the BCS-BEC crossover for various values of the SOC strength $\lambda$ and the Zeeman field $h = 0$ at temperature $T = 0$.

Consequently to second order, the effecitve action $S_{\text{eff}}$ is

$$S_{\text{eff}} = S_{\text{MF}} + S_{\text{fluctuation}} \qquad (5.4.25)$$

with $S_{\text{MF}}$ the mean field action and

$$S_{\text{fluctuation}} = \frac{1}{2} \sum_q \left( \delta\Delta_q^*, \delta\Delta_{-q} \right) M \begin{pmatrix} \delta\Delta_q \\ \delta\Delta_{-q}^* \end{pmatrix} \qquad (5.4.26)$$

where $q$ is actually a four vector $(\mathbf{q}, iq_n)$ and the elements of the two-by-two matrix $M$ are given in Appendix. A.2. Therefore, the thermodynamic

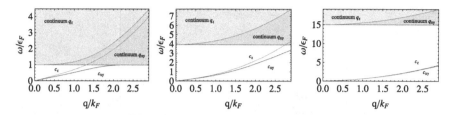

**Fig. 5.4.2.** Dispersions of the collective modes propagating in the $z$ direction (grey line) or in the $xy$ plane (blue line) and their corresponding continuum regions (green area and pink area) for the SOC strength $\lambda/k_F = 2$ on the BCS side ($1/k_F a = -2$, the left graph), at the unitary point ($1/k_F a = 0$, the middle one) and on the BEC side ($1/k_F a = 2$, the right one). Here we plot for $T = 0$ and $h = 0$.

potential up to the Gaussian fluctuation order is similarly given by

$$\Omega = \Omega_{\mathrm{MF}} + \Omega_{\mathrm{fluctuation}}, \qquad (5.4.27)$$

where the Gaussian fluctuation part is given by

$$\Omega_{\mathrm{fluctuation}} = \frac{1}{2\beta} \sum_q \ln\left[\det M(q)\right]. \qquad (5.4.28)$$

The dispersion of the collective modes can be determined by the equation $\det M(\mathbf{q}, \omega_{\mathbf{q}}) = 0$ with $\omega_{\mathbf{q}}$ being smaller than the lower boundary of the continuum region, which is set by $\min_{\mathbf{p}}(E_{\mathbf{p},\beta} + E_{\mathbf{p}-\mathbf{q},\beta'})$. Here $\beta, \beta'$ labels the branches of quasiparticles. Because of the fact that the spin-orbit coupling only affect the spatial motion of single particles in the $xy$-plane, the continuum region varies with the direction of the momentum of the collective modes and so does their dispersion. Therefore, as shown in Fig. 5.4.2, there are two continuum regions: One is for the momentum $\mathbf{q}$ along the $z$ direction (green area) and the other for $\mathbf{q}$ lies in the $xy$-plane (pink area). The sound velocities of the phonon modes also depends on the momentum direction. The one for the phonon mode in the $xy$-plane through the BEC-BCS crossover is given in Fig. 5.4.3. Similar calculation is also done in Ref. 26.

## 5.5. Superfluid and Normal Density of a Spin-orbit Coupled Fermi Gas

For standard superfluids like Helium four, the superfluid density $\rho_s$ equals the total density $\rho_t$ at zero temperature. In other words, the normal density $\rho_n$ is zero. A crucial fact that leads to the above conclusion is that the low

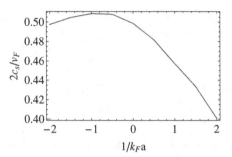

**Fig. 5.4.3.**  The sound velocity for the phonon modes in the $xy$ plane through the BEC-BCS crossover for SOC strength $\lambda/k_F = 2$ and $T = 0, h = 0$.

energy long wave-length excitations are dominated by phonons.[43] Being of longitudinal nature, these excitations are not probed in measuring the normal density which involves transverse perturbation. It turns out that this conclusion is invalid in the case of a spin-orbit coupled boson gas, as was investigated in Ref. 24. There, the existence of a gapped elementary excitation in the long wave-length limit gives rises to a non-zero normal density which in fact exhausts the total density at the transition from the plane wave to the zero momentum phase.

In the case of two-component fermions without spin-orbit interaction, the long wave-length excitations consist of either single quasi-particle excitations or phonon modes. The later, being longitudinal, do not contribute to the normal density. The quasi-particles, while gapped in the long wave-length limit, do not contribute to the normal density either because of vanishing matrix element induced by a transverse current operator [see later]. The presence of spin-orbit coupling, however, changes the picture completely. In the first place, a transverse current operator will induce non-zero matrix elements between the ground and the single quasi-particle excited states, and secondly, the existence of multiple long wave-length elementary excitations should also modify the normal density of the system, much the same way as in the boson case. Thus we are lead to investigate the contribution of both the quasi-particles and the collective excitations to the calculation of the normal density of a spin-orbit coupled system. Unfortunately, at the time of writing, we haven't obtained any concrete result regarding the contribution of the collective excitations to the calculation of the normal density. In the following, we shall nonetheless point out a few equivalent ways that can be used to obtain concrete results. The contribution from the quasi-particles actually has been investigated in several papers.[25-28,30,31]

## 5.A. *Phase twist*

To calculate the superfluid density $\rho_s$ directly, one can employ the method of phase twist.[44,45] Suppose that the superfluid is confined in a cylinder with two ends (A and B). By performing a phase twist at the boundaries, the phases of order parameter on side A and side B becomes different and the phase gradient is characterized by a wave vector $\mathbf{Q}$. Consequently, the superfluid velocity can be given by $\mathbf{v}_s = \hbar\mathbf{Q}/2m$. The twist only affects the superfluid part due to its long range phase coherence. The difference of the free energy $F$ only comes from superfluid part and can be written as

$$\Delta F \equiv \frac{1}{2}\rho_s v_s^2 \tag{5.5.1}$$

which means that the superfluid density $\rho_s$ should satisfies

$$\rho_s = \left(\frac{\partial^2 F}{\partial v_s^2}\right)_{v_s \to 0} = \frac{4m^2}{\hbar^2}\left(\frac{\partial^2 F}{\partial Q^2}\right)_{Q \to 0}. \tag{5.5.2}$$

In the grand canonical ensemble,

$$\left(\frac{\partial^2 F}{\partial Q^2}\right)_{Q \to 0} = \left(\frac{\partial^2 \Omega}{\partial Q^2} + N\frac{\partial^2 \mu}{\partial Q^2}\right)_{Q \to 0}. \tag{5.5.3}$$

When fixing temperature $T$ and volume $V$, the thermodynamic potential $\Omega$ would be just the function of the pairing gap $\Delta$ and the chemical potential $\mu$. By making use of the property $\left(\frac{\partial \mu}{\partial Q}\right)_{Q \to 0} = 0$ and $\left(\frac{\partial \Delta}{\partial Q}\right)_{Q \to 0} = 0$ from the saddle point approximation, we have

$$\left(\frac{\partial^2 F}{\partial Q^2}\right)_{Q \to 0} = \left[\left(\frac{\partial^2 \Omega}{\partial Q^2}\right)_{\Delta,\mu} + \left(\frac{\partial \Omega}{\partial \Delta}\right)_\mu \frac{\partial^2 \Delta}{\partial Q^2}\right.$$
$$\left. + \left(\frac{\partial \Omega}{\partial \mu}\right)_\Delta \frac{\partial^2 \mu}{\partial Q^2} + N\frac{\partial^2 \mu}{\partial Q^2}\right]_{Q \to 0} \tag{5.5.4}$$

Since the total number $N$ satisfies

$$N = -\left(\frac{\partial \Omega}{\partial \mu}\right) = -\left(\frac{\partial \Omega}{\partial \mu}\right)_\Delta - \left(\frac{\partial \Omega}{\partial \Delta}\right)_\mu \frac{\partial \Delta}{\partial \mu}, \tag{5.5.5}$$

we find

$$\left(\frac{\partial^2 F}{\partial Q^2}\right)_{Q \to 0} = \left\{\left(\frac{\partial^2 \Omega}{\partial Q^2}\right)_{\Delta,\mu} + \left(\frac{\partial \Omega}{\partial \Delta}\right)_\mu \left[\frac{\partial^2 \Delta}{\partial Q^2} - \left(\frac{\partial \Delta}{\partial \mu}\right)\frac{\partial^2 \mu}{\partial Q^2}\right]\right\}_{Q \to 0}. \tag{5.5.6}$$

When the Gaussian fluctuation part of the thermodynamic potential is ignored in the gap equation, that is, $\left(\frac{\partial \Omega}{\partial \Delta}\right)_\mu = \left(\frac{\partial \Omega_0}{\partial \Delta}\right)_\mu = 0$, the superfluid density can be simply expressed as

$$\rho_s = \frac{4m^2}{\hbar^2} \left(\frac{\partial^2 \Omega}{\partial Q^2}\right)_{\Delta,\mu,(Q\to0)}. \tag{5.5.7}$$

If one wants to include the contribution from both the quasiparticle and the collective mode into the superfluid density, the thermodynamic potential can be written up to the Gaussian fluctuation term, that is

$$\rho_s = \frac{4m^2}{\hbar^2} \left(\frac{\partial^2 \left[\Omega_0(Q) + \Omega_{\text{fluctuation}}(Q)\right]}{\partial Q^2}\right)_{\Delta,\mu,(Q\to0)}. \tag{5.5.8}$$

### 5.B.  Transverse current-current correlation

To calculate explicitly the normal density $\rho_n$, one can calculate directly the transverse current-current correlation.[3] In this approach, the normal density is defined as the derivative of the mass current $\mathbf{J}$ with respect to the normal velocity, that is (assuming $\mathbf{J} \parallel \mathbf{v}_n$ as will be the case for our discussion below)

$$\rho_n = \frac{\partial \mathbf{J}}{\partial \mathbf{v}_n} \tag{5.5.9}$$

which implies that the normal density is just the linear coefficient of the expansion of $\mathbf{J}$ with respect to $\mathbf{v_n}$. Moreover, the difference of the free energy in the $v_n = 0$ frame and the $v_n \neq 0$ frame is given by

$$\Delta F = \int d\mathbf{r} \mathbf{v}_n \cdot \mathbf{J}(\mathbf{r}). \tag{5.5.10}$$

Consequently, the normal density can be written as the current-current correlation function form

$$\rho_n = \lim_{q_x q_z \to 0} \lim_{q_y \to 0} \gamma_{yy}(\mathbf{q}, \omega = 0) \tag{5.5.11}$$

Here it is crucial to note the order of limit of $q_{x,y,z}$ to zero. The direction along $\mathbf{v}_n$ (assumed to be in $y$ direction) should be taken first, implying the cylindrical tube is infinitely long. Otherwise, the limit with order reversed will simply give the total density.

Within the framework of the transverse current-current correlation function, it is most easy to discuss the various contributions to the normal density $\rho_n$. Let the current operator along $y$-direction be $J_y(\mathbf{q})$, the $\rho_n$ is given by

$$
\begin{aligned}
\rho_{ny} &= \frac{1}{V} \lim_{q_z,q_x \to 0} \lim_{q_y \to 0} \sum_n \left( \frac{|\langle n|J_y(\mathbf{q})|0\rangle|^2}{\omega_{n0}} + \mathbf{q} \to -\mathbf{q} \right) \\
&= \frac{1}{V} \sum_{n'} \left( \frac{2|\langle n'|P_y|0\rangle|^2}{\omega_{n'0}} \right)
\end{aligned}
\tag{5.5.12}
$$

where we have used the fact that in the limit when $\mathbf{q} \to 0$,

$$
J_y(\mathbf{q}=0) = \int d\mathbf{r}\, j_y(\mathbf{r}) = P_y = \sum_i (p_{i,y} + m\lambda\sigma_{i,x})
\tag{5.5.13}
$$

The state $\langle n|$ represents all the possible excitation including quasiparticle excitations and elementary collective excitations, while the states $\langle n'|$ are those excitations except the phonon modes. Within the BdG formalism, the operator $P$ can be expressed in terms of the creation or annihilation operators of the quasi-particles $(\gamma^\dagger, \gamma)$ and then by making use of the properties of these operators the normal density becomes

$$
\rho_{ny} = \frac{m^2\lambda^2}{V} \sum_{\mathbf{p},\delta=\pm 1} \frac{\Delta_0^2 + \xi_{\mathbf{p}}^2 + \xi_{\mathbf{p}}\delta|\gamma_{\mathbf{p}}|}{4\xi_{\mathbf{p}}\delta|\gamma_{\mathbf{p}}|E_{\mathbf{p},\delta}}
\tag{5.5.14}
$$

where $\gamma_{\mathbf{p}} = \lambda(p_y + ip_x)$. This expression is consistent with the result of Refs. 25, 26 and includes only the contributions from the quasi-particle excitations. It is important to recognize that in the intermediate states $|n\rangle$, there also could be states that correspond to collective excitations, which in the case of bosons, are entirely responsible for the suppression of superfluid density in the ground state.

### 5.B.1. Superfluid density from the exact relation $\rho_s/\rho_t = mc^2\kappa$.

Using the definition of the normal density in terms of the transverse current correlation function, one can further derive exact relations relating $\rho_n$ to other thermodynamic quantities. One of these relations is given by

$$
\frac{\rho_s}{\rho_t} = mc^2\kappa
\tag{5.5.15}
$$

where $c$ is the sound velocity of the phonon modes in the $xy$-plane that we have been obtained before. $\kappa$ corresponds to the compressibility of the system, which can be calculated via compressibility sum rule by density-density correlation function

$$\kappa = \frac{1}{n_t} \lim_{\mathbf{q} \to 0} \chi_{nn}(\mathbf{q}, 0). \tag{5.5.16}$$

The proof of this relation is straightforward. Since the normal density is given by the current-current correlation function through

$$\rho_{ny} = \lim_{\mathbf{q} \to 0} \frac{1}{V} \sum_n \left( \frac{|\langle n|J_y^T(\mathbf{q})|0\rangle|^2}{\omega_{n0}} + \mathbf{q} \to -\mathbf{q} \right)$$

$$= \lim_{\mathbf{q} \to 0} \frac{1}{V} \sum_{n'} \left( \frac{|\langle n'|J_y^L(\mathbf{q})|0\rangle|^2}{\omega_{n'0}} + \mathbf{q} \to -\mathbf{q} \right) \tag{5.5.17}$$

where $J^T$ and $J^L$ are the transverse current and the longitudinal current respectively. The reason for the equality is that for gapped quasi-particles and collective excitations, there is no difference between the longitudinal and the transverse probes, while the phonon modes only couple to the longitudinal probe. In addition, the total density is given by the $f$-sum rule

$$\rho_t = \lim_{\mathbf{q} \to 0} \frac{1}{V} \sum_n \left( \frac{|\langle n|J_y^L(\mathbf{q})|0\rangle|^2}{\omega_{n0}} + \mathbf{q} \to -\mathbf{q} \right). \tag{5.5.18}$$

Therefore, the superfluid density is given by

$$\rho_s \equiv \rho_t - \rho_n = \frac{1}{V} \lim_{\mathbf{q} \to 0} \sum_{\text{phonon}} \left( \frac{|\langle n|J_y^L(\mathbf{q})|0\rangle|^2}{\omega_{\text{phonon}}} + \mathbf{q} \to -\mathbf{q} \right). \tag{5.5.19}$$

According to the continuity equation $\mathbf{q} \cdot \mathbf{J} = \omega\rho$, and for the phonon modes $\omega = c\,|\,\mathbf{q}\,|$ with $c$ being the sound velocity, we have

$$\rho_s = c^2 \frac{1}{V} \sum_{\text{phonon}} \lim_{\mathbf{q} \to 0} \left( \frac{|\langle n|\rho(\mathbf{q})|0\rangle|^2}{\omega_{\text{phonon}}} + \mathbf{q} \to -\mathbf{q} \right). \tag{5.5.20}$$

By analyzing the asymptotic form of the numerator at the limitation $\mathbf{q} \to 0$ with the f-sum rule, it can be concluded that the phonon modes exhaust the compressibility sum rule, that is

$$\frac{1}{V} \sum_{\text{phonon}} \lim_{\mathbf{q} \to 0} \left( \frac{|\langle n|\rho(\mathbf{q})|0\rangle|^2}{\omega_{\text{phonon}}} + \mathbf{q} \to -\mathbf{q} \right) = m^2 \lim_{\mathbf{q} \to 0} \chi_{nn}(\mathbf{q}, 0). \tag{5.5.21}$$

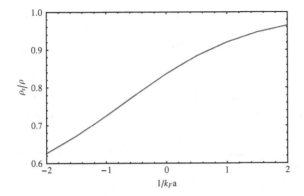

**Fig. 5.5.1.** Superfluid density in BEC-BCS crossover calculated with $\rho_s = \rho_t mc^2 \kappa$ (blue line) and from Eq. (5.5.14) for SOC strength $\lambda/k_F = 2$. The two lines are identical.

Thus the superfluid density can be expressed in terms of the compressibility as

$$\rho_s = mc^2 \rho_t \kappa. \qquad (5.5.22)$$

Calculation based on the above equality for the superfluid density gives identical result as Eq. (5.5.14) within the random phase approximation for $\chi_{nn}$, as shown in Fig. 5.5.1, which takes into account only contributions from the quasiparticles.

### 5.B.2. *Normal density from the spin-spin correlation function*

There is also a relation between the spin-spin correlation function $\chi_{\sigma_x \sigma_x}$ and the normal density. Following the same logic while establishing the relation between $\rho_n$ and $\kappa$ in the previous section, we have

$$\rho_{ny} = \lim_{\mathbf{q} \to 0} \frac{1}{V} \sum_{n'} \left( \frac{|\langle n'|J_y^T(\mathbf{q})|0\rangle|^2}{\omega_{n'0}} + \mathbf{q} \to -\mathbf{q} \right) \qquad (5.5.23)$$

$$= \frac{1}{V} \sum_{n'} \left( \frac{|\langle n'| \sum_i (p_{i,y} + m\lambda\sigma_{i,x})|0\rangle|^2}{\omega_{n'0}} + \mathbf{q} \to -\mathbf{q} \right) \qquad (5.5.24)$$

$$= \frac{(m\lambda)^2}{V} \sum_{n'} \left( \frac{|\langle n'| \sum_i \sigma_{i,x}|0\rangle|^2}{\omega_{n'0}} + \mathbf{q} \to -\mathbf{q} \right). \qquad (5.5.25)$$

We also need to keep in mind that since we are considering the transverse current, it is the transverse spin-spin correlation function that is obtained in the following. The second equality holds because $\langle n' | \sum_i p_{i,y} | 0 \rangle = 0$. Within the mean field approximation, the phonon modes will not be considered in calculating the spin-spin correlation function $\chi_{\sigma_x, \sigma_x}$, therefore we have

$$\rho_{ny} = m^2 \lambda^2 \chi_{\sigma_x, \sigma_x} (\mathbf{q} \to 0, \omega = 0) \qquad (5.5.26)$$

A similar formula has also been obtained in Ref. 28 by comparing the expression of the spin susceptibility with the normal density.

### 5.6. Vortex Structure in a Spin-orbit Coupled Fermion Superfluid

We now continue to study the response of a spin-orbit coupled Fermi superfluid under rotation. To setup the problem that is amenable to numerical calculation, we consider a cylindrical geometry where the central axis lies along $z$-direction. The radius of the cylinder in the $xy$-plane is given by $R$. Now the single particle states are labeled by the good quantum numbers $k_z$ describing the linear momentum in $z$-direction and $m$ describing the angular momentum around the $z$-axis. The single particle wave function can be written as

$$\mathbf{\Phi}_{nmk_z}(\mathbf{r}) = \begin{bmatrix} \Phi_{nmk_z\uparrow}(\mathbf{r}) \\ \Phi_{nmk_z\downarrow}(\mathbf{r}) \end{bmatrix}, \qquad (5.6.1)$$

where $n$ labels the radial wave function. The form of $\Phi_{nmk_z\sigma}(\mathbf{r})$ can be determined by demanding that $\mathbf{\Phi}_{nmk_z}(R) = 0$ and its detailed form is given in the appendix of Ref. 46.

To investigate the structure of the quantized vortex and the associated elementary excitations, we generalize the Bogoliubov-de-Genne equations to the cylindrical geometry using $\mathbf{\Phi}_{nmk_z}(\mathbf{r})$ as the expansion basis (see Appendix for details). For a singly quantized vortex, we can write

$$\Delta(\mathbf{r}) = g \langle \Psi_\downarrow(\mathbf{r}) \Psi_\uparrow(\mathbf{r}) \rangle = \Delta(r) e^{i\theta} \qquad (5.6.2)$$

where we have assumed that attractive interaction only operates between the opposite spin components. The full mean field Hamiltonian $\hat{H}_{\mathrm{MF}}$ thus becomes

$$\hat{H}_{\mathrm{MF}} = \sum_{\sigma, \sigma' = \uparrow, \downarrow} \int d\mathbf{r} \Psi_\sigma^\dagger(\mathbf{r}) K_{\sigma\sigma'} \Psi_{\sigma'}(\mathbf{r}) + \hat{V}_{\mathrm{MF}}, \qquad (5.6.3)$$

where $\hat{K} \equiv \hat{H}_0 - \mu \mathbf{I}$ and $\hat{V}_{\mathrm{MF}} = \int d\mathbf{r} \left[ \Psi_\uparrow^\dagger(\mathbf{r}) \Psi_\downarrow^\dagger(\mathbf{r}) \Delta(\mathbf{r}) + \Delta^*(\mathbf{r}) \Psi_\downarrow(\mathbf{r}) \Psi_\uparrow(\mathbf{r}) \right]$. Explicitly in matrix form, we have, neglecting the constant terms (compare with Eq. (5.4.13))

$$\hat{H}_{MF} = \frac{1}{2} \int d\mathbf{r} \mathbf{\Psi}^\dagger(\mathbf{r}) \begin{bmatrix} K_{\uparrow\uparrow} & K_{\uparrow\downarrow} & \Delta(\mathbf{r}) & 0 \\ K_{\downarrow\uparrow} & K_{\downarrow\downarrow} & 0 & -\Delta(\mathbf{r}) \\ \Delta^*(\mathbf{r}) & 0 & -K_{\downarrow\downarrow} & K_{\uparrow\downarrow} \\ 0 & -\Delta^*(\mathbf{r}) & K_{\downarrow\uparrow} & -K_{\uparrow\uparrow} \end{bmatrix} \mathbf{\Psi}(\mathbf{r}) \quad (5.6.4)$$

$$= \frac{1}{2} \int d\mathbf{r} \mathbf{\Psi}^\dagger(\mathbf{r}) \mathcal{M}(\mathbf{r}) \mathbf{\Psi}(\mathbf{r}), \quad (5.6.5)$$

with $\mathbf{\Psi}^\dagger(\mathbf{r}) = [\Psi_\uparrow^\dagger(\mathbf{r}), \Psi_\downarrow^\dagger(\mathbf{r}), \Psi_\downarrow(\mathbf{r}), \Psi_\uparrow(\mathbf{r})]$. The eigenvectors of $\mathcal{M}(\mathbf{r})$ are denoted as $[\boldsymbol{u}^T, \boldsymbol{v}^T] = [u_\uparrow(\mathbf{r}), u_\downarrow(\mathbf{r}), v_\downarrow(\mathbf{r}), v_\uparrow(\mathbf{r})]$ satisfying

$$\mathcal{M} \begin{bmatrix} \boldsymbol{v} \\ \boldsymbol{u} \end{bmatrix} = E \begin{bmatrix} \boldsymbol{v} \\ \boldsymbol{u} \end{bmatrix}. \quad (5.6.6)$$

Similar to the uniform case, for each positive eigenenergy $E$ there exists a corresponding negative eigenenergy $-E$ whose eigenvector is given by

$$\mathcal{M} \begin{bmatrix} \sigma_x \boldsymbol{v}^* \\ \sigma_x \boldsymbol{u}^* \end{bmatrix} = -E \begin{bmatrix} \sigma_x \boldsymbol{v}^* \\ \sigma_x \boldsymbol{u}^* \end{bmatrix}. \quad (5.6.7)$$

We can also introduce the quasi-particle operators as

$$\gamma_\sigma^\dagger = \sum_{\sigma'} \int d\mathbf{r} [u_{\sigma\sigma'}(\mathbf{r}) \Psi_{\sigma'}^\dagger(\mathbf{r}) + v_{\sigma\sigma'}(\mathbf{r}) \Psi_{\sigma'}(\mathbf{r})]. \quad (5.6.8)$$

that create quasi-particle excitations with spin $-\sigma$ when operated on the mean field ground state. In terms of these quasi-particle operators,

$$\hat{H}_{\text{MF}} = \sum_{\ell\sigma} E_{\ell\sigma} \gamma_{\ell\sigma}^\dagger \gamma_{\ell\sigma}, \quad (5.6.9)$$

up to a term independent of $\gamma_{\ell\sigma}$. The summation runs over all positive eigenenergies $E_{\ell\sigma} > 0$. Here $\ell$ stands for quantum numbers other than the spin.

### 6.A. *Vortex profiles*

We determine the profiles of the gap function, the density distribution and the current density around the core of vortex through the BCS-BEC crossover by solving the BdG equation numerically with parameters $k_F R = 12$, $L = 0.5R$ and the cut-off energy in the expansion $\epsilon_c = 10\epsilon_F$.

6.A.1.  *Gap function and density distribution*

In Fig. 5.6.1(a), We plot $\Delta(r)$ as a function of $r$ for various values of $1/k_F a_s$.
At $r = 0$, the center of the vortex, $\Delta(0)$ vanishes. As $r$ increases, $\Delta(r)$
increases monotonically and approaches its bulk value $\Delta_0$. The bulk gap
$\Delta_0$ as a function of $1/k_F a_s$ is given in Fig. 5.6.1(b), which increases monotonically from the BCS side to the BEC side, similar to the result without
spin-orbit couplings.[32,47] To characterize the core size of the vortex, we
define the healing length $\zeta$ at which the gap function is half of its bulk
value, namely $\Delta(\zeta) = \frac{1}{2}\Delta_0$, as indicated by the cross point of the dashed
lines in Fig. 5.6.1(a). As shown in Fig. 5.6.1(c), the healing length $\zeta$ is a

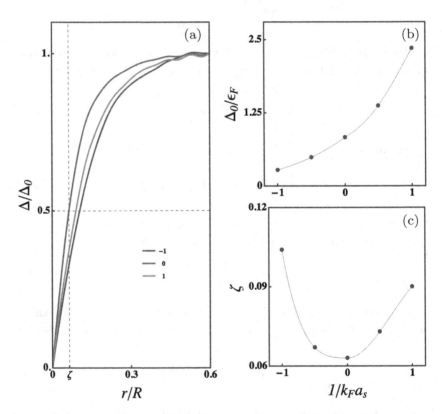

**Fig. 5.6.1.** The gap function $\Delta(r)$ for various $1/k_F a_s$ with taking $\lambda/R = 0.14\epsilon_F$,
$h = 0.32\epsilon_F$. (a) $\Delta/\Delta_0$ as a function of $r/R$ at $1/k_F a_s = -1, 0$ and 1. (b) The bulk
value of the gap function $\Delta_0/\epsilon_F$ as a function of $1/k_F a_s$. (c) The healing length $\zeta$ as a
function of $1/k_F a_s$.

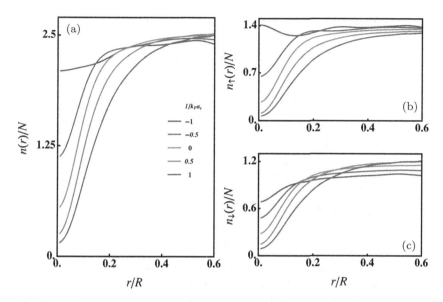

**Fig. 5.6.2.** Density function for $1/k_F a_s = 1, 0.5, 0, -0.5, -1$ with taking $\lambda/R = 0.15\epsilon_F$, $h = 0.32\epsilon_F$. (a) Total density $n(r)$ as a function of $r/R$. (b) Density distribution $n_\uparrow(r)$ for the spin-up component. (c) Density distribution $n_\downarrow(r)$ for the spin-down component.

non-monotonic function of $1/k_F a_s$ which displays a minimum around unitarity for $\lambda/R = 0.15\epsilon_F$ and $h = 0.32\epsilon_F$. Later we will find that the position of the minimum can be shifted to the BEC side.

We show the density function $n(r)$ in Fig. 5.6.2(a). The density is the smallest at the vortex center and increases with $r$. Throughout the crossover, the density configuration at the vortex core evolves from an almost full vortex on the BCS side to an empty vortex on the BEC side; the density is strongly suppressed on the BEC side. The spin polarization in the vortex core is much stronger on the BCS side than on the BEC side as shown in Fig. 5.6.2(b) and (c).

### 6.A.2. *Current density*

Another quantity of interest is the current density circulating the vortex. By the continuity equation

$$\frac{\partial \hat{n}}{\partial t} + \nabla \cdot \hat{\mathbf{j}} = 0 \tag{5.6.10}$$

with $\hat{n} = \sum_{\sigma} \Psi_{\sigma}^{\dagger}(\mathbf{r})\Psi_{\sigma}(\mathbf{r})$, we can derive the circulating current density

$$
\begin{aligned}
\hat{j}_{\theta} =& \frac{1}{2mi} \sum_{\sigma} \left( \Psi_{\sigma}^{\dagger}(\mathbf{r}) \frac{1}{r} \frac{\partial}{\partial \theta} \Psi_{\sigma}(\mathbf{r}) - \frac{1}{r} \frac{\partial}{\partial \theta} \Psi_{\sigma}^{\dagger}(\mathbf{r})\Psi_{\sigma}(\mathbf{r}) \right) \\
&+ \lambda \left( \Psi_{\uparrow}^{\dagger}(\mathbf{r})\Psi_{\downarrow}(\mathbf{r})e^{-i\theta} + \Psi_{\downarrow}^{\dagger}(\mathbf{r})\Psi_{\uparrow}(\mathbf{r})e^{i\theta} \right),
\end{aligned}
\tag{5.6.11}
$$

and the current along the radial direction equals to zero. In terms of the solutions of BdG equations, $\hat{j}_{\theta}$ is given by

$$
\begin{aligned}
\hat{j}_{\theta} =& \frac{1}{i\hbar} \frac{\hbar^2}{2m} \frac{1}{r} \sum_{\kappa\sigma} \left\{ [v_{\sigma}^{\kappa} \frac{\partial}{\partial \theta} v_{\sigma}^{\kappa*} f(-E^{\kappa}) + u_{\sigma}^{\kappa*} \frac{\partial}{\partial \theta} u_{\sigma}^{\kappa} f(E^{\kappa})] - \text{h.c.} \right\} \\
&+ \lambda \sum_{\kappa} \left\{ e^{-i\theta} [v_{\uparrow}^{\kappa} v_{\downarrow}^{\kappa*} f(-E^{\kappa}) + u_{\uparrow}^{\kappa*} u_{\downarrow}^{\kappa} f(E^{\kappa})] + \text{h.c.} \right\},
\end{aligned}
\tag{5.6.12}
$$

where the second term vanishes in the absence of spin-orbit coupling.

As shown in Fig. 5.6.3(a), the current density vanishes at $r = 0$ and reaches its maximum value $j_{\theta max}$ at some finite radius $r$. The maximal current, which is related to the critical velocity beyond which the superfluid

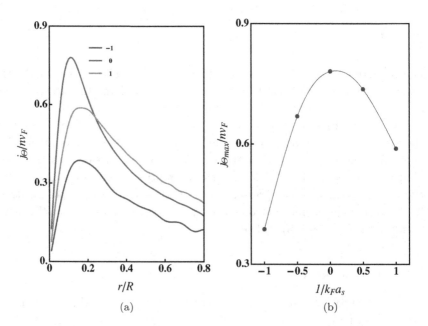

(a)                    (b)

**Fig. 5.6.3.** Circulating current density profiles for $\lambda/R = 0.15\epsilon_F$ and $h = 0.32\epsilon_F$. (a) Circulating current density as a function of $r/R$ for $1/k_F a_s = -1, 0, 1$. (b) Maximal circulating current $j_{\theta max}$ (critical current) as a function of $1/k_F a_s$ through the crossover.

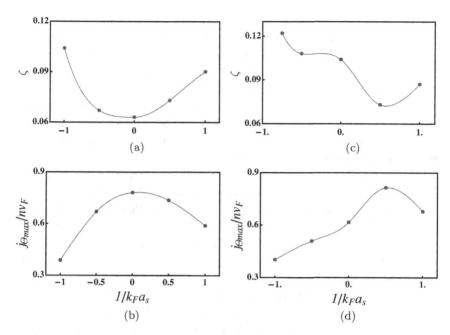

**Fig. 5.6.4.** Healing length and critical current at two different set of parameters. (a)–(b): Strong SOC strength $\lambda/R = 0.15\epsilon_F$, weak spin imbalance $h = 0.32\epsilon_F$. (c)–(d): Weak SOC strength $\lambda/R = 0.03\epsilon_F$, strong spin imbalance $h = 0.19\epsilon_F$.

will be destroyed,[47] displays a non-monotonic behavior through the BCS to BEC crossover as given in Fig. 5.6.3(b). Together with the minimal healing length $\zeta$, the maximal $j_{\theta max}$ shows that the superfluid is the most robust close to the unitarity point $1/k_F a_s = 0$ as shown in Fig. 5.6.4(a)–(b) for $\lambda/R = 0.15\epsilon_F$ and $h = 0.32\epsilon_F$. We find that by tuning to larger $h$ (larger spin imbalance) and smaller $\lambda$ (weaker spin-orbit coupling), the most robust point of the superfluidity can be shifted to the BEC side as shown in Fig. 5.6.4(c)–(d). This change is because, as we mentioned for the uniform case, in the helicity basis, at small $h$, the pairing mostly occurs between inter-band which corresponds to the result of Fig. 5.6.4(a)–(b). The increase of $h$ tends to weaken or destroy the pairing which is weaker on the BCS side; consequently, the most robust point is pushed towards the BEC side.

## 6.B. *Excitations*

Three phases of different topological properties have been predicted for our system in the uniform case at zero temperature:[42] For $0 < |h| < \Delta$, the bulk

gap is fully open and the system is in a topologically trivial phase called $A$;
For $\Delta < |h| < \sqrt{\Delta^2 + \mu^2}$, there are four Fermi points in the bulk excitation
spectrum and the system is in a topologically nontrivial phase called $B$; For
$|h| > \sqrt{\Delta^2 + \mu^2}$, there are two Fermi points instead and the system is in
another topologically nontrivial phase called $C$. It is interesting to see how
the topological properties can affect the excitations in the presence of a
vortex core and the hard wall boundary of the cylinder. By solving the
BdG equation (C.2) self-consistently, we also obtain the spectrum for the
collective excitations $E^\kappa_{mk_z}$. Since $E^\kappa_{mk_z}$ with $k_z \neq 0$ are shifted from those
with $k_z = 0$ only by the amount $k_z^2/2M$, and due to particle-hole symmetry
[Eq. (5.6.7)], we focus on the nonnegative part of $E^\kappa_m \equiv E^\kappa_{m,k_z=0}$.

Figure 5.6.5 shows the excitation spectrum of the system throughout
the crossover with the parameters: $\lambda/R = 0.03\epsilon_F$ and $h = 0.19\epsilon_F$. One
can see that the bulk gap continuously increases from the BCS to BEC
side which is consistent with result of gap function in Fig. 5.6.1(b). On the
BCS side up to the vicinity of the unitary point, the low positive energy
excitations are of nonpositive quantum number $m$. We shall see through
the argument below that this branch of the excitations are the core states
whose wavefunctions are localized around the vortex core. For Fig. 5.6.5(a)
the smallest positive $E^\kappa_m$ is equal to $0.012\epsilon_F$ when $m = 0$. On the BEC
side, this branch is pushed to higher excitation energies; The number of
low energy excitations decreases from the BCS to BEC side.

Figure 5.6.6 shows the excitation spectrum with another set of the
parameters: $\lambda/R = 0.14\epsilon_F$ and $h = 0.32\epsilon_F$. On the BCS side, the low

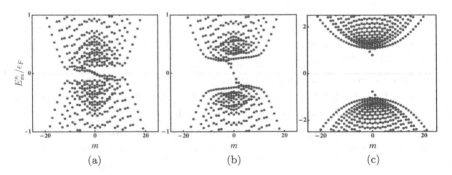

**Fig. 5.6.5.** Excitation spectrum as a function of $m$ when $k_z = 0$ through the BCS to
BEC crossover for $\lambda/R = 0.03\epsilon_F$ and $h = 0.19\epsilon_F$ with various $1/k_F a_s$. (a) Excitation
spectrum on the BCS side where $1/k_F a_s = -0.75$. (b) Excitation spectrum at unitarity
where $1/k_F a_s = 0$. (c) Excitation spectrum on the BEC side where $1/k_F a_s = 1$.

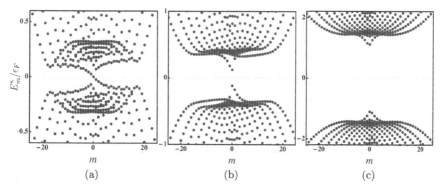

**Fig. 5.6.6.** Excitation spectrum as a function of $m$ when $k_z = 0$ through the BCS to BEC crossover for $\lambda/R = 0.14\epsilon_F$ and $h = 0.32\epsilon_F$ with various $1/k_F a_s$. (a) Excitation spectrum on the BCS side where $1/k_F a_s = -1$. (b) Excitation spectrum at unitarity where $1/k_F a_s = 0$. (c) Excitation spectrum on the BEC side where $1/k_F a_s = 1$.

positive energy excitations seem to include the states of quantum number $m$ both positive and negative. For Fig. 5.6.6(a) the smallest positive $E_m^\kappa$ is equal to $1.5 \times 10^{-4}\epsilon_F$ when $m = 0$. To understand this difference from Fig. 5.6.5, let us look at the corresponding wavefunctions. Both Figs. 5.6.7 and 5.6.8 plot the wavefunctions for fixed $m$ with smallest positive $E_m^\kappa$ for the case of Fig. 5.6.6(a). One can see from Fig. 5.6.7 that the excitations with $m \leq 0$ are states well localized in the vortex core until the angular momentum $|m|$ is so big that the centrifugal force pushes the state to the cylinder edge. On the other hand, Fig. 5.6.8 shows that the excitation states with $m > 0$ well penetrate into the bulk and have substantial weight in the vicinity of the edge. Such a situation usually will not happen until the system is topologically nontrivial and zero energy modes are expected to reside at both the vortex and the edge. Consider that for the case of Fig. 5.6.6(a), since $\Delta/\epsilon_F = 0.26$ and $\sqrt{\Delta^2 + \mu^2}/\epsilon_F = 0.42$, the system is in the topologically nontrivial phase $B$, while for the case of Fig. 5.6.5(a), since $\Delta/\epsilon_F = 0.37$, the system is in the topologically trivial phase $A$. Our results is consistent with this topological classification; the smallness of $|E_{m=0}^\kappa| = 1.5 \times 10^{-4}\epsilon_F$ and the structure of the $m > 0$ excitations are due to the finite size of the system. The zero energy modes which would have result in an infinite system at the vortex core and the edge hybridize to lift the otherwise degenerate modes exactly at zero energy.

J. Yao, S. Ding, Z. Yu, and S. Zhang

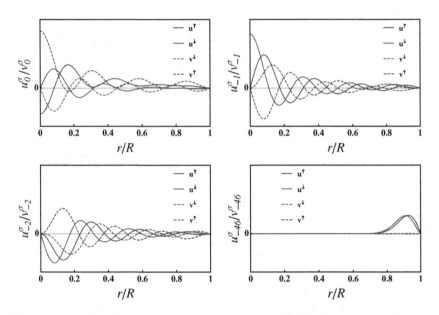

**Fig. 5.6.7.** Excitation wavefunctions with lowest positive $E_m^\kappa$ for fixed $m(\leq 0)$ in the case of Fig. 5.6.6(a).

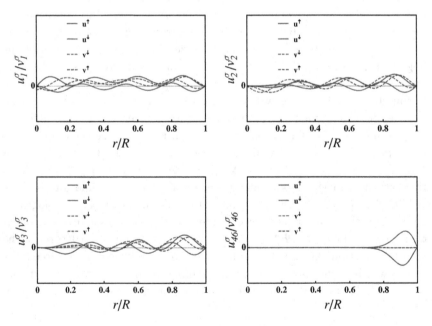

**Fig. 5.6.8.** Excitation wavefunctions with lowest positive $E_m^\kappa$ for fixed $m(> 0)$ in the case of Fig. 5.6.6(a).

## 5.7. Conclusion

In this chapter, we discussed the superfluid properties of a two-component Fermi gas with spin-orbit coupling. We show how the presence of spin-orbit coupling modifies the normal and superfluid density of the system. The structure of the vortex is discussed and the core states are calculated. Further work is required to investigate quantitatively the collective mode contributions to the normal and superfluid densities.

## Acknowlegement

Yao Juan, Ding Shanshan and Shizhong Zhang are supported by Hong Kong Research Grants Council (General Research Fund, HKU 17306414, 17318316 and Collaborative Research Fund, HKUST3/CRF/13G) and the Croucher Foundation under the Croucher Innovation Award. Zhenhua Yu is supported by NSFC Grant NO. 11474179.

## Appendix A. Path Integral Formulation of a Spin-orbit Coupled Fermi Gas

In this appendix, we will use the path integral formalism to set up the general framework for investigating the ground state as well as collective excitations of a spin-orbit coupled Fermi gas.

Within the path integral formalism, the partition function of the system is given by

$$Z = \int D\phi_\sigma^* D\phi_\sigma e^{-S(\phi_\sigma^*, \phi_\sigma)}, \qquad (A.1)$$

where $\phi_\sigma^*$ and $\phi_\sigma$ are Grassmann numbers, and the action takes the form

$$S(\phi_\sigma^*, \phi_\sigma) = \int_0^\beta d\tau \int d^3x \left\{ \sum_\sigma \phi_\sigma^* \left( \frac{\partial}{\partial \tau} - \frac{\nabla_x^2}{2m} - \mu + \lambda \sigma \times p \right) \phi_\sigma \right.$$

$$\left. + g\phi_\uparrow^* \phi_\downarrow^* \phi_\downarrow \phi_\uparrow \right\} \qquad (A.2)$$

where $\beta = \frac{1}{k_B T}$ with $k_B$ being the Boltzmann constant and $T$ the temperature. We will take the limit $T \to 0 (\beta \to \infty)$ in the end of calculation. By performing the Hubbard-Stratonovich transformation, the action can be transformed into the following effective action

$$S = \int_0^\beta d\tau \int d^3x \left\{ -\frac{|\Delta|^2}{g} \right.$$

$$\left. + \frac{1}{2} \begin{pmatrix} \phi_\uparrow^* & \phi_\downarrow^* & \phi_\uparrow & \phi_\downarrow \end{pmatrix} \begin{pmatrix} \frac{\partial}{\partial\tau} + \xi_p & \gamma_p & 0 & \Delta \\ \gamma_p^* & \frac{\partial}{\partial\tau} + \xi_p & -\Delta & 0 \\ 0 & -\Delta^* & \frac{\partial}{\partial\tau} - \xi_p & \gamma_p^* \\ \Delta^* & 0 & \gamma_p & \frac{\partial}{\partial\tau} - \xi_p \end{pmatrix} \begin{pmatrix} \phi_\uparrow \\ \phi_\downarrow \\ \phi_\uparrow^* \\ \phi_\downarrow^* \end{pmatrix} \right\}.$$

(A.3)

Here $\Delta$ and $\Delta^*$ are the auxiliary fields employed in the Hubbard-Stratonovich transformation. After integrating over the Grassmann numbers $\phi_\sigma^*$ and $\phi_\sigma$, we obtain the effective action in terms of the auxiliary fields as

$$S_{\text{eff}}(\Delta, \Delta^*) = -\beta V \frac{|\Delta|^2}{g} - \frac{1}{2}\text{Tr} \ln \det \left(-G^{-1}\right), \qquad (A.4)$$

where

$$G^{-1} = -\begin{pmatrix} \frac{\partial}{\partial\tau} + \xi_p & \gamma_p & 0 & \Delta \\ \gamma_p^* & \frac{\partial}{\partial\tau} + \xi_p & -\Delta & 0 \\ 0 & -\Delta^* & \frac{\partial}{\partial\tau} - \xi_p & \gamma_p^* \\ \Delta^* & 0 & \gamma_p & \frac{\partial}{\partial\tau} - \xi_p \end{pmatrix}. \qquad (A.5)$$

## A.1. Saddle Point Approximation

We can first apply the saddle point approximation to the partition function. In the saddle point approximation, for a uniform auxiliary field $\Delta(x) = \Delta(x)^* = \Delta_0$, the effective action becomes

$$S_{\text{eff0}}(\Delta, \Delta^*) = -\beta V \frac{\Delta_0^2}{g} - \frac{1}{2}\text{Tr} \ln \det \left(-G_0^{-1}\right), \qquad (A.1)$$

where $G_0^{-1}$ equals $G^{-1}$ in which $\Delta$ and $\Delta^*$ are substituted by $\Delta_0$. After summing over the Matsubara frequency, the mean field thermodynamic potential is given by

$$\Omega_0 = -\frac{1}{\beta} \ln Z$$

$$= -V\frac{\Delta_0^2}{g} + \frac{1}{2}\sum_{p,\delta}(\xi_p - E_{p,\delta}) - \frac{1}{\beta}\sum_{p,\delta}\ln\left(1 + e^{-\beta E_{p,\delta}}\right), \qquad (A.2)$$

where $E_{p,\delta} = \sqrt{\left(\xi_p + \delta|\gamma_p|\right)^2 + \Delta_0^2}$ with $|\gamma_p| = \lambda\sqrt{p_x^2 + p_y^2}$ and $\delta = \pm 1$. Given the derived explicit expression for the thermodynamic potential $\Omega_0$ at the mean field level, by setting the first order derivative of $\Omega_0$ with respect to $\Delta_0$ to be zero and with respect to the chemical potential $\mu$ to be $-N$, we obtain the gap and number equations for the BEC-BCS crossover

$$\frac{1}{g} = -\frac{1}{V}\sum_{p,\delta}\frac{\tanh\left[\beta E_{p,\delta}/2\right]}{4E_{p,\delta}}, \tag{A.3a}$$

$$n = \frac{1}{2}\sum_{p,\delta}\left[1 - \tanh\left[\frac{\beta E_{p,\delta}}{2}\right]\frac{\xi_p + \delta|\gamma_p|}{E_{p,\delta}}\right]. \tag{A.3b}$$

The ultraviolet divergence of the gap equation is due to the artificial contact interaction constant $g$ and can be renormalized by the experimental measurable parameter–the $s$-wave scattering length $a_s$ via the following equation

$$\frac{1}{g} = \frac{m}{4\pi\hbar^2 a_s} - \frac{1}{V}\sum_p\frac{m}{p^2}. \tag{A.4}$$

We are thus lead to the two equations given by Eq. (5.4.23).

## A.2. Gaussian Fluctuation

The matrix M given in Section. 4.C is defined by its matrix elements as

$$M_{11} = \frac{1}{2}\frac{1}{\beta V}\sum_p N_{11} - \frac{1}{g} \tag{A.1a}$$

$$M_{12} = \frac{1}{2}\frac{1}{\beta V}\sum_p N_{12} \tag{A.1b}$$

$$M_{21} = \frac{1}{2}\frac{1}{\beta V}\sum_p N_{21} \tag{A.1c}$$

$$M_{22} = \frac{1}{2}\frac{1}{\beta V}\sum_p N_{22} - \frac{1}{g} \tag{A.1d}$$

with

$$N11 = [-G_{034}(p)]\,G_{012}(p-q) + G_{033}(p)G_{022}(p-q)$$
$$+ [-G_{044}(p)]\,[-G_{011}(p-q)] + G_{043}(p)\,[-G_{021}(p-q)] \tag{A.2a}$$
$$N22 = G_{012}(p)\,[-G_{034}(p-q)] + [-G_{011}(p)]\,[-G_{044}(p-q)]$$
$$+ G_{022}(p)G_{033}(p-q) + [-G_{021}(p)]\,G_{043}(p-q) \tag{A.2b}$$

$$N12 = [-G_{014}(p)]\,[-G_{014}(p-q)] + G_{013}(p)\,[-G_{024}(p-q)]$$
$$+ [-G_{024}(p)]\,G_{013}(p-q) + G_{023}(p)G_{023}(p-q) \qquad \text{(A.2c)}$$
$$N21 = G_{032}(p)G_{032}(p-q) + [-G_{031}(p)]\,G_{042}(p-q)$$
$$+ G_{042}(p)\,[-G_{031}(p-q)] + [-G_{041}(p)]\,[-G_{041}(p-q)] \qquad \text{(A.2d)}$$

Here $G_{0ij}$ are the elements of the mean field Green's function and $p$ is the four vector $(\mathbf{p}, i\omega_n)$. By summing over the Matsubara frequency using the residue theorem, we find

$$\left[\frac{1}{\beta V}\sum_p N_{11}\right](\mathbf{q}, iq_n)$$

$$= \frac{1}{V}\sum_{\mathbf{p}}\sum_{\alpha=\pm 1,\beta=\pm 1}\sum_{\alpha'=\pm 1,\beta'=\pm 1}\left\{\left[f(-\alpha' E_{p-q,\beta'}) - f(\alpha E_{p,\beta})\right]\right.$$

$$\times \frac{1}{iq_n - \alpha E_{p,\beta} - \alpha' E_{p-q,\beta'}} \times \left[\frac{\xi_{p-q} + \beta'|\gamma_{p-q}| - \alpha' E_{p-q,\beta'}}{4\alpha' E_{p-q,\beta'}}\right]$$

$$\left.\times \left[\frac{\xi_p + \beta|\gamma_p| - \alpha E_{p,\beta}}{4\alpha E_{p,\beta}}\right] \times \left[\left(1 + \frac{\gamma_p\gamma_{p-q}^*}{\gamma_p^*\gamma_{p-q}}\right)\frac{\gamma_{p-q}\gamma_p^*}{|\gamma_{p-q}||\gamma_p|}\beta'\beta + 2\right]\right\}$$

$$\text{(A.3a)}$$

$$\left[\frac{1}{\beta V}\sum_p N_{22}\right](\mathbf{q}, iq_n) = \left[\frac{1}{\beta V}\sum_p N_{11}\right](\mathbf{q}, -iq_n) \qquad \text{(A.3b)}$$

$$\left[\frac{1}{\beta V}\sum_p N_{12}\right](\mathbf{q}, iq_n)$$

$$= \frac{1}{V}\sum_{\mathbf{p}}\sum_{\alpha=\pm 1,\beta=\pm 1}\sum_{\alpha'=\pm 1,\beta'=\pm 1}\left\{\left[f(\alpha E_{p,\beta}) - f(-\alpha' E_{p-q,\beta'})\right]\right.$$

$$\times \frac{\Delta_0}{4\alpha E_{p,\beta}} \times \frac{\Delta_0}{4\alpha' E_{p-q,\beta'}}\left[\frac{2}{iq_n - \alpha E_{p,\beta} - \alpha' E_{p-q,\beta'}} + \frac{\gamma_{p-q}^*\gamma_p}{|\gamma_{p-q}||\gamma_p|}\right.$$

$$\left.\left.\times \beta'\beta\left(\frac{1}{iq_n - \alpha E_{p,\beta} - \alpha' E_{p-q,\beta'}} - \frac{1}{iq_n + \alpha E_{p,\beta} + \alpha' E_{p-q,\beta'}}\right)\right]\right\}$$

$$\text{(A.3c)}$$

$$\left[\frac{1}{\beta V}\sum_p N_{21}\right](\mathbf{q}, iq_n) = \left[\frac{1}{\beta V}\sum_p N_{12}\right](\mathbf{q}, iq_n) \qquad \text{(A.3d)}$$

In the above expression, $f(x) = \frac{1}{e^{\beta x}+1}$ is the Fermi distribution function. Similar to the divergence of the gap equation in section A.1, the interaction constant $g$ in the expression of $M$ also should be renormalized.

## Appendix B. Correlation Function

### B.1. Density-Density Correlation Function

When $\mathbf{q}$ is taken along the $\mathbf{x}$ direction, the spin fluctuation can be decoupled from that of the density and the pairing fields $\Delta$ and $\Delta^*$. Consequently the random phase approximation gives the density-density correlation function as

$$\chi_{nn} = -\frac{\Pi_{11}}{2} + \left(\Pi_{12}/2 \ \Pi_{13}/2\right) \begin{pmatrix} M_{22} & M_{21} \\ M_{12} & M_{11} \end{pmatrix}^{-1} \begin{pmatrix} \Pi_{21}/2 \\ \Pi_{31}/2 \end{pmatrix} \tag{B.1}$$

where

$$\Pi_{11} = 2\Pi_{11A} - 4\Pi_{11B} + \Pi_{11C} + 2\Pi_{11D} + 2\Pi_{11E} + \Pi_{11F} \tag{B.2a}$$

$$\Pi_{12} = -2\Pi_{12A} + \Pi_{12B} + 2\Pi_{12C} \tag{B.2b}$$

$$\Pi_{13} = \Pi_{21} = \Pi_{31} = \Pi_{12} \tag{B.2c}$$

and

$$\Pi_{11A} = \frac{1}{V} \sum_{\mathbf{p}} \sum_{\alpha=\pm1,\beta=\pm1} \sum_{\alpha'=\pm1,\beta'=\pm1} \left\{ \left[-n(-\alpha'E_{p-q,\beta'}) + n(\alpha E_{p,\beta})\right] \right.$$

$$\times \frac{1}{iq_n - \alpha E_{p,\beta} - \alpha' E_{p-q,\beta'}} \times \left[\frac{\xi_{p-q} + \beta'|\gamma_{p-q}| - \alpha' E_{p-q,\beta'}}{4\alpha' E_{p-q,\beta'}}\right]$$

$$\left. \times \left[\frac{\xi_p + \beta|\gamma_p| + \alpha E_{p,\beta}}{4\alpha E_{p,\beta}}\right] \right\} \tag{B.3a}$$

$$\Pi_{11B} = \frac{1}{V} \sum_{\mathbf{p}} \sum_{\alpha=\pm1,\beta=\pm1} \sum_{\alpha'=\pm1,\beta'=\pm1} \left\{ \left[-n(-\alpha'E_{p-q,\beta'}) + n(\alpha E_{p,\beta})\right] \right.$$

$$\left. \times \frac{1}{iq_n - \alpha E_{p,\beta} - \alpha' E_{p-q,\beta'}} \times \left[\frac{\Delta_0}{4\alpha' E_{p-q,\beta'}}\right] \times \left[\frac{\Delta_0}{4\alpha E_{p,\beta}}\right] \right\} \tag{B.3b}$$

$$\Pi_{11C} = \frac{1}{V} \sum_{\mathbf{p}} \sum_{\alpha=\pm1,\beta=\pm1} \sum_{\alpha'=\pm1,\beta'=\pm1} \left\{ 2\Re\left[\gamma_{p-q}\gamma_p^*\right]\left[-n(-\alpha'E_{p-q,\beta'})\right.\right.$$

$$\left.\left. + n(\alpha E_{p,\beta})\right] \times \frac{\beta\beta'}{iq_n - \alpha E_{p,\beta} - \alpha' E_{p-q,\beta'}} \times \left[\frac{\xi_p + \beta|\gamma_p| + \alpha E_{p,\beta}}{4\alpha E_{p,\beta}|\gamma_p|}\right] \right.$$

$$\times \left[ \frac{\xi_{p-q} + \beta'|\gamma_{p-q}| - \alpha' E_{p-q,\beta'}}{4\alpha' E_{p-q,\beta'} |\gamma_{p-q}|} \right] \right\} \tag{B.3c}$$

$$\Pi_{11D} = -\frac{1}{V} \sum_{\mathbf{p}} \sum_{\alpha=\pm1,\beta=\pm1} \sum_{\alpha'=\pm1,\beta'=\pm1} \left\{ \left[ -n(-\alpha' E_{p-q,\beta'}) + n(\alpha E_{p,\beta}) \right] \right.$$

$$\times \left[ \frac{1}{iq_n - \alpha E_{p,\beta} - \alpha' E_{p-q,\beta'}} - \frac{1}{iq_n + \alpha E_{p,\beta} + \alpha' E_{p-q,\beta'}} \right]$$

$$\left. \times \left[ \frac{\Delta_0}{4\alpha' E_{p-q,\beta'}} \right] \times \left[ \frac{\Delta_0}{4\alpha E_{p,\beta}} \right] \times \left[ \frac{\gamma_p \gamma_{p-q}^*}{|\gamma_p||\gamma_{p-q}|} \right] \beta\beta' \right\} \tag{B.3d}$$

$$\Pi_{11E} = \frac{1}{V} \sum_{\mathbf{p}} \sum_{\alpha=\pm1,\beta=\pm1} \sum_{\alpha'=\pm1,\beta'=\pm1} \left\{ \left[ -n(-\alpha' E_{p-q,\beta'}) + n(\alpha E_{p,\beta}) \right] \right.$$

$$\times \frac{1}{iq_n - \alpha E_{p,\beta} - \alpha' E_{p-q,\beta'}} \times \left[ \frac{\xi_{p-q} + \beta'|\gamma_{p-q}| + \alpha' E_{p-q,\beta'}}{4\alpha' E_{p-q,\beta'}} \right]$$

$$\left. \times \left[ \frac{\xi_p + \beta|\gamma_p| - \alpha E_{p,\beta}}{4\alpha E_{p,\beta}} \right] \right\} \tag{B.3e}$$

$$\Pi_{11F} = \frac{1}{V} \sum_{\mathbf{p}} \sum_{\alpha=\pm1,\beta=\pm1} \sum_{\alpha'=\pm1,\beta'=\pm1} \left\{ 2\Re \left[ \gamma_{p-q}^* \gamma_p \right] \left[ -n(-\alpha' E_{p-q,\beta'}) \right. \right.$$

$$+ n(\alpha E_{p,\beta}) \left] \times \frac{\beta\beta'}{iq_n - \alpha E_{p,\beta} - \alpha' E_{p-q,\beta'}} \times \left[ \frac{\xi_p + \beta|\gamma_p| - \alpha E_{p,\beta}}{4\alpha E_{p,\beta}|\gamma_p|} \right] \right.$$

$$\left. \times \left[ \frac{\xi_{p-q} + \beta'|\gamma_{p-q}| + \alpha' E_{p-q,\beta'}}{4\alpha' E_{p-q,\beta'}|\gamma_{p-q}|} \right] \right\} \tag{B.3f}$$

$$\Pi_{12A} = \frac{1}{V} \sum_{\mathbf{p}} \sum_{\alpha=\pm1,\beta=\pm1} \sum_{\alpha'=\pm1,\beta'=\pm1} \left\{ \left[ -n(-\alpha' E_{p-q,\beta'}) + n(\alpha E_{p,\beta}) \right] \right.$$

$$\times \frac{1}{iq_n - \alpha E_{p,\beta} - \alpha' E_{p-q,\beta'}} \times \left[ \frac{\xi_p + \beta|\gamma_p| + \alpha E_{p,\beta}}{4\alpha E_{p,\beta}} \right]$$

$$\left. \times \left[ \frac{\Delta_0}{4\alpha' E_{p-q,\beta'}} \right] \right\} \tag{B.3g}$$

$$\Pi_{12B} = \frac{1}{V} \sum_{\mathbf{p}} \sum_{\alpha=\pm1,\beta=\pm1} \sum_{\alpha'=\pm1,\beta'=\pm1} \left\{ 4\Re \left[ \gamma_{p-q} \gamma_p^* \right] \right.$$

$$\left[ n(-\alpha' E_{p-q,\beta'}) - n(\alpha E_{p,\beta}) \right] \times \frac{\beta\beta'}{iq_n - \alpha E_{p,\beta} - \alpha' E_{p-q,\beta'}}$$

$$\times \left[ \frac{\Delta_0}{4\alpha' E_{p-q,\beta'} |\gamma_{p-q}|} \right] \times \left[ \frac{\xi_p + \beta|\gamma_p| + \alpha E_{p,\beta}}{4\alpha E_{p,\beta} |\gamma_p|} \right] \Bigg\} \tag{B.3h}$$

$$\Pi_{12C} = \frac{1}{V} \sum_{p} \sum_{\alpha=\pm 1, \beta=\pm 1} \sum_{\alpha'=\pm 1, \beta'=\pm 1} \Bigg\{ \left[ n(-\alpha' E_{p-q,\beta'}) - n(\alpha E_{p,\beta}) \right]$$

$$\times \frac{1}{iq_n - \alpha E_{p,\beta} - \alpha' E_{p-q,\beta'}} \times \left[ \frac{\xi_{p-q} + \beta'|\gamma_{p-q}| + \alpha' E_{p-q,\beta'}}{4\alpha' E_{p-q,\beta'}} \right]$$

$$\times \left[ \frac{\Delta_0}{4\alpha E_{p,\beta}} \right] \Bigg\} \tag{B.3i}$$

## B.2. Spin-Spin Correlation Function

In this part of the appendix, we will study the spin-spin correlation function of the system. We take $\mathbf{q}$ along the $\mathbf{x}$ direction, and find by the random phase approxmation

$$\chi_{ss}^{RPA}(q_x, iq_n) = \chi_{ss}^{MF}(q_x, iq_n) = -\frac{\Pi_{44}}{2} \tag{B.1}$$

where, for $\chi_{\sigma_z \sigma_z}$,

$$\Pi_{44} = \frac{1}{\beta V} \sum_k \left[ 2G_{011}(k)G_{011}(k-q) + 4G_{014}(k)G_{014}(k-q) \right.$$

$$- G_{012}(k-q)G_{021}(k) - G_{012}(k)G_{021}(k-q)$$

$$+ 2G_{013}(k-q)G_{024}(k) + 2G_{013}(k)G_{024}(k-q)$$

$$+ 2G_{033}(k)G_{033}(k-q) - G_{034}(k-q)G_{043}(k)$$

$$\left. - G_{034}(k)G_{043}(k-q) \right] \tag{B.2}$$

while for $\chi_{\sigma_x \sigma_x}$,

$$\Pi_{44} = \frac{1}{\beta V} \sum_k \left[ 2G_{011}(k)G_{011}(k-q) + G_{012}(k)G_{012}(k-q) \right.$$

$$+ 4G_{014}(k)G_{014}(k-q) + G_{021}(k)G_{021}(k-q)$$

$$+ 2G_{013}(k)G_{013}(k-q) + 2G_{024}(k)G_{024}(k-q)$$

$$+ 2G_{033}(k)G_{033}(k-q) + G_{034}(k)G_{034}(k-q)$$

$$\left. + G_{043}(k)G_{043}(k-q) \right] \tag{B.3}$$

After summing over Matsubara frequency $i\omega_n$ of four vector $k = (\mathbf{k}, i\omega_n)$ and taking the zero temperature limitation, we can obtain that

$$
\chi_{\sigma_z\sigma_z}(\mathbf{q}, iq_n) = -\frac{1}{2}\frac{1}{V}\sum_{\mathbf{k}}\sum_{\beta\beta'}\left\{\left[2 - \frac{\gamma_k\gamma_{k-q}^* + \gamma_k^*\gamma_{k-q}}{|\gamma_k||\gamma_{k-q}|}\beta\beta'\right]\right.
$$
$$
\left[\frac{\xi_k + \beta|\gamma_k| - E_{k\beta}}{4E_{k\beta}}\frac{\xi_{k-q} + \beta'|\gamma_{k-q}| + E_{k-q\beta'}}{4E_{k-q\beta'}}\right.
$$
$$
+ \frac{2\Delta^2}{4E_{k\beta}4E_{k-q\beta'}}
$$
$$
\left.+ \frac{\xi_k + \beta|\gamma_k| + E_{k\beta}}{4E_{k\beta}}\frac{\xi_{k-q} + \beta'|\gamma_{k-q}| - E_{k-q\beta'}}{4E_{k-q\beta'}}\right]
$$
$$
\left.\times\left[\frac{1}{iq_n + E_{k\beta} + E_{k-q\beta'}} - \frac{1}{iq_n - E_{k\beta} - E_{k-q\beta'}}\right]\right\}
$$

$$\tag{B.4a}$$

$$
\chi_{\sigma_x\sigma_x}(\mathbf{q}, iq_n) = -\frac{1}{2}\frac{1}{V}\sum_{\mathbf{k}}\sum_{\beta\beta'}\left\{\left[2 + \frac{\gamma_k\gamma_{k-q} + \gamma_k^*\gamma_{k-q}^*}{|\gamma_k||\gamma_{k-q}|}\beta\beta'\right]\right.
$$
$$
\left[\frac{\xi_k + \beta|\gamma_k| - E_{k\beta}}{4E_{k\beta}}\frac{\xi_{k-q} + \beta'|\gamma_{k-q}| + E_{k-q\beta'}}{4E_{k-q\beta'}}\right.
$$
$$
+ \frac{2\Delta^2}{4E_{k\beta}4E_{k-q\beta'}}
$$
$$
\left.+ \frac{\xi_k + \beta|\gamma_k| + E_{k\beta}}{4E_{k\beta}}\frac{\xi_{k-q} + \beta'|\gamma_{k-q}| - E_{k-q\beta'}}{4E_{k-q\beta'}}\right]
$$
$$
\left.\left[\frac{1}{iq_n + E_{k\beta} + E_{k-q\beta'}} - \frac{1}{iq_n - E_{k\beta} - E_{k-q\beta'}}\right]\right\}
$$

$$\tag{B.4b}$$

From this we can verify the relation given in the main text between the normal density and the correlation function. In addition, we also obtain analytically

$$
\chi_{\sigma_x\sigma_x}(\mathbf{q} = 0, iq_n) = \frac{1}{2}\chi_{\sigma_z\sigma_z}(\mathbf{q} = 0, iq_n) \tag{B.5}
$$

and when the momentum transfer $\mathbf{q}$ is infinitely large

$$
\chi_{\sigma_x\sigma_x}(\mathbf{q} \to \infty, iq_n) = \chi_{\sigma_z\sigma_z}(\mathbf{q} \to \infty, iq_n). \tag{B.6}
$$

## Appendix C. Details of Bogoliubov-de Gennes Equation in a Cylindrical Goemetry

To solve the eigenfunctions of Eq. (5.6.6), we expand $\mathbf{u}(\mathbf{r})$ and $\mathbf{v}(\mathbf{r})$ with respect to the complete basis $\boldsymbol{\Phi}_{nmk_z}(\mathbf{r})$ which are solutions to Eq. (5.6.1). Due to the assumed form $\Delta(\mathbf{r}) = \Delta(r)e^{i\theta}$, the component $\boldsymbol{\Phi}_{nmk_z}(\mathbf{r})$ of $\mathbf{u}$ couples to the component $\boldsymbol{\Phi}_{n',-m,-k_z}(\mathbf{r})$ of $\mathbf{v}$; namely, for each eigenfunction, we can write

$$
\mathbf{u}(\mathbf{r}) = \sum_n a_n \begin{bmatrix} \Phi_{nmk_z\uparrow}(\mathbf{r}) \\ \Phi_{nmk_z\downarrow}(\mathbf{r}) \end{bmatrix} \equiv \sum_n a_n \begin{bmatrix} \Phi_{nm\uparrow}(r)e^{im\theta}e^{ik_z z} \\ \Phi_{nm\downarrow}(r)e^{i(m+1)\theta}e^{ik_z z} \end{bmatrix},
$$

$$
\mathbf{v}(\mathbf{r}) = \sum_n b_n \begin{bmatrix} \Phi^*_{n,-m,-k_z\downarrow}(\mathbf{r}) \\ \Phi^*_{n,-m,-k_z\uparrow}(\mathbf{r}) \end{bmatrix} \equiv \sum_n b_n \begin{bmatrix} \Phi^*_{n,-m\downarrow}(r)e^{i(m-1)\theta}e^{ik_z z} \\ \Phi^*_{n,-m\uparrow}(r)e^{im\theta}e^{ik_z z} \end{bmatrix}.
$$

$$(\text{C.1})$$

The coefficients $a_n$ and $b_n$ are to be determined self-consistently using the gap equation to be shown below. For each given $m$ and $k_z$, the BdG equation, Eq. (5.6.6), is now reduced to a matrix form

$$
\begin{bmatrix} T^{mk_z} & \mathcal{B}^m \\ \mathcal{B}^{m\dagger} & -T^{-m,-k_z} \end{bmatrix}_{nn'} \begin{bmatrix} a_{n'}^{mk_z} \\ b_{n'}^{-m,-k_z} \end{bmatrix} = E_{mk_z} \begin{bmatrix} a_n^{mk_z} \\ b_n^{-m,-k_z} \end{bmatrix}, \qquad (\text{C.2})
$$

where the matrix elements $T^{mk_z}$ and $\mathcal{B}^m$ in terms of the complete basis Eq. (5.6.1) are given by

$$
[T^{mk_z}]_{nn'} = (\epsilon_{nmk_z} - \mu)\delta_{nn'},
$$

$$
[\mathcal{B}^m]_{nn'} = \int dr r \Delta(r)[\Phi_{nm\uparrow}(r)\Phi_{n',-m\downarrow}(r) - \Phi_{nm\downarrow}(r)\Phi_{n',-m\uparrow}(r)].
$$

$$(\text{C.3})$$

On the other hand, the gap function $\Delta(r)$ at temperature $T = 1/k_B\beta$ is given by

$$
\Delta(r) = g \sum_{\kappa m k_z} \sum_{n_1 n_2} a_{n_1}^{\kappa,m,k_z} b_{n_2}^{\kappa,-m,-k_z}
$$

$$
[\Phi_{n_1 m\uparrow}(r)\Phi_{n_2,-m\downarrow}(r)f(E_{mk_z}^\kappa) + \Phi_{n_1 m\downarrow}(r)\Phi_{n_2,-m\uparrow}(r)f(-E_{mk_z}^\kappa)],
$$

$$(\text{C.4})$$

where $f(E)$ is the Fermi distribution and the summation takes over only positive energies $E_{mk_z}^\kappa > 0$. The particle number density has the expression

$$
n_\sigma(r) = 2\pi L \Bigg\{ \sum_{\kappa m k_z} \sum_{n_1 n_2} [f(E_{mk_z}^\kappa) a_{n_1}^{\kappa,m,k_z} a_{n_2}^{\kappa,m,k_z *} \Phi_{n_1 m\sigma}(r)\Phi^*_{n_2 m\sigma}(r)
$$

$$+ f(-E^\kappa_{mk_z}) b^{\kappa,l-m-1,-k_z}_{n_1} b^{\kappa,,l-m-1,-k_z*}_{n_2}$$

$$\times \Phi^*_{n_1,l-m-1,\sigma}(r) \Phi_{n_2,l-m-1,\sigma}(r)] \Big\} \tag{C.5}$$

where $\sigma = \uparrow$ and $\downarrow$. Note that the normalization of single particle wave-function $\Phi_{n,m,\sigma}(r)$ is given by

$$\int_0^R dr r [\Phi^*_{n_1 m\uparrow}(r) \Phi_{n_2 m\uparrow}(r) + \Phi^*_{n_1 m\downarrow}(r) \Phi_{n_2 m\downarrow}(r)] = \delta_{n_1 n_2}. \tag{C.6}$$

Thus the total number of particles is

$$N = 2\pi L \sum_{\kappa m k_z n} \left[ f(E^\kappa_{mk_z}) a^{\kappa,m,k_z}_n a^{\kappa,m,k_z*}_n \right. \tag{C.7}$$

$$\left. + f(-E^\kappa_{mk_z}) b^{\kappa,l-m-1,-k_z}_n b^{\kappa,,l-m-1,-k_z*}_n \right].$$

By numerically calculating $\Delta(r)$ and $\mu$ self-consistently via Eqs. (C.2), (C.4) and (C.7), we can study properties of the vortex line in a gas of $N$ fermions through the BEC-BCS crossover.

## References

1. C. J. Pethick and H. Smith, *Bose-Einstein Condensation in Dilute Gases.* Cambridge university press (2002).
2. A. Griffin, D. W. Snoke, and S. Stringari, *Bose-Einstein Condensation.* Cambridge University Press (1996).
3. A. J. Leggett, *Quantum Liquids: Bose Condensation and Cooper Pairing in Condensed-Matter Systems.* Oxford University Press (2006).
4. J. Bardeen, L. N. Cooper, and J. R. Schrieffer, Theory of superconductivity, *Physical Review*, **108**(5), 1175 (1957).
5. S. Giorgini, L. P. Pitaevskii, and S. Stringari, Theory of ultracold atomic fermi gases. *Reviews of Modern Physics*, **80**(4), 1215 (2008).
6. C. Regal, M. Greiner, and D. S. Jin, Observation of resonance condensation of fermionic atom pairs. *Physical Review Letters*, **92**(4), 040403 (2004).
7. M. Zwierlein, C. Stan, C. Schunck, S. Raupach, A. Kerman, and W. Ketterle, Condensation of pairs of fermionic atoms near a feshbach resonance. *Physical Review Letters*, **92**(12), 120403 (2004).
8. J. Kinast, S. Hemmer, M. Gehm, A. Turlapov, and J. Thomas, Evidence for superfluidity in a resonantly interacting fermi gas. *Physical Review Letters*, **92**(15), 150402 (2004).
9. Y.-J. Lin, K. Jiménez-García, and I. Spielman, Spin-orbit-coupled bose-einstein condensates. *Nature*, **471**(7336), 83–86 (2011).
10. J.-Y. Zhang, S.-C. Ji, Z. Chen, L. Zhang, Z.-D. Du, B. Yan, G.-S. Pan, B. Zhao, Y.-J. Deng, H. Zhai, et al., Collective dipole oscillations of a spin-orbit coupled bose-einstein condensate. *Physical Review Letters*, **109**(11), 115301 (2012).

11. S.-C. Ji, J.-Y. Zhang, L. Zhang, Z.-D. Du, W. Zheng, Y.-J. Deng, H. Zhai, S. Chen, and J.-W. Pan, Experimental determination of the finite-temperature phase diagram of a spin-orbit coupled bose gas. *Nature physics*, **10**(4), 314–320 (2014).

12. L. W. Cheuk, A. T. Sommer, Z. Hadzibabic, T. Yefsah, W. S. Bakr, and M. W. Zwierlein, Spin-injection spectroscopy of a spin-orbit coupled fermi gas. *Physical Review Letters*, **109**(9), 095302 (2012).

13. P. Wang, Z.-Q. Yu, Z. Fu, J. Miao, L. Huang, S. Chai, H. Zhai, and J. Zhang, Spin-orbit coupled degenerate fermi gases. *Physical Review Letters*, **109**(9), 095301 (2012).

14. Z. Fu, L. Huang, Z. Meng, P. Wang, X.-J. Liu, H. Pu, H. Hu, and J. Zhang, Radio-frequency spectroscopy of a strongly interacting spin-orbit-coupled fermi gas. *Physical Review A*, **87**(5), 053619 (2013).

15. Z. Fu, L. Huang, Z. Meng, P. Wang, L. Zhang, S. Zhang, H. Zhai, P. Zhang, and J. Zhang, Production of feshbach molecules induced by spin-orbit coupling in fermi gases. *Nature Physics*, **10**(2), 110–115 (2014).

16. M. Khamehchi, Y. Zhang, C. Hamner, T. Busch, and P. Engels, Measurement of collective excitations in a spin-orbit-coupled bose-einstein condensate. *Physical Review A*, **90**(6), 063624 (2014).

17. A. J. Olson, S.-J. Wang, R. J. Niffenegger, C.-H. Li, C. H. Greene, and Y. P. Chen, Tunable landau-zener transitions in a spin-orbit-coupled bose-einstein condensate. *Physical Review A*, **90**(1), 013616 (2014).

18. S.-C. Ji, L. Zhang, X.-T. Xu, Z. Wu, Y. Deng, S. Chen, and J.-W. Pan, Softening of roton and phonon modes in a bose-einstein condensate with spin-orbit coupling. *Physical Review Letters*, **114**(10), 105301 (2015).

19. Z. Wu, L. Zhang, W. Sun, X.-T. Xu, B.-Z. Wang, S.-C. Ji, Y. Deng, S. Chen, X.-J. Liu, and J.-W. Pan, Realization of two-dimensional spin-orbit coupling for bose-einstein condensates. *Science*, **354**(6308), 83–88 (2016).

20. L. Huang, Z. Meng, P. Wang, P. Peng, S.-L. Zhang, L. Chen, D. Li, Q. Zhou, and J. Zhang, Experimental realization of two-dimensional synthetic spin-orbit coupling in ultracold fermi gases. *Nature Physics*, **12** (6), 540–544 (2016).

21. Z. Meng, L. Huang, P. Peng, D. Li, L. Chen, Y. Xu, C. Zhang, P. Wang, and J. Zhang, Experimental observation of a topological band gap opening in ultracold fermi gases with two-dimensional spin-orbit coupling. *Physical Review Letters*, **117**(23), 235304 (2016).

22. D. Campbell, R. Price, A. Putra, A. Valdés-Curiel, D. Trypogeorgos, and I. Spielman, Magnetic phases of spin-1 spin-orbit-coupled bose gases. *Nature communications*, **7** (2016).

23. S. Kolkowitz, S. Bromley, T. Bothwell, M. Wall, G. Marti, A. Koller, X. Zhang, A. Rey, and J. Ye, Spin–orbit-coupled fermions in an optical lattice clock. *Nature* (2016).

24. Y.-C. Zhang, Z.-Q. Yu, T. K. Ng, S. Zhang, L. Pitaevskii, and S. Stringari, Superfluid density of a spin-orbit-coupled bose gas. *Physical Review A*, **94**(3), 033635 (2016).

25. K. Zhou and Z. Zhang, Opposite effect of spin-orbit coupling on condensation and superfluidity. *Physical Review Letters*, **108**(2), 025301 (2012).

26. L. He and X.-G. Huang, Superfluidity and collective modes in rashba spin–orbit coupled fermi gases. *Annals of Physics*, **337**, 163–207 (2013).

27. L. He and X.-G. Huang, Bcs-bec crossover in 2d fermi gases with rashba spin-orbit coupling. *Physical Review Letters*, **108**(14), 145302 (2012).

28. L. He and X.-G. Huang, Bcs-bec crossover in three-dimensional fermi gases with spherical spin-orbit coupling. *Physical Review B*, **86** (1), 014511 (2012).

29. J. P. Vyasanakere and V. B. Shenoy, Collective excitations across the bcs-bec crossover induced by a synthetic rashba spin-orbit coupling. arXiv preprint arXiv:1201.5332 (2012).

30. L. He and X.-G. Huang, Unusual zeeman-field effects in two-dimensional spin-orbit-coupled fermi superfluids. *Physical Review A*, **86**(4), 043618 (2012).

31. Y. Cao, X.-J. Liu, L. He, G.-L. Long, and H. Hu, Superfluid density and berezinskii-kosterlitz-thouless transition of a spin-orbit-coupled fulde-ferrell superfluid. *Physical Review A*, **91**(2), 023609 (2015).

32. A. J. Leggett. Diatomic molecules and cooper pairs. In *Modern trends in the theory of condensed matter*, pp. 13–27. Springer (1980).

33. X. Cui, Mixed-partial-wave scattering with spin-orbit coupling and validity of pseudopotentials. *Physical Review A*, **85**(2), 022705 (2012).

34. P. Zhang, L. Zhang, and Y. Deng, Modified bethe-peierls boundary condition for ultracold atoms with spin-orbit coupling. *Physical Review A*, **86**(5), 053608 (2012).

35. Z. Yu, Short-range correlations in dilute atomic fermi gases with spin-orbit coupling. *Physical Review A*, **85**(4), 042711 (2012).

36. L. Zhang, Y. Deng, and P. Zhang, Scattering and effective interactions of ultracold atoms with spin-orbit coupling. *Physical Review A*, **87**(5), 053626 (2013).

37. H. Duan, L. You, and B. Gao, Ultracold collisions in the presence of synthetic spin-orbit coupling. *Physical Review A*, **87**(5), 052708 (2013).

38. S.-J. Wang and C. H. Greene, General formalism for ultracold scattering with isotropic spin-orbit coupling. *Physical Review A*, **91**(2), 022706 (2015).

39. Q. Guan and D. Blume, Scattering framework for two particles with isotropic spin-orbit coupling applicable to all energies. *Physical Review A*, **94**(2), 022706 (2016).

40. Y. Wu and Z. Yu, Short-range asymptotic behavior of the wave functions of interacting spin-1/2 fermionic atoms with spin-orbit coupling: A model study. *Physical Review A*, **87**(3), 032703 (2013).

41. Q. Guan and D. Blume, Analytical coupled-channel treatment of two-body scattering in the presence of three-dimensional isotropic spin-orbit coupling. *Physical Review A*, **95**(2), 020702 (2017).

42. M. Gong, S. Tewari, and C. Zhang, Bcs-bec crossover and topological phase transition in 3d spin-orbit coupled degenerate fermi gases. *Physical Review Letters*, **107**(19), 195303 (2011).

43. J. Gavoret and P. Nozieres, Structure of the perturbation expansion for the bose liquid at zero temperature. *Annals of Physics*, **28**(3), 349–399 (1964).

44. M. E. Fisher, M. N. Barber, and D. Jasnow, Helicity modulus, superfluidity, and scaling in isotropic systems. *Physical Review A*, **8**(2), 1111 (1973).

45. E. Taylor, A. Griffin, N. Fukushima, and Y. Ohashi, Pairing fluctuations and the superfluid density through the bcs-bec crossover. *Physical Review A*, **74**(6), 063626 (2006).

46. Z. Xu, Z. Yu, and S. Zhang, Evidence for correlated states in a cluster of bosons with rashba spin-orbit coupling. *New Journal of Physics*, **18**(2), 025002 (2016).

47. R. Sensarma, M. Randeria, and T.-L. Ho, Vortices in superfluid fermi gases through the bec to bcs crossover. *Physical Review Letters*, **96**(9), 090403 (2006).

# Chapter 6

# Spin-Orbit Coupling in Three-Component Bose Gases

Wei Han[1]* and Wei Zhang[2]†

[1]*Key Laboratory of Time and Frequency Primary Standards,
National Time Service Center, Chinese Academy of Sciences,
Xi'an 710600, China*
[2]*Department of Physics, Renmin University of China,
Beijing 100872, China*

We review our recent work on three-component Bose-Einstein condensates with spin-orbit coupling. In contrast to the conventional spin-orbit coupled systems, where the spin is represented by the SU(2) Pauli matrices, our work mainly focuses on the case of SU(3) spin-orbit coupling, where the spin is described by the Gell-Mann matrices, i.e., the generators of SU(3) group. We begin with a general discussion on the recent progress in three-component Bose-Einstein condensates, and indicate that SU(3) spin system is a potential direction of research in cold atom physics. Then we discuss a typical form of SU(3) spin-orbit coupling, which may be realized by the Raman laser dressing technology. We investigate the single-particle spectrum, many-body ground states and vortex configurations. Interesting findings involve the triple-well dispersion relation, lattice phase and time-reversal symmetry broken stripe phase. More surprisingly, it is demonstrated that the SU(3) spin-orbit coupling breaks the ordinary phase rule of spinor Bose gases and allows the spontaneous emergence of double-quantum spin vortices, which is in stark contrast to the singly quantized spin vortices observed in existing experiments and can be readily observed by the current magnetization-sensitive phase-contrast imaging technique.

*Correspondence addressed to: weihan@ntsc.ac.cn.
†Correspondence addressed to: wzhangl@ruc.edu.cn.

## 6.1. Introduction

### 1.A. *SU(2) spin-orbit coupling*

Ultracold atoms have become an ideal platform for simulating and implementing quantum systems with a variety of gauge fields. One of the most popular topic is synthetic spin-orbit (SO) coupling in ultracold atomic gases. The experimental realization of both one-dimensional (1D) and 2D SO coupling[1–6] not only paves the way towards the emulation of electronic behavior in solids with highly controllable degree of freedom,[7,8] but also provides opportunities for designing new Hamiltonian never happened in conventional quantum systems.[9–11]

With the development of SO coupling in ultra-cold atomic gases, intense theoretical and experimental interest goes towards to large spin systems.[4,12–15] In stark contrast to the extensively studied spin-1/2 case, where only two hyperfine states are coupled, large spin systems involve more spin states, thus make it possible to realize more complicated SO coupling unattainable in electronic materials. This opens an interesting avenue for exploring SO-related exotic states of matter and quantum phenomena.

The feasibility of realizing SO coupling in large spin systems is attributed to the highly controllable platform of the interaction between laser fields and ultracold atomic gases, which provide the opportunities for designing sophisticated Hamiltonian desired by physicists. The first example of large spin systems is three-component condensates, which allow the internal coupling between three hyperfine spin states. So far, in three-component Bose-Einstein condensates (BECs) a type of one-dimensional SO coupling, which can be considered as an equally weighted superposition of Rashba and Dresselhaus (ERD) SO couplings, has been realized in experiments by using Ramman laser dressing[4] or modulating gradient magnetic field.[12] Theory interests focus on not only the already realized form of 1D ERD SO coupling,[14–18] but also many other types of SO coupling, involving the spin-1 generalization of the pure 2D Rashba or Dresselhaus SO coupling[13,19–27] and the angular SO coupling.[28,29]

It should be noted that all most all of existing works on SO coupling focus on the case of SU(2) spin systems, in which the spin is described by the Pauli matrices

$$\sigma_x = \begin{pmatrix} 0 & 1 \\ 1 & 0 \end{pmatrix}, \sigma_y = \begin{pmatrix} 0 & -i \\ i & 0 \end{pmatrix}, \sigma_z = \begin{pmatrix} 1 & 0 \\ 0 & -1 \end{pmatrix}, \tag{6.1.1}$$

Table 1.   Common spin-orbit coupling forms in cold atomic systems.

| Type of SOC | Formula | Dimension | Experiment | Reference |
|---|---|---|---|---|
| NIST SOC | $p_x \sigma_y$ | 1D | Yes | Ref. 1 |
| Rashba or Dresselhaus SOC | $p_x \sigma_x \pm p_y \sigma_y$ | 2D | Yes | Ref. 5 |
| Raman Lattice | $M_x(x,z)\sigma_x + M_y(x,z)\sigma_y$ | 2D | Yes | Ref. 6 |
| Weyl SOC | $p_x \sigma_x + p_y \sigma_y + p_z \sigma_z$ | 3D | No | Ref. 30 |
| Angular SOC | $\frac{1}{r^2} L_z \sigma_z$ | 1D | No | Refs. 31, 32 |
| Cavity-assisted SOC | $\cos(k_0 x)\sigma_x$ | 2D | No | Ref. 33 |
| Position-dependent SOC | $[1 - \epsilon(x,y)]p_x \sigma_x - [1 + \epsilon(x,y)]p_y \sigma_y$ | 2D | No | Ref. 34 |
| SU(3) SOC | $p_x(\lambda_2 - \lambda_5 + \lambda_7) + p_y(\lambda_3 + \sqrt{3}\lambda_8)$ | 2D | No | Ref. 35 |

Here $p_{x,y,z}$ and $L_z$ represent the linear momentum and angular momentum operators respectively. While $\sigma_{x,y,z}$ describe the Pauli matrices, i.e., the generators of SU(2) group, $\lambda_i$ with $i = 1, 2, ..., 8$ represent the Gell-Mann matrices, which are the generators of SU(3) group.

i.e., the generators of SU(2) group [see Table 1]. The three-component SO coupling is just an extension of the SU(2) Pauli matrices in the $3 \times 3$ representation[4,13,19,36,37]

$$f_x = \frac{1}{\sqrt{2}} \begin{pmatrix} 0 & 1 & 0 \\ 1 & 0 & 1 \\ 0 & 1 & 0 \end{pmatrix}, f_y = \frac{1}{\sqrt{2}} \begin{pmatrix} 0 & -i & 0 \\ i & 0 & -i \\ 0 & i & 0 \end{pmatrix}, f_z = \begin{pmatrix} 1 & 0 & 0 \\ 0 & 0 & 0 \\ 0 & 0 & -1 \end{pmatrix}. \quad (6.1.2)$$

As the spin has the same symmetry, many similar properties, such as the emergence of plan-wave and stripe phases, as well as chiral magnetism induced by Rashba SO coupling, are discovered both in two- and three-component systems.[19,20] On the other hand, the three-component system with SO coupling has its own characteristics. In contrast to the two-state coupling, it introduces more internal states and establishes the coupling among three states by the $3 \times 3$ spin matrices. This makes it possible to simulating many new Hamiltonians and discovering more exotic phases and quantum phenomena.

In the experiment of Refs. 4, 12, the 1D SO coupling represented by equal Rashba and Dresselhaus contributions in three-component systems is generated together with a spin-tensor potential, which has fundamentally different rotation properties and acts only in spin-1 (or higher-spin) space.[18] This enables exploration of spin-tensor-related physics in the SO-coupled superfluid. It has been revealed that the competition between SO coupling and spin-tensor potential leads to rich single-particle energy band structures,[14] which enriches the phase diagram and make it more accessible and easy to understand many important scientific problems, such as itinerant magnetism,[4] tricriticalities[15] and quantum phase transitions.[4,15,18] In addition, there exists a double maxon-roton structure in the Bogoliubov-excitation spectrum, which is attributed to the three band minima of the SO-coupled spin-1 BEC, and absent in a spin-1/2 system.[18] The spin-1 SO coupled Bose gases also offer unique insight into the interplay between competing orders, which are ubiquitous in strongly correlated systems. It is found that the SO coupled spin-1 Bose gas supports an translational-symmetry-broken ferronematic phase,[16] which is distinct from the conventional translational-symmetry-conserved ferronematic phase occurred in higher spin systems, such as spin-3 chromium Bose condensate[38] or dipolar fermions,[39] in the absence of SO coupling.

The experimental feasibility of realizing the generalized 2D Rashba-Dresselhaus-type SO coupling in a three-component system is also explored by using stimulated Raman couplings between Zeeman sublevels of the ground state of alkali-metal atoms.[13] In contrast to the spin-1/2 case, a significant difference of the Rashba-Dresselhaus-type SO coupling in three-component systems is that there exists an extra branch of the energy band with a flat dispersion around zero momentum. The formation of this branch leads to interesting phenomena, such as the possibility to have a negative refraction at a potential step, characterized by a larger amplitude as compared to the spin-1/2 case.[40–42]

Several recent theoretical proposals also focus on systems with angular SO coupling.[31,32,43,44] In contrast to the usual discussed manner of coupling by which the spin is coupled to the linear momentum, the angular SO coupling establishes the coupling between the spin and orbit angular momentum. With the presence of angular SO coupling, a many-body Rabi oscillation between two different striped phases is observed in three-component Bose gases by a sudden quench of the quadratic Zeeman shift.[29]

## 1.B. *SU(3) spin-orbit coupling*

Besides the SU(2) spin described by the $3 \times 3$ Pauli matrices, a three-component system allows designing more complicated forms of SO coupling,[10,35,45-48] in which the description of spin requires the Gell-Mann matrices,

$$\lambda_1 = \begin{pmatrix} 0 & 1 & 0 \\ 1 & 0 & 0 \\ 0 & 0 & 0 \end{pmatrix}, \quad \lambda_2 = \begin{pmatrix} 0 & -i & 0 \\ i & 0 & 0 \\ 0 & 0 & 0 \end{pmatrix}, \quad \lambda_3 = \begin{pmatrix} 1 & 0 & 0 \\ 0 & -1 & 0 \\ 0 & 0 & 0 \end{pmatrix},$$

$$\lambda_4 = \begin{pmatrix} 0 & 0 & 1 \\ 0 & 0 & 0 \\ 1 & 0 & 0 \end{pmatrix}, \quad \lambda_5 = \begin{pmatrix} 0 & 0 & -i \\ 0 & 0 & 0 \\ i & 0 & 0 \end{pmatrix},$$

$$\lambda_6 = \begin{pmatrix} 0 & 0 & 0 \\ 0 & 0 & 1 \\ 0 & 1 & 0 \end{pmatrix}, \quad \lambda_7 = \begin{pmatrix} 0 & 0 & 0 \\ 0 & 0 & -i \\ 0 & i & 0 \end{pmatrix}, \quad \lambda_8 = \frac{1}{\sqrt{3}} \begin{pmatrix} 1 & 0 & 0 \\ 0 & 1 & 0 \\ 0 & 0 & -2 \end{pmatrix},$$

$$(6.1.3)$$

i.e., the generators of SU(3) group.[49] This goes beyond the traditional electron systems in condensed matter physics, in which the spin is just described by the Pauli matrices. Recalling the $3 \times 3$ Pauli matrices in describing the coupling among the three hyperfine spin states, it is found that only the states $|1\rangle$ and $|0\rangle$ (or $|0\rangle$ and $|-1\rangle$) can be directly coupled as shown in Fig. 6.1.1(a). The Pauli matrices with SU(2) symmetry is incapable of describing the direct coupling between the states $|1\rangle$ and $|-1\rangle$. A complete description of the coupling between the three spin states requires the SU(3) Gell-Mann matrices as shown in Fig. 6.1.1(b). The SU(3) SO coupling works with different algebraic structure, geometry and topology from the familiar spin case, thus will bring potentially more interesting physics never happened in traditional condensed matter or any other matter. The SO coupling with new symmetry of the spin will bring more exotic and nontrivial many-body phenomena. As the gauge group becomes larger, it is possible to contain more subgroups which have nontrivial structures. Therefore, more kinds of spontaneous symmetry breaking scenarios will occur in a interaction many-body system.

Experimental schemes for realizing a SU(3) SO coupling have been preliminarily discussed.[10,35,45-48] It is suggested that an SU(3) SO coupling can be readily realized using the very flexible N-pod scheme, when the detuning from resonance induced by quadratic Zeeman shift is eliminated.[9,13,45] It is also proposed to generate an SU(3) SO coupling by using a so-called

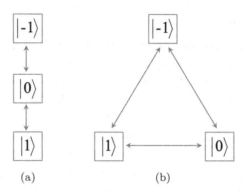

**Fig. 6.1.1.** Possible coupling between the hyperfine states in three-component BECs with (a) SU(2) spin-orbit coupling and (b) SU(3) spin-orbit coupling. The SU(2) Pauli matrices can only directly couple the states $|1\rangle$ and $|0\rangle$ (or $|0\rangle$ and $|-1\rangle$)), but is incapable of describing the direct coupling between the states $|1\rangle$ and $|-1\rangle$. A complete description of the coupling between the three spin states requires the SU(3) spin described in terms of the Gell-Mann matrices.

two-tripod scheme, in which two tripod configurations share a common ground-state level and only one-photon resonant transition is utilized.[46] Besides, it is also possible to realize an SU(N) SO coupling in the cold atomic systems.[35,45]

Theoretical interest in SU(3) SO coupled systems have focused on the topological properties and spin textures.[35,45,48] An exciting finding is that even a homogeneous SU(3) field on a simple square lattice can lead nontrivial topological states.[35] This is in distinct contrast to the SU(2) systems, which always have trivial topology. By consider a strip geometry with SU(3) SO coupling, one can observe three gapless topological edge states, implying the existence of an SU(3) topological insulator.[35] The properties of the SU(3) SO system in the strongly may-body interacting regime is also discussed. A quantum order states with periodicity controlled spiral spin textures is predicted, which is distinct from the SU(2) system due to the coexistence of the ferromagnetic and spiral orders.[45]

It should be noted that the existing works on SU(3) SO coupling mainly focus on the properties in the insulator phase, while there is a lack of investigation in the superfluid case. It has been shown that an SU(2) SO coupling can significantly change the superfluid properties, such as the spontaneous emergence of spin stripe,[19,50–53] vortex lattice[22,54–56] and quantum quasicrystal.[57] An interesting question is how does the SU(3) SO coupling influence the properties of a superfluid?

Here we review our recent work on SU(3) SO coupled Bose gases, in which we mainly focus on the superfluid properties influenced by SO coupling. We first establish a model to simulate an SU(3) SO coupled cold atom system. Compared with the SU(2) spin case, we analyze the single-particle energy spectrum with the presence of SU(3) SO coupling. Based on this we predict the possible cases of multi-particle occupation. In the Gross-Pitaevskii mean-field approximation, we obtained the zero-temperature ground-state phase diagram.

As is known to all, quantized vortex is a remarkable feature of superfluid. We specifically discuss the influences of the SU(3) SO coupling on the vortex configurations. A surprising finding is that a SU(3) SO coupling allows the existence of stable double-quantum spin vortex, which is usually considered to be topologically unstable and has never been discovered in the experiments.[58] By means of a variational approach, we discuss the change of the phase requirements caused by SO coupling, then predict some possible vortex configurations. According to numerical simulation, a stable lattice composed of double-quantum spin vortices is observed in the ground states, which confirm our prediction. The discovered double-quantum spin vortex in SU(3) SO coupled systems add a new member in the family of topological defects, thus will attract wide interest in range from condensed matter physics to ultra-cold quantum gases and magnetic materials. Finally, we consider the experimental relevance, and suggest how to realized such an SU(3) SO coupling.

## 6.2. Single-Particle Spectrum

We consider an SU(3) SO coupling with the form

$$\mathcal{V}_{so} = \kappa \boldsymbol{\lambda} \cdot \mathbf{p}, \qquad (6.2.1)$$

where $\boldsymbol{\lambda} = (\lambda_x, \lambda_y)$ is spanned by the Gell-Mann matrices $\lambda^{(i)}(i = 1, ...8)$ as shown in Eq. (6.1.3), i.e., the generators of the SU(3) group,[49] and can be expressed as

$$\lambda_x = \lambda^{(1)} + \lambda^{(4)} + \lambda^{(6)}, \qquad (6.2.2a)$$

$$\lambda_y = \lambda^{(2)} - \lambda^{(5)} + \lambda^{(7)}. \qquad (6.2.2b)$$

The vector $\mathbf{p} = (p_x, p_y)$ represents the 2D momentum, and the SO coupling strength is denoted as $\kappa$. In contrast to the famous SU(2) Rashba SO coupling in a three-component systems,[13] where only the hyperfine spin states $\Psi_1(\mathbf{r})$ and $\Psi_0(\mathbf{r})$ (or $\Psi_0(\mathbf{r})$ and $\Psi_{-1}(\mathbf{r})$) are directly coupled, the SU(3)

SO coupling considered here involves the direct coupling between $\Psi_1(\mathbf{r})$ and $\Psi_{-1}(\mathbf{r})$, thus supports all pairwise couplings between the three states $\Psi_1(\mathbf{r})$, $\Psi_0(\mathbf{r})$ and $\Psi_{-1}(\mathbf{r})$.

The single-particle Hamiltonian involving this kind of SO coupling then is given as

$$H = -\frac{\hbar^2\nabla^2}{2M}\mathbb{1} + \hbar\kappa \begin{pmatrix} 0 & -i\partial_x - \partial_y & -i\partial_x + \partial_y \\ -i\partial_x + \partial_y & 0 & -i\partial_x - \partial_y \\ -i\partial_x - \partial_y & -i\partial_x + \partial_y & 0 \end{pmatrix} \quad (6.2.3)$$

with $\mathbb{1}$ being the $3 \times 3$ unit matrix. The Hamiltonian with such SU(3) SO coupling may be realized using an Raman laser dressing technology as similar as those used in an SU(2) case.[1,5] A detail discussion on the experimental relevance will be given in Section 6.5.

By diagonalizing the Hamiltonian in Eq. (6.2.3), one can obtain the single-particle dispersion relation. It is found that there exist three discrete minima of the energy bands residing on the vertices of an equilateral triangle in the $k_x$-$k_y$ plane of momentum space, as shown in Figs. 6.2.1(a) and 6.2.1(b). This kind of single-particle dispersion relation is distinctly different from those of conventional SU(2) SO coupling, such as the NIST or the Rashba types. In the NIST type of SO coupling, the single-particle spectrum

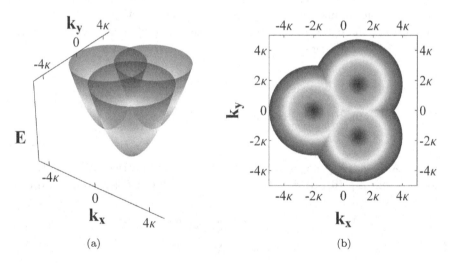

(a)                                     (b)

**Fig. 6.2.1.** Single-particle energy spectrum with SU(3) spin-orbit coupling. (a) A triple-well dispersion relation is observed with three discrete minima of the single-particle energy band locating on the vertices of an equilateral triangle in the $k_x$-$k_y$ plane. (b) Projection of the first energy band on a 2D plane. Units with $\hbar = M = 1$ are used for simplicity.

has either a single or two minima, depending on the strength of the Raman coupling,[1] while in the case of Rashba type the minima in the energy bands locate on a continuous ring in momentum space and form a Mexican-hat band structure.[59]

The single-particle spectrum can provide us some useful informations about the many-body ground states of Bose condensates. When a number of particles are condensated in the zero-temperature ground states, it is preferred for them to occupy the energy minima of the single-particle bands. From this point of view, it is reasonable to predict that there exist three kinds of possible many-body ground states, which respectively corresponds to three forms of particle occupations. The choices among the three minima are made by the interatomic interactions. When only one of the energy minima is occupied, the many-body ground state is expect to be a three-fold degenerate plane-wave state, which is in contrast with the double-fold degenerate plane-wave state discovered in a system with SU(2) SO coupling.[60] When two of the energy minima are occupied, the ground state is expect to be a three-fold degenerate stripe state with time-reversal symmetry broken, where the two minima are randomly chosen. When all the three energy minima are occupied, the ground state is expect to be a spontaneous lattice state with spatial translational symmetry broken in the two directions of a 2D plane.

## 6.3. Many-Body Phase Diagram

The many-body Hamiltonian for a three-component BECs with SU(3) SO coupling can be written in the Gross-Pitaevskii mean-field approximation as

$$\mathcal{H} = \int d\mathbf{r} \left[ \mathbf{\Psi}^\dagger \left( -\frac{\hbar^2 \mathbf{\nabla}^2}{2m} + \mathcal{V}_{\text{so}} \right) \mathbf{\Psi} + \frac{c_0}{2} n^2 + \frac{c_2}{2} |\mathbf{F}|^2 \right], \qquad (6.3.1)$$

where the three-component order parameters $\mathbf{\Psi} = [\Psi_1(\mathbf{r}), \Psi_0(\mathbf{r}), \Psi_{-1}(\mathbf{r})]^\top$ are normalized with $\int d\mathbf{r} \mathbf{\Psi}^\dagger \mathbf{\Psi} = N$. Here $N$ is the total particle number. $n$ describes the total particle density of the three components, and is given as $n = \sum_{m=1,0,-1} \Psi_m^*(\mathbf{r})\Psi_m(\mathbf{r})$. The spin density vector $\mathbf{F} = (F_x, F_y, F_z)$ is defined by $F_\nu(\mathbf{r}) = \mathbf{\Psi}^\dagger f_\nu \mathbf{\Psi}$ with $\mathbf{f} = (f_x, f_y, f_z)$ being the spin-1 Pauli matrix vectors given in the irreducible representation as shown in Eq. (6.1.2). Here we only consider two types of hard-core interactions involving the density-density and spin-exchange ones, whose strengths are described by the parameters $c_0$ and $c_2$ respectively.

### 3.A.  Variational analysis

When the interatomic interactions are weak, a variational analysis is effective by assuming that the many-body ground state is written as an linear superposition of the three degenerate single-particle ground states. We suppose the many-body ground state is given with a trial wave function $\Psi = \alpha_1 \Psi_1 + \alpha_2 \Psi_2 + \alpha_3 \Psi_3$, where

$$\Psi_1 = \frac{1}{\sqrt{3}} \begin{pmatrix} 1 \\ 1 \\ 1 \end{pmatrix} e^{-i2\kappa x}, \tag{6.3.2a}$$

$$\Psi_2 = \frac{1}{\sqrt{3}} \begin{pmatrix} e^{-i\frac{\pi}{3}} \\ e^{i\frac{\pi}{3}} \\ e^{i\pi} \end{pmatrix} e^{i\kappa(x-\sqrt{3}y)}, \tag{6.3.2b}$$

$$\Psi_3 = \frac{1}{\sqrt{3}} \begin{pmatrix} e^{i\frac{\pi}{3}} \\ e^{-i\frac{\pi}{3}} \\ e^{i\pi} \end{pmatrix} e^{i\kappa(x+\sqrt{3}y)}, \tag{6.3.2c}$$

correspond to the many-body states with all particles condensing on one of the three minima of the single-particle energy bands, and $\alpha_{i=1,2,3}$ are the expansion coefficients. Substituting Eqs. (6.3.2a)–(6.3.2c) into the interaction energy functional

$$E = \int d\mathbf{r} \left( \frac{c_0}{2} n^2 + \frac{c_2}{2} |\mathbf{F}|^2 \right), \tag{6.3.3}$$

one obtains the energy expression written in terms of the variational parameters $|\alpha_i|^2$ as

$$\frac{E}{N} = \left( \frac{c_0}{2} + \frac{4c_2}{9} \right) \bar{n} - \frac{7c_2}{9\bar{n}} \sum_{i \neq j} |\alpha_i|^2 |\alpha_j|^2, \tag{6.3.4}$$

where $\bar{n} = |\alpha_1|^2 + |\alpha_2|^2 + |\alpha_3|^2$ describes the mean particle density. By minimizing the interaction energy with respect to the variation of $|\alpha_i|^2$, one can obtain the many-body ground states. It is found that the spin-exchange interaction plays an important role in determining the ground-state phase diagram.

When the spin-exchange interaction is negative, i.e., $c_2 < 0$, the system favors a state with either $|\alpha_1|^2 = \bar{n}$, $|\alpha_2|^2 = |\alpha_3|^2 = 0$, or $|\alpha_2|^2 = \bar{n}$, $|\alpha_1|^2 = |\alpha_3|^2 = 0$, or $|\alpha_3|^2 = \bar{n}$, $|\alpha_1|^2 = |\alpha_2|^2 = 0$. In this case, all the particles in the ground state only occupy one of the three minima in the single-particle spectrum and form a time-reversal symmetry broken plane-wave phase. At the same time, the choice among the three minima is random, thus we in fact obtained a three-fold degenerate many-body states. This is in stark contrast to the case of SU(2) SO coupling, where the plane-wave state is double-fold degenerate.[60]

When the spin-exchange interaction is positive, i.e., $c_2 > 0$, the system prefers a state with $|\alpha_1|^2 = |\alpha_2|^2 = |\alpha_3|^2 = \bar{n}/3$. In this case, the many-body ground state is an equally weighted superposition of the three degenerate states $\Psi_1$, $\Psi_2$ and $\Psi_3$ in the three minima of the energy spectrum. The interference of these three states in space lead to the formation of a triangular lattice phase with spatial translational symmetry broken.

It is noticed that under the present consideration of weak interatomic interactions, only the single-well or triple-well occupation is chosen. The many-body states with double-well occupation are not favored. However, the variational wave function Eqs. (8a)–(8c) is a good starting point only when the interatomic interactions are weak and the SO coupling is strong enough to dominate the chemical potential. For the case with weak SO coupling but strong interatomic interactions, the many-body ground state must be determined relying on numerical simulations. In such a situation, we find a double-well occupied stripe state with two random minima are chosen for $c_2 \gg \kappa^2$, which will be discussed latter.

### 3.B. *Numerical simulation*

The many-body ground states can be numerically obtained by minimizing the energy functional associated with the Hamiltonian Eq. (6.3.1) via the imaginary time evolution method. in the weak interaction region with $c_2 \lesssim \kappa^2$, it is found that the numerical results are consistent with the analytical analysis discussed above. The ground state of the SU(3) SO coupled Bose gases prefers a single-well occupied plane-wave state or a triple-well occupied lattice state depending on the sign of the spin-exchange interaction strength $c_2$.

When $c_2 < 0$, the three components are miscible and all the particles occupy one of the three minima of the triple-well dispersion by spontaneous symmetry breaking. In this case, the system forms a plane-wave state with spatial transitional symmetry preserved but time-reversal symmetry broken [See Figs. 6.3.1(a)–6.3.1(b)]. This plane-wave state is three-fold degenerate instead of doubly degenerate in the SU(2) case.[52,60] It has been shown

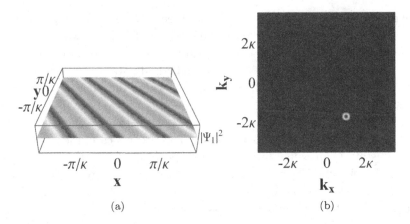

**Fig. 6.3.1.** Three-fold degenerate plane-wave phase with time-reversal symmetry broken. (a) The density and phase distributions of $\Psi_1$ represented by heights and colors respectively. (b) The corresponding momentum distributions. The spin interaction is chosen as $c_2 < 0$.

that in the case of double-well dispersion induced by SU(2) SO coupling, a resonant quantum oscillation can be observed under external perturbation which drives resonant couplings between the two degenerate many-body states. An interesting question for future study is what nontrivial quantum many-body dynamics can be expected on the triple-well oscillation with the SU(3) SO coupling. This may offer us insights into nonequilibrium quantum dynamical phase transitions.[60]

When $c_2 > 0$, the three components are immiscible with all the three minima of the triple-well dispersion are equally weighted occupied. The components are arranged as an interlaced triangular vortex lattice with the spatial transitional symmetry broken [See Figs. 6.3.2(a)–6.3.2(d)]. This phase simultaneously possesses a diagonal long-range order of solid and an off-diagonal long-range order of superfluid, thus can be considered as a supersolid phase.

The supersolidity of the stripe phase with spontaneously broken translational invariance in an SU(2) SO coupled Bose gases recent has been theoretically predicted[19,51–53] and experimentally observed.[50] It should be indicated that actually the stripe state is spatially periodic only in one direction and can be considered as superfluid nematic liquid crystals. In contrast, the vortex lattice state discovered here possesses spatially translational symmetry broken in both directions of the two-dimensional plane, thus is much closer to the general concept of supersolid.[61]

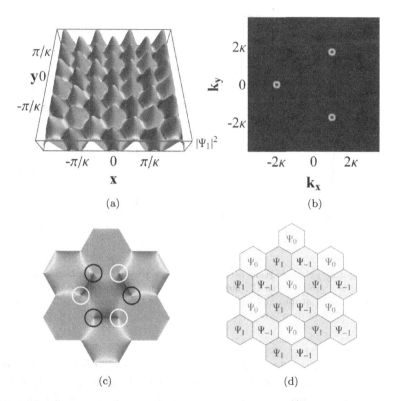

**Fig. 6.3.2.** Lattice phase with nontrivial topology. (a) The density and phase distributions of $\Psi_1$ represented by heights and colors respectively. (b) The corresponding momentum distributions. (c) The phase distribution within one unit cell. The positions of vortices and antivortices are signed by white and black circles respectively. (d) The structural schematic drawing of the phase separation between the three components. The spin interaction is weak and chosen as $c_2 > 0$.

Recalling the case of SU(2) SO coupling that only the plane-wave state and stripe state can be stabilized in a homogeneous system.[19,52,62] Although the Rashba SO coupling provides infinite degenerate minima in the single-particle spectrum, a many-body ground state condensed in one or two points in momentum space is always energetically favorable due to the presence of spin-exchange interaction.[19] As a result, a lattice state with the condensates occupying three or more momentum points for SU(2) SO coupling is unstable, unless a strong harmonic trap is introduced.[21,54,56] Here we reveal that a stable lattice phase with triple-well occupation can exist in a uniform SU(3) SO coupled BEC. In addition, this lattice phase is

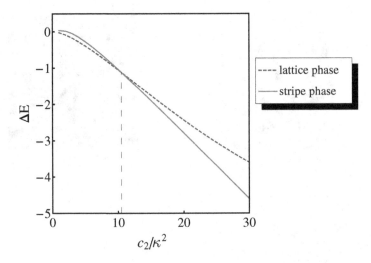

**Fig. 6.3.3.** Energy comparison between the stripe phase and the lattice phase. The energy difference $\Delta E$ between the numerical simulation and the variational calculation are shown by solid (stripe) and dashed (lattice) lines. The density-density interatomic interaction is fixed at $c_0 = 20\kappa^2$.

topologically nontrivial and embedded by vortices and antivortices as shown in Fig. 6.3.2(c). More details on the structure of vortices as well as their unique spin configurations will be investigated later. For strong antiferromagnetic spin interaction with $c_2 \gg \kappa^2$, numerical simulations indicate that the many-body ground state prefers a stripe phase with two of the three minima occupied in the momentum space. We take the states with two or three minima occupied in the momentum space as trial wave functions, and perform imaginary time evolution to find their respective optimized ground state energy. Figure 6.3.3 summarizes the results and presents the energy comparison with different values of interatomic interactions. It is found that while the lattice phase is energy favored with weak antiferromagnetic spin interaction strength $c_2$, the stripe phase will has lower energy than the lattice phase when the interatomic interaction $c_2$ exceeds a critical value. At the same time, the two minima are randomly chosen with spontaneous symmetry breaking. As the particles has finite momentum in vertical direction of the stripe [See Fig. 6.3.4(c)], both the spatial translational and time-reversal symmetries are broken [See Figs. 6.3.4(a)–6.3.4(b)]. This is distinct from the conventional stripe phase induced by SU(2) SO coupling, where the time-reversal symmetry is preserved.[19,50–53]

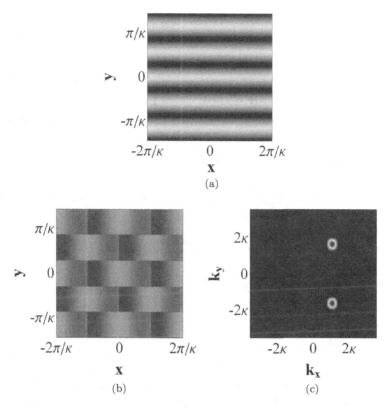

**Fig. 6.3.4.** Stripe phase with time-reversal symmetry breaking. The ground-state density, phase and momentum distributions with the parameters $c_2 = 20\kappa^2$ and $c_0 = 10c_2$ are shown in (a), (b) and (c).

## 6.4. Vortices with SU(3) Spin-orbit Coupling

### 4.A. *Phase requirement*

Vortex plays a critical role on the understanding of superfluidity. In order to demonstrate the influence of the SU(3) SO coupling on the properties of superfluid, we discuss the change of vortex configurations caused by the SO coupling. The vortex configuration in a spinor BEC directly depends on the phase relation between the three hyperfine spin states. Thus it is necessary to make clear the phase requirement of the vortex configuration in the presence of SO coupling. We assume that the spinor order parameter of a vortex in the polar coordinate $(r, \theta)$ is described as

$$\psi_j(r, \theta) = \phi_j e^{i w_j \theta + \alpha_j}, \tag{6.4.1}$$

where $j = 0, \pm 1$ and $\phi_j \geq 0$.

### 4.A.1. *Without spin-orbit coupling*

Without SO coupling, the phase requirement and possible vortex configurations have been previously studied.[37,63,64] In the Hamiltonian of a spinor BECs, only the phase-dependent terms can influence the vortex configurations, and are written as

$$
H^{\text{phase}} = E_{\text{kin}}^{\text{phase}} + E_{\text{int}}^{\text{phase}}
$$

$$
= -\frac{1}{2} \int \mathbf{\Psi}^* \frac{1}{r^2} \frac{\partial^2}{\partial \theta^2} \mathbf{\Psi} d\mathbf{r} + 2c_2 \int \Re(\psi_1^* \psi_{-1}^* \psi_0^2) d\mathbf{r}, \qquad (6.4.2)
$$

where the first term results from the kinetic energy and the second from the spin-exchange interaction. Substituting Eq. (6.4.1) into (6.4.2), one obtains

$$
E_{\text{kin}}^{\text{phase}} = \sum_{j=1,0,-1} \mathrm{w}_j^2 \int \frac{\pi \phi_j^2}{r} dr, \qquad (6.4.3)
$$

$$
E_{\text{int}}^{\text{phase}} = 2c_2 \int \phi_1 \phi_{-1} \phi_0^2 r \cos\left[(\mathrm{w}_1 - 2\mathrm{w}_0 + \mathrm{w}_{-1})\theta + (\alpha_1 - 2\alpha_0 + \alpha_{-1})\right] dr d\theta. \qquad (6.4.4)
$$

From Eq. (6.4.3), it is easy to read that the system prefers small winding numbers energetically. Moreover, from Eq. (6.4.4) the energy minimization requires the winding number and phase satisfy the following relations

$$
\mathrm{w}_1 - 2\mathrm{w}_0 + \mathrm{w}_{-1} = 0, \qquad (6.4.5a)
$$

$$
\alpha_1 - 2\alpha_0 + \alpha_{-1} = n\pi, \qquad (6.4.5b)
$$

where $n$ is even for $c_2 < 0$ and odd for $c_2 > 0$. Under the phase requirement of Eq. (6.4.5a), the following types of winding combination, such as $\langle \pm 1, \times, 0 \rangle$, $\langle 0, \times, \pm 1 \rangle$, $\langle \pm 1, 0, \mp 1 \rangle$, $\langle \pm 1, \pm 1, \pm 1 \rangle$, $\langle \pm 2, \pm 1, 0 \rangle$ and $\langle 0, \pm 1, \pm 2 \rangle$ are allowed in a spinor BEC, where the symbol "$\times$" denotes the absence of the $\Psi_0$ component. This result is consistent with the existing investigations.

### 4.A.2. *With SU(2) spin-orbit coupling*

For the case of SU(2) SO coupling, we take the Rashba type as an example. As the SO coupling term in the Hamiltonian is phase-dependent, thus will significantly change the phase requirement. We write the Rashba SO coupling term of the Hamiltonian as

$$E_{\text{soc}} = \int \kappa \psi^\dagger \begin{pmatrix} 0 & -i\partial_x - \partial_y & 0 \\ -i\partial_x + \partial_y & 0 & -i\partial_x - \partial_y \\ 0 & -i\partial_x + \partial_y & 0 \end{pmatrix} \psi \, d\mathbf{r}, \qquad (6.4.6)$$

where $\psi = [\psi_1, \psi_0, \psi_{-1}]^\mathsf{T}$. Substituting the trial spinor order parameters Eq. (6.4.1) into the Hamiltonian Eq. (6.4.6), one can obtain

$$\begin{aligned}
E_{\text{soc}} = \int dr d\theta \Big[ &(\phi_0 r \partial_r \phi_1 - \text{w}_1 \phi_0 \phi_1) e^{i[(\text{w}_1 - \text{w}_0 + 1)\theta + (\alpha_1 - \alpha_0 - \frac{\pi}{2})]} \\
&- (\phi_1 r \partial_r \phi_0 + \text{w}_0 \phi_1 \phi_0) e^{-i[(\text{w}_1 - \text{w}_0 + 1)\theta + (\alpha_1 - \alpha_0 - \frac{\pi}{2})]} \\
&+ (\phi_0 r \partial_r \phi_{-1} + \text{w}_{-1} \phi_0 \phi_{-1}) e^{i[(\text{w}_{-1} - \text{w}_0 - 1)\theta + (\alpha_{-1} - \alpha_0 - \frac{\pi}{2})]} \\
&- (\phi_{-1} r \partial_r \phi_0 - \text{w}_0 \phi_{-1} \phi_0) e^{-i[(\text{w}_{-1} - \text{w}_0 - 1)\theta + (\alpha_{-1} - \alpha_0 - \frac{\pi}{2})]} \Big]. \quad (6.4.7)
\end{aligned}$$

From Eq. (6.4.7), in order to minimize the SO coupling energy, it is preferred that the winding number and phase satisfy the following relations

$$\text{w}_1 - \text{w}_0 + 1 = 0, \qquad (6.4.8\text{a})$$

$$\text{w}_{-1} - \text{w}_0 - 1 = 0, \qquad (6.4.8\text{b})$$

$$\alpha_1 - \alpha_0 - \frac{\pi}{2} = m\pi, \qquad (6.4.8\text{c})$$

$$\alpha_{-1} - \alpha_0 - \frac{\pi}{2} = n\pi. \qquad (6.4.8\text{d})$$

After the phase requirement Eq. (6.4.8) is satisfied, the SO coupling energy can be rewritten as

$$\begin{aligned}
E_{\text{soc}} = 2\pi \int [\phi_0 r \partial_r \phi_1 - \phi_1 r \partial_r \phi_0 - (\text{w}_1 + \text{w}_0)\phi_0 \phi_1] dr \cos m\pi \\
+ 2\pi \int [\phi_0 r \partial_r \phi_{-1} - \phi_{-1} r \partial_r \phi_0 + (\text{w}_{-1} + \text{w}_0)\phi_0 \phi_{-1}] dr \cos n\pi,
\end{aligned}$$

$$(6.4.9)$$

where $m$ and $n$ are odd or even, which can be determined by minimizing the energy expressed in Eq. (6.4.9). It is found that the SU(2) SO coupling does not violate the ordinary requirement on the winding combination in Eq. (6.4.5a), but introduces further requirements in Eqs. (6.4.8a)–(6.4.8b). As a result, vortex configurations with $\langle -1, 0, 1 \rangle$, $\langle -2, -1, 0 \rangle$ and $\langle 0, 1, 2 \rangle$ are still allowed, but some winding combinations such as $\langle \pm 1, \pm 1, \pm 1 \rangle$, $\langle \pm 1, \times, 0 \rangle$, $\langle 0, \times, \pm 1 \rangle$, $\langle 1, 0, -1 \rangle$, $\langle 2, 1, 0 \rangle$ and $\langle 0, -1, -2 \rangle$ are forbidden. It is noticed that while $\langle -1, 0, 1 \rangle$, $\langle -2, -1, 0 \rangle$ and $\langle 0, 1, 2 \rangle$ are allowed, $\langle 1, 0, -1 \rangle$, $\langle 2, 1, 0 \rangle$ and $\langle 0, -1, -2 \rangle$ are forbidden. This implies that the chiral symmetry of the system is broken, and may lead to interesting chiral spin textures.[20]

4.A.3. *With SU(3) spin-orbit coupling*

The effective Hamiltonian of the SU(3) SO coupling can be written as

$$E_{\text{soc}} = \int \kappa \psi^\dagger \begin{pmatrix} 0 & -i\partial_x - \partial_y & -i\partial_x + \partial_y \\ -i\partial_x + \partial_y & 0 & -i\partial_x - \partial_y \\ -i\partial_x - \partial_y & -i\partial_x + \partial_y & 0 \end{pmatrix} \psi d\mathbf{r}. \qquad (6.4.10)$$

Substituting the trial spinor order parameters Eq. (6.4.1) into (6.4.10), we obtain

$$\begin{aligned}
E_{\text{soc}} = \int dr d\theta \Big[ & (\phi_0 r \partial_r \phi_1 - w_1 \phi_0 \phi_1) e^{i[(w_1 - w_0 + 1)\theta + (\alpha_1 - \alpha_0 - \frac{\pi}{2})]} \\
& - (\phi_1 r \partial_r \phi_0 + w_0 \phi_1 \phi_0) e^{-i[(w_1 - w_0 + 1)\theta + (\alpha_1 - \alpha_0 - \frac{\pi}{2})]} \\
& + (\phi_0 r \partial_r \phi_{-1} + w_{-1} \phi_0 \phi_{-1}) e^{i[(w_{-1} - w_0 - 1)\theta + (\alpha_{-1} - \alpha_0 - \frac{\pi}{2})]} \\
& - (\phi_{-1} r \partial_r \phi_0 - w_0 \phi_{-1} \phi_0) e^{-i[(w_{-1} - w_0 - 1)\theta + (\alpha_{-1} - \alpha_0 - \frac{\pi}{2})]} \\
& + (\phi_{-1} r \partial_r \phi_1 + w_1 \phi_{-1} \phi_1) e^{i[(w_1 - w_{-1} - 1)\theta + (\alpha_1 - \alpha_{-1} - \frac{\pi}{2})]} \\
& - (\phi_1 r \partial_r \phi_{-1} - w_{-1} \phi_1 \phi_{-1}) e^{-i[(w_1 - w_{-1} - 1)\theta + (\alpha_1 - \alpha_{-1} - \frac{\pi}{2})]} \Big].
\end{aligned}$$

$$(6.4.11)$$

By minimizing the SO coupling energy, it is found that the winding number and phase satisfy the following relations

$$w_1 - w_0 + 1 = 0, \qquad (6.4.12a)$$

$$w_{-1} - w_0 - 1 = 0, \qquad (6.4.12b)$$

$$w_1 - w_{-1} - 1 = 0, \qquad (6.4.12c)$$

$$\alpha_1 - \alpha_0 - \frac{\pi}{2} = m\pi, \qquad (6.4.12d)$$

$$\alpha_{-1} - \alpha_0 - \frac{\pi}{2} = n\pi, \qquad (6.4.12e)$$

$$\alpha_1 - \alpha_{-1} - \frac{\pi}{2} = l\pi. \qquad (6.4.12f)$$

After the phase requirement Eq. (6.4.12) is satisfied, the SO coupling energy can be rewritten as

$$\begin{aligned}
E_{\text{soc}} = & \, 2\pi \int [\phi_0 r \partial_r \phi_1 - \phi_1 r \partial_r \phi_0 - (w_1 + w_0)\phi_0 \phi_1] dr \cos m\pi \\
& + 2\pi \int [\phi_0 r \partial_r \phi_{-1} - \phi_{-1} r \partial_r \phi_0 + (w_{-1} + w_0)\phi_0 \phi_{-1}] dr \cos n\pi \\
& + 2\pi \int [\phi_{-1} r \partial_r \phi_1 - \phi_1 r \partial_r \phi_{-1} + (w_1 + w_{-1})\phi_{-1} \phi_1] dr \cos l\pi,
\end{aligned}$$

$$(6.4.13)$$

where $m$, $n$ and $l$ are odd or even, which can be determined by minimizing the energy expressed in Eq. (6.4.13). However, it is obvious that the three winding requirements Eqs. (6.4.12a)–(6.4.12c) can not be satisfied simultaneously. As a result, the SU(3) SO coupling may choose two out of the three winding requirements for the following three cases:

Case I:

$$w_1 - w_0 + 1 = 0, \tag{6.4.14a}$$

$$w_{-1} - w_0 - 1 = 0, \tag{6.4.14b}$$

$$\alpha_1 - \alpha_0 - \frac{\pi}{2} = m\pi, \tag{6.4.14c}$$

$$\alpha_{-1} - \alpha_0 - \frac{\pi}{2} = n\pi. \tag{6.4.14d}$$

Case II:

$$w_1 - w_0 + 1 = 0, \tag{6.4.15a}$$

$$w_1 - w_{-1} - 1 = 0, \tag{6.4.15b}$$

$$\alpha_1 - \alpha_0 - \frac{\pi}{2} = m\pi, \tag{6.4.15c}$$

$$\alpha_1 - \alpha_{-1} - \frac{\pi}{2} = l\pi. \tag{6.4.15d}$$

Case III:

$$w_{-1} - w_0 - 1 = 0, \tag{6.4.16a}$$

$$w_1 - w_{-1} - 1 = 0, \tag{6.4.16b}$$

$$\alpha_{-1} - \alpha_0 - \frac{\pi}{2} = n\pi, \tag{6.4.16c}$$

$$\alpha_1 - \alpha_{-1} - \frac{\pi}{2} = l\pi. \tag{6.4.16d}$$

For case I, while $\langle -1, 0, 1 \rangle$ is allowed, the winding combination $\langle 1, 0, -1 \rangle$ is forbidden, indicating the chiral symmetry broken. For case II and case III, one can find that the SU(3) SO coupling breaks the ordinary requirement on the winding combination in Eq. (6.4.5a), thus new winding combinations, such as $\langle 0, 1, -1 \rangle$ and $\langle 1, -1, 0 \rangle$, are possible. Given the above, the phase requirements and possible vortex configurations are summarized in Table 2.

## 4.B. *Vortex configurations*

The vortex configurations of spinor BECs can be classified according to the combination of winding numbers and the magnetization of vortex

Table 2.    Phase requirements and possible vortex configurations in spinor BECs.

| Type of SOC | Phase requirements | Possible vortex configurations |
|---|---|---|
| Without SOC | $w_1 - 2w_0 + w_{-1} = 0,$ | $\langle\pm1, \times, 0\rangle$, $\langle0, \times, \pm1\rangle$, $\langle\pm1, 0, \mp1\rangle$, |
| | $\alpha_1 - 2\alpha_0 + \alpha_{-1} = n\pi.$ | $\langle\pm1, \pm1, \pm1\rangle$, $\langle\pm2, \pm1, 0\rangle$, $\langle0, \pm1, \pm2\rangle.$ |
| Rashba SOC | $w_1 - w_0 + 1 = 0,$ | $\langle-1, 0, 1\rangle$, $\langle-2, -1, 0\rangle$, $\langle0, 1, 2\rangle.$ |
| | $w_{-1} - w_0 - 1 = 0,$ | |
| | $\alpha_1 - \alpha_0 - \frac{\pi}{2} = m\pi,$ | |
| | $\alpha_{-1} - \alpha_0 - \frac{\pi}{2} = n\pi.$ | |
| SU(3) SOC[*] | $w_1 - w_0 + 1 = 0,$ | $\langle-1, 0, 1\rangle$, $\langle0, 1, -1\rangle$, $\langle1, -1, 0\rangle.$ |
| | $w_{-1} - w_0 - 1 = 0,$ | |
| | $w_1 - w_{-1} - 1 = 0,$ | |
| | $\alpha_1 - \alpha_0 - \frac{\pi}{2} = m\pi,$ | |
| | $\alpha_{-1} - \alpha_0 - \frac{\pi}{2} = n\pi,$ | |
| | $\alpha_1 - \alpha_{-1} - \frac{\pi}{2} = l\pi.$ | |

[*]The three winding requirements of the SU(3) spin-orbit coupling can not be satisfied simultaneously, thus two of them may be chosen.

core.[37,63,64] For example, a Mermin-Ho vortex has winding combination $\langle\pm2, \pm1, 0\rangle$ with a ferromagnetic core, where the plus and minus signs represent different chirality of the vortices,[65] and the expression of $\langle w_1, w_0, w_{-1}\rangle$ indicates that the components of $\Psi_1$, $\Psi_0$ and $\Psi_{-1}$ in the wave function acquire winding numbers of $w_1$, $w_0$ and $w_{-1}$, respectively. Using this notation, a polar-core vortex has winding combination $\langle\pm1, 0, \mp1\rangle$ with an antiferromagnetic core, and a half-quantum vortex has winding combination $\langle\pm1, \times, 0\rangle$ with a ferromagnetic core, where the symbol "$\times$" denotes the absence of the $\Psi_0$ component.

According to the discussion in Section 4.1.3, three kinds of vortex configurations with winding combination $\langle-1, 0, 1\rangle$, $\langle0, 1, -1\rangle$ and $\langle1, -1, 0\rangle$ are predicted in the SU(3) SO coupled spinor BECs. At the same time, numerical simulations indicate that vortices exist in the ground state of the lattice phase with antiferromagnetic spin interaction. Thus it is important to make a detailed analysis on the vortex configurations obtained by numerical simulations to verify the previous prediction. Figure 6.4.1 presents the vortex arrangement among the three-component condensates. It is found that there exists three kinds of vortices: one is a polar-core vortex with

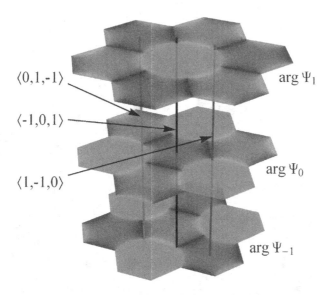

**Fig. 6.4.1.** Vortex configurations in the lattice phase. Vortex arrangement among the three components of the condensates is present, and one can identify three types of vortices, including a polar-core vortex with winding combination $\langle -1, 0, 1 \rangle$ and two ferromagnetic-core vortices with winding combinations $\langle 1, -1, 0 \rangle$ and $\langle 0, 1, -1 \rangle$.

winding combination $\langle -1, 0, 1 \rangle$, and the other two are ferromagnetic-core vortices with winding combinations $\langle 1, -1, 0 \rangle$ and $\langle 0, 1, -1 \rangle$. This is exactly consistent with the analytical prediction. Meanwhile, the vortex configurations with opposite chirality of each type, such as $\langle 1, 0, -1 \rangle$, $\langle -1, 1, 0 \rangle$ and $\langle 0, -1, 1 \rangle$, are forbidden in the present system. This can be understood by noting that the chiral symmetry is intrinsically broken in SU(3) SO coupled systems, as demonstrated in Section 4.1.3.

A surprising observation is that the two types of ferromagnetic-core vortices $\langle 1, -1, 0 \rangle$ and $\langle 0, 1, -1 \rangle$ violate the conventional phase requirement $2w_0 = w_1 + w_{-1}$ of an ordinary spinor BECs.[37,63,64] This can be understood by noting that the relative phase among different wave function components are no longer uniquely determined by the spin-exchange interaction but also affected by the SU(3) SO coupling, as qualitatively explained in Section 4.1.3. As the $\langle 1, -1, 0 \rangle$ and $\langle 0, 1, -1 \rangle$ is lacking in an ordinary spinor BECs, the interlaced arrangement of the three types of vortices forms a new class of vortex lattice which has no analogue in systems without SO coupling.

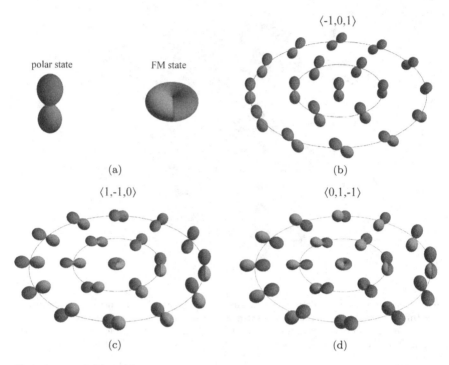

**Fig. 6.4.2.** Spherical-harmonic representation of the vortices in antiferromagnetic spinor BECs with SU(3) spin-orbit coupling. (a) Spherical-harmonic representation of the order parameters for the polar state and ferromagnetic state. (b)–(d) The surface plots of $|\Phi(\theta,\phi)|^2$ for (b) the polar-core vortex $\langle -1,0,1 \rangle$, (c) the ferromagnetic-core vortex $\langle 1,-1,0 \rangle$, and (d) the ferromagnetic-core vortex $\langle 0,1,-1 \rangle$ with the colors representing the phase of $\Phi(\theta,\phi)$. Here, $\Phi(\theta,\phi) = \sum_{m=-1}^{1} Y_{1m}(\theta,\phi)\Psi_m$ and $Y_{1m}$ is the rank-1 spherical-harmonic function.

The spherical-harmonic representation is usually used to describe the vortex structures.[37] Figure 6.4.2 shows the configurations of the three types of vortices induced by the SU(3) SO coupling with antiferromagnetic interaction. One can see that the vortices are essentially different from those usually observed in ferromagnetic spinor BECs.[37] For the polar-core vortex, the antiferromagnetic order parameter varies continuously everywhere, while for the ferromagnetic-core vortex, the magnetic order parameter acquires a singularity at the vortex core. In contrast, in the ordinary ferromagnetic spinor BECs, the ferromagnetic order parameter varies continuously everywhere for the ferromagnetic-core vortex, but has a singularity at the core for the polar-core vortex.[37]

## 4.C. Double-quantum spin vortices

In a spinor BEC, the conception of vortex is generalized. Besides the well-known mass vortex, which characterize the circulating flow of the mass current around the core, there also exists spin vortex, which characterize the circulating flow of the spin current. Spin vortex is a complex topological defect resulting from symmetry breaking, and is usually characterized by zero net mass current and quantized spin current around an unmagnetized core.[36,37,66-68] It is not only different from the magnetic vortex found in magnetic thin films,[69-71] but also from the 2D skyrmion[72,73] due to the existence of singularity in the spin textures.[74] Single-quantum spin vortex with the spin current showing one quantum of circulation has been experimentally observed in ferromagnetic spinor BECs by magnetization-sensitive phase-contrast imaging technique.[58] Multi-quantum spin vortices with $l$ ($l \geq 2$) quanta circulating spin current, however, are considered to be topologically unstable and have not been discovered yet.[58]

A particularly important finding in a SU(3) SO coupled system is that the polar-core vortex of the lattice phase has a spin current with two quanta of circulation around the unmagnetized core, hence can be identified as a double-quantum spin vortex. Figure 6.4.3 presents the transverse magnetization $F_+ = F_x + iF_y$, longitudinal magnetization $F_z$, and amplitude of the total magnetization $|\mathbf{F}|$ in the lattice phase, which are experimentally observable by magnetization-sensitive phase-contrast imaging technique.[58] From these results, one can find two distinct types of topological defects, double-quantum spin vortex (DSV) and half skyrmion (HS),[31,75] which correspond to the polar-core vortex with winding combinations $\langle -1, 0, 1 \rangle$ and the ferromagnetic-core vortex with winding combinations $\langle 1, -1, 0 \rangle$ or $\langle 0, 1, -1 \rangle$, respectively. In particular, for the double-quantum spin vortex, the core is unmagnetized and the orientation of the magnetization along a closed path surrounding the core acquires a rotation of $4\pi$. This finding indicates that a regular lattice of multi-quantum spin vortices can emerge spontaneously in antiferromagnetic spinor BECs with SU(3) SO coupling. By exploring the effect of a small but finite temperature, we confirm that the double-quantum spin vortices are robust against thermal fluctuations and hence are observable in experiments.

The emergence of spin current with two quanta of circulation can be analytically understood by expanding the wave function obtained by the variational methods around the center of a double-quantum spin vortex.

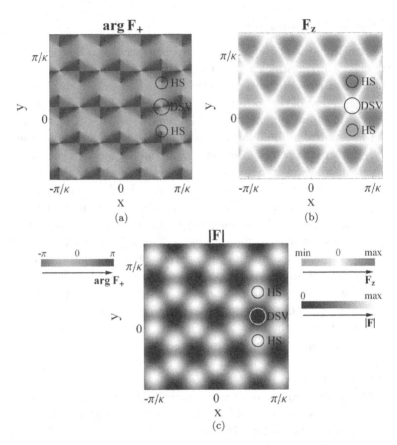

**Fig. 6.4.3.** Double-quantum spin vortex in antiferromagnetic spinor BECs with SU(3) spin-orbit coupling. (a) Spatial maps of the transverse magnetization with colors indicating the magnetization orientation. (b) Longitudinal magnetization. (c) Amplitude of the total magnetization $|\mathbf{F}|$. Two kinds of topological defects, double-quantum spin vortex (DSV) and half skyrmion (HS) are marked by big and small circles, respectively. The transverse magnetization orientation $\arg F_+$ along a closed path (indicated by big circles) surrounding the unmagnetized core shows a net winding of $4\pi$, revealing the presence of a double-quantum spin vortex.

We suppose that the wave function of the lattice phase is written as

$$\psi = \frac{1}{3}\begin{pmatrix}1\\1\\1\end{pmatrix}e^{-i2\kappa x} + \frac{1}{3}\begin{pmatrix}e^{-i\frac{\pi}{3}}\\e^{i\frac{\pi}{3}}\\e^{i\pi}\end{pmatrix}e^{i\kappa(x-\sqrt{3}y)} + \frac{1}{3}\begin{pmatrix}e^{i\frac{\pi}{3}}\\e^{-i\frac{\pi}{3}}\\e^{i\pi}\end{pmatrix}e^{i\kappa(x+\sqrt{3}y)}. \quad (6.4.17)$$

Then one can expand $\psi$ around the center of a vortex with winding number $\langle -1, 0, 1\rangle$, e.g., at the location of $(x, y) = (0, \pi/(3\sqrt{3}\kappa))$. Substituting $x =$

$\epsilon \cos \theta$ and $y = \pi/(3\sqrt{3}\kappa) + \epsilon \sin \theta$ into $\psi$ and expanding with respect to the infinitesimal $\epsilon$, we obtain

$$\psi = \begin{pmatrix} -i\kappa e^{-i\theta}\epsilon - \frac{1}{2}\kappa^2 e^{i2\theta}\epsilon^2 \\ 1 - \kappa^2\epsilon^2 \\ -i\kappa e^{i\theta}\epsilon - \frac{1}{2}\kappa^2 e^{-i2\theta}\epsilon^2 \end{pmatrix} + O\left(\epsilon^3\right). \qquad (6.4.18)$$

Notice that the second-order terms with $e^{\pm i2\theta}$ have no essential influence on the phases, thus the winding number for each component can still be represented as $\langle -1, 0, 1 \rangle$ [See Figs. 6.4.4(a)–6.4.4(c)]. However, since the first-order terms are canceled out when calculating the transverse magnetization $F_+ = \sqrt{2}\left[\psi_1^*\psi_0 + \psi_0^*\psi_{-1}\right]$, the second-order terms play a dominant role, leading to the emergence of spin current with two quanta of circulation around an unmagnetized core

$$F_+ \propto \epsilon^2 e^{-i2\theta}, \qquad (6.4.19)$$

as illustrated in Fig. 6.4.4(d).

## 6.5. Experimental Relevance

The SU(3) SO coupling in Eq. (6.2.1) may be realized by Raman laser dressing technology. As shown in Fig. 6.5.1(a), three laser beams with different polarizations and frequencies, intersecting at an angle of $2\pi/3$, illuminate the three-component BECs. Each of the three Raman lasers dresses one hyperfine spin state from the $F = 1$ manifold ($|F = 1, m_F = 1\rangle$, $|F = 1, m_F = 0\rangle$ and $|F = 1, m_F = -1\rangle$) to the excited state $|e\rangle$ [See Fig. 6.5.1(b)]. The internal dynamics of a single particle under this scheme can be described by the Hamiltonian

$$H = \sum_{j=1}^{3} \left(\frac{\hbar^2 k^2}{2m} + \varepsilon_j\right) |j\rangle\langle j| + \sum_{l=1}^{n} E_l |l\rangle\langle l|$$
$$+ \sum_{j=1}^{3} \sum_{l=1}^{n} \left[\Omega_j e^{i(\mathbf{K}_j \cdot \mathbf{r} + \omega_j t)} M_{lj} |l\rangle\langle j| + h.c.\right], \qquad (6.5.1)$$

where $\hbar\mathbf{k}$ is the momentum of the particles, and $\varepsilon_j$ and $E_l$ are the energies of the ground and excited states, respectively. In the atom-light coupling term, $\mathbf{K}_j$ and $\omega_j$ are the wave vectors and frequencies of the three Raman lasers with $\Omega_j$ the corresponding Rabi frequencies, and $M_{lj}$ is the matrix element of the dipole transition. One can see that this Hamiltonian is similar to that used in the scheme for creating 2D SO coupling in ultracold

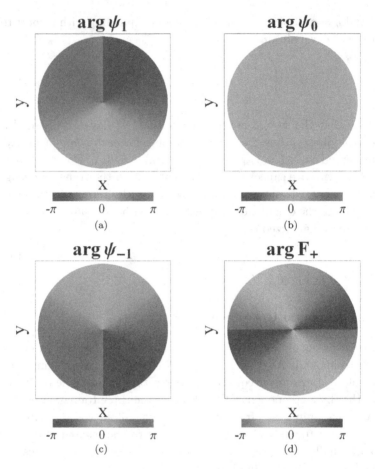

**Fig. 6.4.4.** (a)–(c) Phases of the polar-core vortex described by the wave function in Eq. (6.4.18), displaying the winding combination $\langle -1, 0, 1 \rangle$. (d) Direction of the transverse magnetization, indicating the emergence of spin current with two quanta of circulation.

Fermi gases,[5] thus may be readily realized in Bose gases. Taking the standard rotating wave approximation to get rid of the time dependence of the Hamiltonian, and adiabatically eliminating the excited states for far detuning, the Hamiltonian can be rewritten as

$$
H = \begin{pmatrix} \frac{\hbar^2 (\mathbf{k}+\mathbf{K}_1)^2}{2m}+\delta_1 & \Omega_{12} & \Omega_{13} \\ \Omega_{21} & \frac{\hbar^2 (\mathbf{k}+\mathbf{K}_2)^2}{2m}+\delta_2 & \Omega_{23} \\ \Omega_{31} & \Omega_{32} & \frac{\hbar^2 (\mathbf{k}+\mathbf{K}_3)^2}{2m}+\delta_3 \end{pmatrix}, \quad (6.5.2)
$$

**Laser geometry**          **Level diagram**

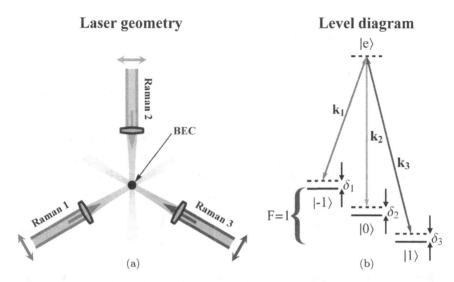

**Fig. 6.5.1.** Scheme for creating SU(3) spin-orbit coupling in spinor BECs. (a) Laser geometry. Three laser beams with different frequencies and polarizations, intersecting at an angle of $2\pi/3$, illuminate the cloud of atoms. (b) Level diagram. Each of the three Raman lasers dresses one hyperfine Zeeman level from $|F = 1, m_F = 1\rangle$, $|F = 1, m_F = 0\rangle$ and $|F = 1, m_F = -1\rangle$ of the $^{87}$Rb $5S_{1/2}$, $F = 1$ ground state. $\delta_1$, $\delta_2$ and $\delta_3$ correspond to the detuning in the Raman transitions.

where $\delta_1$, $\delta_2$ and $\delta_3$ are the two-photon detunings, and the real parameters $\Omega_{jj'} = \Omega_{j'j}$ describe the Raman coupling strength between hyperfine ground states $|j\rangle$ and $|j'\rangle$, which can be expressed as[5,76]

$$\Omega_{jj'} = -\frac{\sqrt{I_j I_{j'}}}{\hbar^2 c \epsilon_0} \sum_{m'} \frac{\langle j'|er_q|m'\rangle \langle m'|er_q|j\rangle}{\Delta}. \tag{6.5.3}$$

Here, $I_j$ is the intensity of each Raman laser, and $\Delta$ denotes the one-photon detuning. Other parameters $c$, $\epsilon_0$ and $e$ in Eq. (6.5.3) are the speed of light, permittivity of vacuum and elementary charge, respectively. In Eq. (6.5.3), $q = x, y, z$ is an index labeling the components of $r$ in the spherical basis, and $|m'\rangle$ describes the middle excited hyperfine spin state in the Raman process. For simplicity, we assume $\Omega = \Omega_{12} = \Omega_{13} = \Omega_{23}$, which can always be satisfied by adjusting the system parameters, such as the laser intensity.

Introducing a unitary transformation

$$U = \frac{1}{\sqrt{3}} \begin{pmatrix} 1 & 1 & 1 \\ -e^{-i\frac{\pi}{3}} & -e^{i\frac{\pi}{3}} & 1 \\ -e^{i\frac{\pi}{3}} & -e^{-i\frac{\pi}{3}} & 1 \end{pmatrix} \tag{6.5.4}$$

and a time-dependent unitary transformation $U(t) = e^{i\left(\frac{\hbar^2 K_0^2}{2m} + \delta_2 - \Omega\right)t}$, the effective Hamiltonian becomes

$$H = \begin{pmatrix} \frac{\hbar^2 \mathbf{k}^2}{2m} + \delta_1 - \delta_2 & 0 & 0 \\ 0 & \frac{\hbar^2 \mathbf{k}^2}{2m} & 0 \\ 0 & 0 & \frac{\hbar^2 \mathbf{k}^2}{2m} + \delta_3 - \delta_2 + 3\Omega \end{pmatrix} + \mathcal{V}_{so}, \qquad (6.5.5)$$

where the laser vectors $K_1 = -K_0 \hat{e}_y$, $K_2 = \frac{\sqrt{3}K_0}{2}\hat{e}_x + \frac{K_0}{2}\hat{e}_y$ and $K_3 = -\frac{\sqrt{3}K_0}{2}\hat{e}_x + \frac{K_0}{2}\hat{e}_y$ are defined with $K_0 = 2m\kappa/\hbar$. The spin-dependent uniform potential induced by the Raman detuning $\delta_i$ and Raman coupling strength $\Omega$ can be eliminated by applying a Zeeman field, leading to

$$H = \begin{pmatrix} \frac{\hbar^2 \mathbf{k}^2}{2m} + \epsilon_1 & 0 & 0 \\ 0 & \frac{\hbar^2 \mathbf{k}^2}{2m} & 0 \\ 0 & 0 & \frac{\hbar^2 \mathbf{k}^2}{2m} + \epsilon_2 \end{pmatrix} + \mathcal{V}_{so}, \qquad (6.5.6)$$

where $\epsilon_1 = \delta_1 - \delta_2 + \Delta_1 + \Delta_2$ and $\epsilon_2 = \delta_3 - \delta_2 - \Delta_1 + \Delta_2 + 3\Omega$ with $\Delta_1$ and $\Delta_2$ denoting the linear and quadratic Zeeman energy respectively. By tuning the detuning, the Zeeman energy and the Raman coupling strength, one can reach the regime $\Delta_1 = \frac{\delta_3 - \delta_1 + 3\Omega}{2}$ and $\Delta_2 = \delta_2 - \frac{\delta_1 + \delta_3 + 3\Omega}{2}$ which satisfying $\epsilon_1 = \epsilon_2 = 0$. Then we have

$$H = \frac{\hbar^2 \mathbf{k}^2}{2m} + \mathcal{V}_{so}, \qquad (6.5.7)$$

which is the single-particle Hamiltonian with SU(3) SO coupling considered in our work.

It should be indicated that the study on SU(3) SO coupling is not limited to the present form, other types of possible SO coupling with SU(3) symmetry are also interesting and their experimental feasibility has been explored.[35,45,46] The new ingredient of the SU(3) spin degrees of freedom will greatly enrich the physics of SO coupling. With the development of artificial gauge potentials in cold atomic gases,[9,11] the accession of SU(3) SO coupling may bring new breakthroughs on discovering new states of matter and quantum simulations.

## 6.6. Conclusion

We have briefly reviewed the recent progress of three-component Bose-Einstein condensates with spin-orbit coupling, and put our emphasis on the ground-state properties with the presence of an SU(3) spin-orbit coupling. It is demonstrated that the SU(3) spin-orbit coupling can lead to a triple-well dispersion relation in the two-dimentional momentum space, which is distinctly different from the ordinary double-well dispersion in the NIST scheme and the Mexican-hat structure of the Rashba spin-orbit coupling. When the interatomic interactions are considered, there exist three different kinds of ground-state occupies on the three minima of the energy bands, depending on the type and strength of the spin-exchange interaction. The different choices of occupy lead three many-body phases: threefold-degenerate plane-wave phase, time-reversal symmetry broken stripe phase and topologically nontrivial lattice phase.

Especially, the supersolid lattice phase possesses the spatially translational symmetry breaking in both directions of the two-dimensional plane. This is in stark contrast to the supersolid stripe observed in recent experiments, in which the translational symmetry is broken only in one direction. From this point of view, the supersolid lattice phase is much closer to the general concept of supersolid. In addition, the topological nontrivial properties of the supersolid lattice phase is illustrated. It is found that double-quantum spin vortices can be stabilized with the presence of SU(3) spin-orbit coupling, which goes beyond the ordinary knowledge that a spin vortex with multiple quanta spin circulating is topologically unstable. We reveal the generation mechanism and suggest possible observation in future experiments by magnetization-sensitive phase-contrast imaging technique. The exotic phases discovered here open new perspectives for the investigation of supersolid phenomena and topological defects in ultracold atomic gases.

## Acknowledgments

This work was supported by the NMFSEID under Grant No. 61127901; the NKRDP under Grants No. 2013CB922000; the NSFC under Grants No. 11434011, No. 11522436, No. 11547126, No. 11704383; the Research Funds of Renmin University of China under Grants No. 10XNL016 and No. 16XNLQ03.

## References

1. Y. J. Lin, K. Jiménez-García, and I. B. Spielman, Spin-orbit-coupled Bose-Einstein condensates. *Nature*, **471**, 83 (2011).
2. P. Wang, Z. Q. Yu, Z. Fu, J. Miao, L. Huang, S. Chai, H. Zhai, and J. Zhang, Spin-orbit coupled degenerate Fermi gases. *Phys. Rev. Lett.*, **109**, 095301 (2012).

3. L. W. Cheuk, A. T. Sommer, Z. Hadzibabic, T. Yefsah, W. S. Bakr, and M. W. Zwierlein, Spin-injection spectroscopy of a spin-orbit coupled Fermi gas. *Phys. Rev. Lett.*, **109**, 095302 (2012).

4. D. L. Campbell, R. M. Price, A. Putra, A. Valdés-Curiel, D. Trypogeorgos, and I. B. Spielman, Magnetic phases of spin-1 spin-orbit-coupled Bose gases. *Nat. Commun.*, **7**, 10897 (2016).

5. L. Huang, Z. Meng, P. Wang, P. Peng, S.-L. Zhang, L. Chen, D. Li, Q. Zhou, and J. Zhang, Experimental realization of two-dimensional synthetic spin-orbit coupling in ultracold Fermi gases. *Nat. Phys.*, **12**, 540 (2016).

6. Z. Wu, L. Zhang, W. Sun, X.-T. Xu, B.-Z. Wang, S.-C. Ji, Y. Deng, S. Chen, X.-J. Liu, J.-W. Pan, Realization of two-dimensional spin-orbit coupling for Bose-Einstein condensates. *Science*, **354**, 83 (2016).

7. D. Xiao, M.-C. Chang, and Q. Niu, Berry phase effects on electronic properties. *Rev. Mod. Phys.*, **82**, 1959 (2010).

8. X.-L. Qi and S.-C. Zhang, Topological insulators and superconductors. *Rev. Mod. Phys.*, **83**, 1057 (2011).

9. J. Dalibard, F. Gerbier, G. Juzeliūnas, and P. Öhberg, Colloquium: Artificial gauge potentials for neutral atoms. *Rev. Mod. Phys.*, **83**, 1523 (2011).

10. V. Galitski and I. B. Spielman, Spin-orbit coupling in quantum gases. *Nature*, **494**, 49 (2013).

11. N. Goldman, G. Juzeliūnas, P. Öhberg, and I B Spielman, Light-induced gauge fields for ultracold atoms. *Rep. Prog. Phys.*, **77**, 126401 (2014).

12. X. Luo, L. Wu, J. Chen, Q. Guan, K. Gao, Z.-F. Xu, L. You, and R. Wang, Tunable atomic spin-orbit coupling synthesized with a modulating gradient magnetic field. *Sci. Rep.*, **6**, 18983 (2016).

13. G. Juzeliūnas, J. Ruseckas, and J. Dalibard, Generalized Rashba-Dresselhaus spin-orbit coupling for cold atoms. *Phys. Rev. A*, **81**, 053403 (2010).

14. Z. Lan, and P. Öhberg, Raman-dressed spin-1 spin-orbit-coupled quantum gas. *Phys. Rev. A*, **89**, 023630 (2014).

15. G. I. Martone, F. V. Pepe, P. Facchi, S. Pascazio, and S. Stringari, Tricriticalities and quantum phases in spin-orbit-coupled spin-1 Bose gases. *Phys. Rev. Lett.*, **117**, 125301 (2016).

16. S. S. Natu, X. Li, and W. S. Cole, Striped ferronematic ground states in a spin-orbit-coupled $S = 1$ Bose gas. *Phys. Rev. A*, **91**, 023608 (2015).

17. Z.-Q. Yu, Phase transitions and elementary excitations in spin-1 Bose gases with Raman-induced spin-orbit coupling. *Phys. Rev. A*, **93**, 033648 (2016).

18. K. Sun, C. Qu, Y. Xu, Y. Zhang, and C. Zhang, Interacting spin-orbit-coupled spin-1 Bose-Einstein condensates. *Phys. Rev. A*, **93**, 023615 (2016).

19. C. Wang, C. Gao, C.-M. Jian, and H. Zhai, Spin-orbit coupled spinor Bose-Einstein condensates. *Phys. Rev. Lett.*, **105**, 160403 (2010).

20. X.-Q. Xu and J. H. Han, Emergence of chiral magnetism in spinor Bose-Einstein condensates with Rashba coupling. *Phys. Rev. Lett.*, **108**, 185301 (2012).

21. Z. F. Xu, Y. Kawaguchi, L. You, and M. Ueda, Symmetry classification of spin-orbit-coupled spinor Bose-Einstein condensates. *Phys. Rev. A*, **86**, 033628 (2012).

22. S.-W. Su, I.-K. Liu, Y.-C. Tsai, W. M. Liu, and S.-C. Gou, Crystallized half-skyrmions and inverted half-skyrmions in the condensation of spin-1 Bose gases with spin-orbit coupling. *Phys. Rev. A*, **86**, 023601 (2012).

23. L. Wen, Q. Sun, H. Q. Wang, A. C. Ji, and W. M. Liu, Ground state of spin-1 Bose-Einstein condensates with spin-orbit coupling in a Zeeman field. *Phys. Rev. A*, **86**, 043602 (2012).

24. S.-W. Song, Y.-C. Zhang, H. Zhao, X. Wang, and W.-M. Liu, Fragmentation of spin-orbit-coupled spinor Bose-Einstein condensates. *Phys. Rev. A*, **89**, 063613 (2014).

25. M. Kato, X.-F. Zhang, D. Sasaki, and H. Saito, Twisted spin vortices in a spin-1 Bose-Einstein condensate with Rashba spin-orbit coupling and dipole-dipole interaction. *Phys. Rev. A*, **94**, 043633 (2016).

26. C.-F. Liu and W. M. Liu, Spin-orbit-coupling-induced half-skyrmion excitations in rotating and rapidly quenched spin-1 Bose-Einstein condensates. *Phys. Rev. A*, **86**, 033602 (2012).

27. S.-W. Song, Y.-C. Zhang, L. Wen and H. Wang, Spin-orbit coupling induced displacement and hidden spin textures in spin-1 Bose-Einstein condensates. *J. Phys. B: At. Mol. Opt. Phys.*, **46**, 145304 (2013).

28. T. Oshima and Y. Kawaguchi, Spin Hall effect in a spinor dipolar Bose-Einstein condensate. *Phys. Rev. A*, **93**, 053605 (2016).

29. L. Chen, H. Pu, and Y. Zhang, Spin-orbit angular momentum coupling in a spin-1 Bose-Einstein condensate. *Phys. Rev. A*, **93**, 013629 (2016).

30. B. M. Anderson, G. Juzeliūnas, V. M. Galitski, and I. B. Spielman, Synthetic 3D spin-orbit coupling. *Phys. Rev. Lett.*, **108**, 235301 (2012).

31. M. DeMarco and H. Pu, Angular spin-orbit coupling in cold atoms. *Phys. Rev. A*, **91**, 033630 (2015).

32. K. Sun, C. Qu, and C. Zhang, Spin-orbital-angular-momentum coupling in Bose-Einstein condensates. *Phys. Rev. A*, **91**, 063627 (2015).

33. J.-S. Pan, X.-J. Liu, W. Zhang, W. Yi, and G.-C. Guo, Topological superradiant states in a degenerate Fermi gas. *Phys. Rev. Lett.*, **115**, 045303 (2015).

34. S.-W. Su, S.-C. Gou, I.-K. Liu, I. B. Spielman, L. Santos, A. Acus, AMekys, J. Ruseckas and G. Juzeliūnas, Position-dependent spin-orbit coupling for ultracold atoms. *New J. Phys.*, **17**, 033045 (2015).

35. R. Barnett, G. R. Boyd, and V. Galitski, SU(3) Spin-orbit coupling in systems of ultracold atoms. *Phys. Rev. Lett.*, **109**, 235308 (2012).

36. D. M. Stamper-Kurn and M. Ueda, Spinor Bose gases: Symmetries, magnetism, and quantum dynamics. *Rev. Mod. Phys.*, **85**, 1191 (2013).

37. Y. Kawaguchi and M. Ueda, Spinor Bose–Einstein condensates. *Phys. Rep.*, **520**, 253 (2012).

38. R. B. Diener and T.-L. Ho, $^{52}$Cr Spinor condensate: A biaxial or uniaxial spin nematic. *Phys. Rev. Lett.*, **96**, 190405 (2006).

39. B. M. Fregoso and E. Fradkin, Ferronematic ground state of the dilute dipolar Fermi gas. *Phys. Rev. Lett.*, **103**, 205301 (2009).

40. G. Juzeliūnas, J. Ruseckas, M. Lindberg, L. Santos, and P. Öhberg, Quasirelativistic behavior of cold atoms in light fields. *Phys. Rev. A*, **77**, 011802(R) (2008).

41. G. Juzeliūnas, J. Ruseckas, A. Jacob, L. Santos, and P. Öhberg, Double and negative reflection of cold atoms in non-Abelian gauge potentials. *Phys. Rev. Lett.*, **100**, 200405 (2008).

42. V. Teodorescu and R. Winkler, Spin angular impulse due to spin-dependent reflection off a barrier. *Phys. Rev. B*, **80**, 041311(R) (2009).

43. C. Qu, K. Sun, and C. Zhang, Quantum phases of Bose-Einstein condensates with synthetic spin-orbital-angular-momentum coupling. *Phys. Rev. A*, **91**, 053630 (2015).

44. Y.-X. Hu, C. Miniatura, and B. Grémaud, Half-skyrmion and vortex-antivortex pairs in spinor condensates. *Phys. Rev. A*, **92**, 033615 (2015).

45. T. GraßSS, R. W. Chhajlany, C. A. Muschik, and M. Lewenstein, Spiral spin textures of a bosonic Mott insulator with SU(3) spin-orbit coupling. *Phys. Rev. B*, **90**, 195127 (2014).

46. Y.-X. Hu, C. Miniatura, D. Wilkowski, and B. Grémaud, U(3) artificial gauge fields for cold atoms. *Phys. Rev. A*, **90**, 023601 (2014).

47. I. Mandal and A. Bhattacharya, Cold atoms in U(3) gauge potentials. *Condens. Matter*, **1**, 2 (2016).

48. S. Ray, A. Ghatak, and T. Das, Photo-induced SU(3) topological material of spinless fermions, arXiv:1701.06319.

49. G. B. Arfken, H. J. Weber, and F. E. Harris, *Mathematical Methods for Physicists*, 7th ed. (Academic Press, New York, 2000).

50. J.-R. Li, J. Lee, W. Huang, S. Burchesky, B. Shteynas, F. Ç. Top, A. O. Jamison, and W. Ketterle, A stripe phase with supersolid properties in spin-orbit-coupled Bose-Einstein condensates. *Nature*, **543**, 91 (2017).

51. T.-L. Ho and S. Zhang, Bose-Einstein condensates with spin-orbit interaction. *Phys. Rev. Lett.*, **107**, 150403 (2011).

52. Y. Li, L. P. Pitaevskii, and S. Stringari, Quantum tricriticality and phase transitions in spin-orbit-coupled Bose-Einstein condensates. *Phys. Rev. Lett.*, **108**, 225301 (2012).

53. Y. Li, G. I. Martone, L. P. Pitaevskii, and S. Stringari, Superstripes and the excitation spectrum of a spin-orbit-coupled Bose-Einstein condensate. *Phys. Rev. Lett.*, **110**, 235302 (2012).

54. S. Sinha, R. Nath, and L. Santos, Trapped two-dimensional condensates with synthetic spin-orbit coupling. *Phys. Rev. Lett.*, **107**, 270401 (2011).

55. Z. F. Xu, R. Lü, and L. You, Emergent patterns in a spin-orbit-coupled spin-2 Bose-Einstein condensate. *Phys. Rev. A*, **83**, 053602 (2011).

56. H. Hu, B. Ramachandhran, H. Pu, and X. J. Liu, Spin-orbit coupled weakly interacting Bose-Einstein condensates in Harmonic Traps. *Phys. Rev. Lett.*, **108**, 010402 (2012).

57. S. Gopalakrishnan, I. Martin, and E. A. Demler, Quantum quasicrystals of Spin-orbit-coupled dipolar bosons. *Phys. Rev. Lett.*, **111**, 185304 (2013).

58. L. E. Sadler, J. M. Higbie, S. R. Leslie, M. Vengalattore, and D. M. Stamper-Kurn, Spiral spin textures of a bosonic Mott insulator with SU(3) spin-orbit coupling. *Nature*, **443**, 312 (2006).

59. T. D. Stanescu, B. Anderson, and V. Galitski, Spin-orbit coupled Bose-Einstein condensates. *Phys. Rev. A*, **78**, 023616 (2008).
60. T. F. J. Poon and X. J. Liu, Quantum spin dynamics in a spin-orbit-coupled Bose-Einstein condensate. *Phys. Rev. A*, **93**, 063420 (2016).
61. M. Boninsegni and N. V. Prokof'ev, Colloquium: Supersolids: What and where are they? *Rev. Mod. Phys.*, **84**, 759 (2012).
62. S. C. Ji, J. Y. Zhang, L. Zhang, Z. D. Du, W. Zheng, Y. J. Deng, H. Zhai, S. Chen, and J. W. Pan, Experimental determination of the finite-temperature phase diagram of a spin-orbit coupled Bose gas. *Nat. Phys.*, **10**, 314 (2014).
63. T. Isoshima, K. Machida, and T. Ohmi, Quantum vortex in a spinor Bose-Einstein condensate. *J. Phys. Soc. Jpn.*, **70**, 1604 (2001).
64. T. Mizushima, N. Kobayashi, and K. Machida, Coreless and singular vortex lattices in rotating spinor Bose-Einstein condensates. *Phys. Rev. A*, **70**, 043613 (2004).
65. H. Saito, Y. Kawaguchi, and M. Ueda, Breaking of chiral symmetry and spontaneous rotation in a spinor Bose-Einstein condensate. *Phys. Rev. Lett.*, **96**, 065302 (2006).
66. H. Saito, Y. Kawaguchi, and M. Ueda, Topological defect formation in a quenched ferromagnetic Bose-Einstein condensates. *Phys. Rev. A*, **75**, 013621 (2007).
67. Y. Kawaguchi, H. Saito, K. Kudo, and M. Ueda, Spontaneous magnetic ordering in a ferromagnetic spinor dipolar Bose-Einstein condensate. *Phys. Rev. A*, **82**, 043627 (2010).
68. J. Lovegrove, M. O. Borgh, and J. Ruostekoski, Energetically stable singular vortex cores in an atomic spin-1 Bose-Einstein condensate. *Phys. Rev. A*, **86**, 013613 (2012).
69. A. Hubert and R. Schafer, *Magnetic Domains* (Springer, Berlin, 1998).
70. T. Shinjo, T. Okuno, R. Hassdorf, K. Shigeto, and T. Ono, Magnetic vortex core observation in circular dots of permalloy. *Science*, **289**, 930 (2000).
71. A. Wachowiak, J. Wiebe, M. Bode, O. Pietzsch, M. Morgenstern, and R. Wiesendanger, Direct observation of internal spin structure of magnetic vortex cores. *Science*, **298**, 577 (2002).
72. L. S. Leslie, A. Hansen, K. C. Wright, B. M. Deutsch, and N. P. Bigelow, Creation and detection of skyrmions in a Bose-Einstein condensate. *Phys. Rev. Lett.*, **103**, 250401 (2009).
73. Jae-yoon Choi, Woo Jin Kwon, and Yong-il Shin, Observation of topologically stable 2D skyrmions in an antiferromagnetic spinor Bose-Einstein condensate. *Phys. Rev. Lett.*, **108**, 035301 (2012).
74. S. Yi and H. Pu, Spontaneous spin textures in dipolar spinor condensates. *Phys. Rev. Lett.*, **97**, 020401 (2006).
75. G. E. Brown and M. Rho, *The Multifaceted Skyrmion* (World Scientific, Singapore, 2010).
76. T. A. Savard, S. R. Granade, K. M. O'Hara, M. E. Gehm, and J. E. Thomas, Raman-induced magnetic resonance imaging of atoms in a magneto-optical trap. *Phys. Rev. A*, **60**, 4788 (1999).

# Chapter 7

# Dynamical Spin-Orbit Coupling in Cold Atoms Induced by Cavity Field

Chuanzhou Zhu, Lin Dong, and Han Pu

*Department of Physics and Astronomy, and Rice Center for Quantum Materials, Rice University, Houston, TX 77251-1892, USA*

We consider ultracold atoms inside a ring optical cavity that supports a single plane-wave mode. The cavity field, together with an external coherent laser field, drives a two-photon Raman transition between two internal pseudo-spin states of the atom. This gives rise to the so-called spin-orbit coupling in cold atoms — an effective coupling between atom's pseudo-spin and external center-of-mass (COM) motion. Due to the back action from the atom to the cavity field, this spin-orbit coupling is dynamical in nature. For the case of a single atom inside the cavity, We show how the spin-orbit coupling modifies the static and dynamic properties of the Jaynes-Cummings (JC) model. In the case of many atoms in thermodynamic limit, we show that the spin-orbit coupling modifies the Dicke superradiance phase transition boundary and the non-superradiant normal phase may become reentrant in some regimes.

## 7.1. Introduction

When an atom interacts with a quantized light field supported by an optical cavity, the atom and the light field mutually affect each other. A *self-consistent* solution for the light field and the atom is thus required. This has been a major theme in CQED.[1-3] In traditional CQED settings, only the internal dynamics of the atom is taken into account: the cavity photons induce quantum transitions among different atomic internal states, and the atom affects the cavity field by emitting/absorbing cavity photons. The link between "Cavity photon" and "Atomic internal states" in Fig. 7.1.1 illustrates the relationship between the cavity and the atom. In recent years,

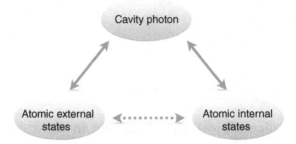

**Fig. 7.1.1.** A new frontier in both CQED and cold atoms research, in which one needs to take into account the interplay among the cavity photons, the atomic external states, and also the atomic internal degrees of freedom.

ultracold atoms have been put inside optical cavities and in such a situation, one can no longer neglect the external degrees of freedom, i.e. the center-of-mass (COM) motion of the atom, as emitting or absorbing even a single photon can significantly change the motional states of the atom. This situation is represented by the link between "Cavity photon" and "Atomic external states" in Fig. 7.1.1. In fact, the mutual influence of the cavity field and the atomic COM motion can be put in the broader context of optomechanics. Here the atoms can be regarded as a mechanical system whose dynamics is controlled by, and in the same time provides a back-action to, the cavity field. A variety of interesting phenomena in this "ultracold atom + cavity" atomic optomechanical system has been explored both experimentally[4–10] and theoretically.[11,12]

As is quite obvious from discussions above, in a most general setting, the same cavity photon can affect both the internal states (via inducing a transition between different states of the atom) and the external COM motion (via photon recoil) of the atom. As a result, it naturally induces an effective coupling between the two atomic degrees of freedom, as is represented by the dashed link between "Atomic external states" and "Atomic internal states" in Fig. 7.1.1. One often regards the internal states of the atom pseudospins, as a result the coupling between the internal and the external atomic states is termed spin-orbit coupling (SOC). SOC in cold atoms induced by coherent laser fields has been realized in both bosonic[13] and fermionic atomic systems without cavity,[14,15] and has attracted tremendous attention in recent years.[16–18] Due to its non-Abelian nature, SOC not only significantly affects the physics of a single atom, but, perhaps more importantly, also profoundly changes the properties of a many-body system. It is an essential ingredient underlying such diverse phenomena as topological superconductors/insulators, Majorana and Weyl fermions, quantum spin-Hall effects, etc.[19–23]

So far, in all the experimental realization, the SOC in cold atoms is induced by external coherent laser fields. These laser fields can be treated as classical fields (i.e., classical electromagnetic waves) which are not involved in the dynamics of the system. By contrast, a *dynamical* SOC can be realized using the scheme depicted in Fig. 7.1.1, where the cavity field induces the SOC in atoms and, in the mean time, the atoms exert a back-action to the cavity photons. As such, the photon field cannot be treated as a static field, but instead represents an intrinsic part of the system dynamics. It is in this sense that the resulting SOC becomes dynamical. In this article, we will consider a simple system[24–26] which clearly illustrates some of the novel features of an atomic system subjected to such dynamical SOC.

## 7.2. Model

We consider an atomic cloud with two relevant internal states (denoted as $|\uparrow\rangle$ and $|\downarrow\rangle$) confined inside a unidirectional optical ring cavity, depicted schematically in Fig. 7.2.1. The cavity is pumped by a coherent laser field with frequency $\omega_p$ and pumping rate $\varepsilon_p$. It supports a single mode traveling wave and has an intrinsic angular frequency $\omega_c$. An additional coherent laser beam with frequency $\omega_R$ shines on the atom, which together with the cavity field provides the Raman transition between $|\uparrow\rangle$ and $|\downarrow\rangle$ states. During the Raman transition, a recoil momentum of $\pm 2\hbar q_r \hat{z}$ is transferred to the atom. We treat the leakage of cavity photon phenomenologically by introducing a cavity decay rate $\kappa$. The model Hamiltonian is thus written as (we take

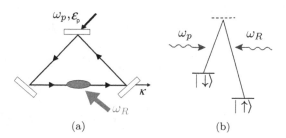

(a)                              (b)

**Fig. 7.2.1.** (a) Schematic diagram of the cavity-assisted spin-orbit coupled system; (b) Level diagram of atom photon/light field interaction.

$\hbar = 1$),

$$H = \int d\mathbf{r} \left[ \Psi^\dagger(\mathbf{r}) \left( \frac{\hat{\mathbf{k}}^2}{2m} + \epsilon^0_\sigma + V(\mathbf{r}) \right) \Psi(\mathbf{r}) \right]$$

$$+ \frac{\Omega}{2} \int d\mathbf{r} \left[ e^{+2iq_r z} \Psi^\dagger_\uparrow(\mathbf{r}) \Psi_\downarrow(\mathbf{r}) \, c \, e^{+i\omega_R t} + h.c. \right]$$

$$+ i\varepsilon_p(c^\dagger e^{-i\omega_p t} - c e^{+i\omega_p t}) + \omega_c c^\dagger c, \qquad (7.2.1)$$

where $\hat{\mathbf{k}}$ is the atomic momentum operator, $\Psi = (\Psi_\uparrow, \Psi_\downarrow)^T$, and $\Psi_\sigma(\mathbf{r})$ ($\sigma = \uparrow, \downarrow$) is the atomic annihilation field operator for spin state $|\sigma\rangle$, $\epsilon^0_\sigma$ is the corresponding bare atomic energy, $V(\mathbf{r})$ is an external trapping potential which is assumed to be spin-independent, and $\tilde{c}$ represents the photon annihilation operator. $\Omega$ characterizes the Raman coupling strength. However, the true Raman coupling strength also includes the cavity photon amplitude of $\tilde{c}$ or $\tilde{c}^\dagger$ which is coupled to the atomic operators. It is this coupling that renders the resulting SOC dynamic. Here we have neglected the inter-atomic interaction. As one shall see, due to the coupling between the cavity field and the atoms, this non-interacting system already exhibits quite rich physics.

It is convenient to work in a frame rotating at pump laser frequency $\omega_p$. This is equivalent to performing a unitary transformation $U = e^{+i\omega_p c^\dagger c t}$ to the Hamiltonian (7.2.1), resulting a new Hamiltonian $H' = UHU^{-1} + i\frac{dU}{dt}U^{-1}$. From $H'$, we perform another unitary transformation $\tilde{U} = e^{i\delta_R t(\Psi^\dagger_\uparrow \Psi_\uparrow - \Psi^\dagger_\downarrow \Psi_\downarrow)/2}$, with $\delta_R = \omega_p - \omega_R$, to obtain the Hamiltonian $H''$. Finally, after a gauge transformation to atomic operators

$$\psi_\uparrow = \Psi_\uparrow e^{-iq_r z}, \qquad \psi_\downarrow = \Psi_\downarrow e^{+iq_r z}, \qquad (7.2.2)$$

we arrive at the following effective Hamiltonian $\mathcal{H}_{\text{eff}}$:

$$\mathcal{H}_{\text{eff}} = \int d\mathbf{r} \left[ \psi^\dagger(\mathbf{r}) \left( \frac{\hat{\mathbf{k}}^2}{2m} + \frac{q_r}{m}\hat{\sigma}_z \hat{k}_z + \hat{\sigma}_z\frac{\tilde{\delta}}{2} + V(\mathbf{r}) \right) \psi(\mathbf{r}) \right]$$

$$+ \frac{\Omega}{2} \int d\mathbf{r} \left[ \psi^\dagger_\uparrow(\mathbf{r})\psi_\downarrow(\mathbf{r})c + c^\dagger \psi^\dagger_\downarrow(\mathbf{r})\psi_\uparrow(\mathbf{r}) \right] + i\varepsilon_p(c^\dagger - c) - \delta_c c^\dagger c,$$

$$(7.2.3)$$

where $\psi = (\psi_\uparrow, \psi_\downarrow)^T$, $\tilde{\delta} = \epsilon^0_\uparrow - \epsilon^0_\downarrow - \delta_R$ represents the two-photon Raman detuning, and $\delta_c = \omega_p - \omega_c$ is the cavity-pump detuning. Note that in Hamiltonian (7.2.3), the operator $\hat{\mathbf{k}}$ has a different meaning as in (7.2.1) and represents the *quasi-momentum* of the atom, which differs from the real atomic momentum by a spin-dependent shift $\sigma_z q_r$ as the result of the unitary transformation in (7.2.2).

We also want to remark that, in the absence of SOC, a situation that can be achieved by setting $q_r = 0$, the atomic COM motion is completely decoupled from the cavity field, our model is then reduces to the Jaynes-Cummings (when a single atom is involved)[27] or Tavis-Cummings (in the case of many atoms)[28] models.

## 7.3. Semi-classical vs. Quantum Approach

From the Hamiltonian (7.2.3), one can easily obtain the following equations of motion:

$$i\frac{d}{dt}c = i\varepsilon_p - (\delta_c + i\kappa)c + \frac{\Omega}{2}\int d\mathbf{r}\,\psi_\downarrow^\dagger(\mathbf{r},t)\,\psi_\uparrow(\mathbf{r},t)\,, \qquad (7.3.1)$$

$$i\frac{\partial}{\partial t}\psi(\mathbf{r},t) = \begin{pmatrix} \frac{\mathbf{k}^2}{2m} + \frac{q_r}{m}\hat{k}_z + V(\mathbf{r}) + \frac{\tilde{\delta}}{2} & \frac{\Omega}{2}c \\ \frac{\Omega}{2}c^\dagger & \frac{\mathbf{k}^2}{2m} - \frac{q_r}{m}\hat{k}_z + V(\mathbf{r}) - \frac{\tilde{\delta}}{2} \end{pmatrix}\psi(\mathbf{r},t)\,, \qquad (7.3.2)$$

where in the equation for the photon operator Eq. (7.3.1), we have included the photon decay term proportional to the cavity decay rate $\kappa$.

In the widely adopted semi-classical approach, the photon operator $c$ is treated as a $c$-number and represents the photon field amplitude, with $|c|^2$ being the cavity photon number. In the bad cavity limit, one assumes that the cavity field reaches steady state in a time scale much faster than the atoms, and hence the cavity field is slaved to the atoms. This amounts to taking the left hand side of Eq. (7.3.1) to be zero and arrive at

$$c(t) = \frac{i\varepsilon_p + \frac{\Omega}{2}\int d\mathbf{r}\,\psi_\downarrow^\dagger(\mathbf{r},t)\,\psi_\uparrow(\mathbf{r},t)}{\delta_c + i\kappa}\,. \qquad (7.3.3)$$

Inserting Eq. (7.3.3) to the atomic equation of motion (7.3.2), we obtain a nonlinear equation for the atoms. For a single atom, or a atomic Bose-Einstein condensate in the mean-field treatment, $\psi(\mathbf{r},t)$ is regarded as the atomic wave function, the resulting Eq. (7.3.2) then takes the form of a nonlinear Schrödinger equation.

The validity of the semi-classical approximation lies on two assumptions: (1) the cavity field can be described as a coherent state; (2) the atomic field and the cavity field have negligible correlation. When these conditions are not met, we have to bo beyond the semi-classical approach. One standard beyond-semi-classical method is the quantum Master equation method. Here we consider the density matrix of the coupled atom-cavity system $\rho$, which obeys the following equation of motion:

$$\dot{\rho} = \frac{1}{i\hbar}[\mathcal{H}_{\text{eff}}, \rho] + \mathcal{L}[\rho] \,. \qquad (7.3.4)$$

When the only dissipation arises from the cavity decay, it can be modeled by the Liouvillean term in the standard form of Lindblad super-operator,[29,30]

$$\mathcal{L}[\rho] = \kappa(2c\rho\hat{c}^\dagger - \hat{c}^\dagger\hat{c}\rho - \rho\hat{c}^\dagger\hat{c}) \,. \qquad (7.3.5)$$

In the following, we will use both the semi-classical and the quantum Master equation approaches, and show that, under certain circumstances, there can be quite significant differences between these two approaches.

## 7.4. Single Atom without Trap

Let us first focus on perhaps the idealized case where a single atom is inside the cavity without any trapping potential with $V(\mathbf{r}) = 0$. In this case, the atomic quasi-momentum is a good quantum number as $[\hat{\mathbf{k}}, \mathcal{H}_{\text{eff}}] = 0$. Exploiting this features greatly simplifies the calculation as we can focus on a single atomic plane-wave $\mathbf{k}$ state. Furthermore, we only consider the atom moving on the $z$-axis as the motion along the perpendicular directions is not affected by the photon field. In this case, we consider that the atomic wave function takes the plane-wave form $\psi_\sigma(\mathbf{r}) = e^{ik_z z}\varphi_\sigma$ with the normalization condition $|\varphi_\uparrow|^2 + |\varphi_\downarrow|^2 = 1$.

### 4.A. *Semi-classical approach*

In the semi-classical approach, the photon operator is treated as a $c$-number whose steady-state value is given by

$$c = \frac{\varepsilon_p - \frac{i}{2}\Omega\varphi_\downarrow^*\varphi_\uparrow}{\kappa - i\delta_c} \,. \qquad (7.4.1)$$

and the corresponding equations for the atomic modes are

$$i\dot{\varphi}_\uparrow = \left(\frac{k_z^2}{2m} + \frac{q_r k_z}{m} + \frac{\tilde{\delta}}{2}\right)\varphi_\uparrow + \frac{\Omega}{2}\frac{\varepsilon_p - \frac{i\Omega}{2}\varphi_\downarrow^*\varphi_\uparrow}{\kappa - i\delta_c}\varphi_\downarrow \,, \qquad (7.4.2)$$

$$i\dot{\varphi}_\downarrow = \left(\frac{k_z^2}{2m} - \frac{q_r k_z}{m} - \frac{\tilde{\delta}}{2}\right)\varphi_\downarrow + \frac{\Omega}{2}\frac{\varepsilon_p + \frac{i\Omega}{2}\varphi_\uparrow^*\varphi_\downarrow}{\kappa + i\delta_c}\varphi_\uparrow \,. \qquad (7.4.3)$$

It is clear from the above two equations that, for an atomic plane-wave state, the SOC term proportional to $q_r$ can be regarded as a momentum-dependent detuning such that the effective detuning becomes $\delta_{k_z}^{\text{eff}} = \tilde{\delta} +$

$2q_r k_z$. To obtain the atomic dispersion, we replace the left hand sides of Eqs. (7.4.2) and (7.4.3) by $\epsilon(k_z)\varphi_\sigma$, which can be shown to satisfy a quartic equation[24,25] in the form of

$$4\epsilon^4 + B\epsilon^3 + C\epsilon^2 + D\epsilon + E = 0, \qquad (7.4.4)$$

where the coefficients are given by

$$B = -(8k_z^2 + 2w),$$
$$C = 6k_z^4 + 3k_z^2 w - 4(q_r k_z + \tilde{\delta}/2)^2 + |v|^2 - 4|u|^2,$$
$$D = -2k_z^6 - \frac{3}{2}wk_z^4 + 4k_z^2(q_r k_z + \tilde{\delta}/2)^2 - |v|^2 k_z^2 + 2w(q_r k_z + \tilde{\delta}/2)^2,$$
$$E = \left[\frac{k_z^4}{4} - (q_r k_z + \tilde{\delta}/2)^2\right](k_z^4 + |v|^2 + wk_z^2) - |u|^2 k_z^4.$$

with $u = \frac{\Omega}{2}\frac{\varepsilon_p}{\kappa - i\delta_c}$, $v = \frac{\Omega}{2}\frac{-\frac{i}{2}\Omega}{\kappa - i\delta_c}$, and $w = v + v^*$.

For a given $k_z$, the quartic equation (7.4.4) either has 2 or 4 real roots. By examining the solution to Eq. (7.4.4), we obtain the "phase diagram" in the parameter space spanned by the atom-photon coupling strength $\Omega$ and the cavity pumping rate $\epsilon_p$, as shown in Fig. 7.4.1(a), where the 4 phases labeled as I, II, III and IV display distinct characters in their dispersion curves, which are displayed in Fig. 7.4.1(b). Note that if the SOC is induced by a classical laser field, instead of a cavity field, Regions III and IV would not exist, and the corresponding dispersion spectrum $\epsilon$ satisfies a quadratic equation with always 2 roots. In fact, we can reach the classic limit by assuming that the pumping and the decay rates of the cavity to be very large, whereas their ratio is finite. Under this condition, the cavity photon amplitude is simply given by $\varepsilon_p/(\kappa - i\delta_c)$, which becomes independent of the atom. This situation corresponds to the bottom region of the phase diagram in Fig. 7.4.1(a), where only Phase I and II are present.

Once $\epsilon(k_z)$ is obtained by solving Eq. (7.4.4), it is straightforward to obtain $\phi_\sigma$ and hence the steady-state cavity photon number $|c|^2$ from Eq. (7.4.1). The solid curves in the upper panels of Fig. 7.4.2 show the cavity photon number as functions of $k_z$. The color of the lines represent the dynamical stability of the corresponding state: black curves correspond to dynamically stable states, while colored curves correspond to unstable states with different colors representing different decay rate of the state, which can be obtained using the standard linear stability analysis.[24]

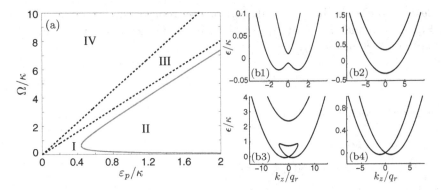

**Fig. 7.4.1.** Single particle eigen-energy spectrum "phase diagram". The dispersion curve is generally catagorized by four regions, represented from I to IV in (a). From (b1) to (b4), we fix $\varepsilon_p = \kappa$. In region I, the dispersion has double minima as shown in (b1) with $\Omega = 0.03\kappa$; region II is enclosed by the red solid curve in (a) and we show the typical dispersion curve in (b2) ($\Omega = \kappa$) where only a single minimum exists in the lower helicity branch; region III is enclosed by the black dashed lines in (a) and it is a region where loop structure emerges in the dispersion curve, as shown in (b3) with $\Omega = 5\kappa$; finally, in region IV we recover the double minimum dispersion although it is different from region I by closing the gap at $k = 0$, as in (b4) with $\Omega = 8\kappa$. Here we have used $\delta_c = \kappa$ and $\tilde{\delta} = 0$, and adopt a dimensionless unit system with $\hbar = m = \kappa = 1$. A typical value for $\kappa$ is $2\pi \times 1$ MHz, and we choose $q_r = 0.22$ in our dimensionless units (based on a realistic experimental parameter estimate).

## 4.B. *Quantum master equation approach*

In the above, we have investigated the steady-state properties of the system using the semi-classical treatment where the cavity field is treated as a classical field whose intensity is dynamically coupled to the atom. Now we want to go beyond this approximation and examine the validity of this approach. To this end, we adopt the quantum Master equation approach, which is a powerful tool for the study of quantum systems with dissipation. Denote $\rho$ as the total density operator for the coupled atom-cavity system, which obeys the Master equation (7.3.4). Again focusing on a single atomic quasi-momentum state $k_z$, we choose our basis states as the direct product states of photon Fock state $|n\rangle$ and atomic internal state $|\sigma = \uparrow, \downarrow\rangle$: $|n; \sigma\rangle \equiv |n\rangle \otimes |\sigma\rangle$, where non-negative integer $n$ denotes cavity photon number. The density matrix element is defined as $\langle m; \sigma|\rho|n; \sigma'\rangle \equiv \rho_{mn}^{\sigma\sigma'}$ whose equation of motion can be straightforwardly obtained from the Master equation as:

$$\frac{d}{dt}\rho_{mn}^{\sigma\sigma'} = -i\left(\frac{k_z^2}{2m} + \frac{q_r k_z}{m} + \frac{\tilde{\delta}}{2}\right)(\delta_{\sigma\uparrow} - \delta_{\sigma'\uparrow})\rho_{mn}^{\sigma\sigma'}$$

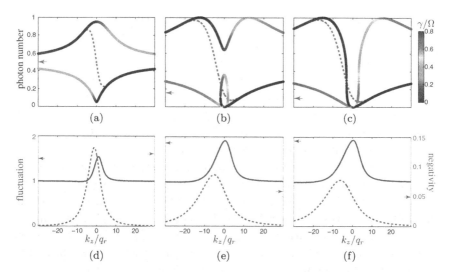

**Fig. 7.4.2.** (a)–(c) Photon number obtained from the semi-classical approach (solid curves) and from quantum Master equation approach (red dashed curves). From (a) to (c), $\Omega = 3\kappa$, $5.6\kappa$, $6\kappa$. The color on the solid curves represents the normalized decay rate $\gamma/\Omega$ of unstable mean-field states. The black color represents stable mean-field states. We have used $\varepsilon_p = \kappa$, and other parameters are the same as in Fig. 7.4.1. (d)–(f): Corresponding photon number fluctuation (blue solid curve) and negativity (green dashed line) obtained from the quantum approach. The parameters used here are the same in (a)–(c), respectively. The horizontal arrows indicate results from the JC model by taking $q_r = 0$.

$$-i\left(\frac{k_z^2}{2m} - \frac{q_r k_z}{m} - \frac{\tilde{\delta}}{2}\right)(\delta_{\sigma\downarrow} - \delta_{\sigma'\downarrow})\rho_{mn}^{\sigma\sigma'}$$

$$+\frac{\Omega}{2i}(\delta_{\sigma\uparrow}\sqrt{m+1}\rho_{m+1n}^{\bar{\sigma}\sigma'} + \delta_{\sigma\downarrow}\sqrt{m}\rho_{m-1n}^{\bar{\sigma}\sigma'} - \delta_{\sigma'\uparrow}\sqrt{n+1}\rho_{mn+1}^{\sigma\bar{\sigma}'}$$

$$- \delta_{\sigma'\downarrow}\sqrt{n}\rho_{mn-1}^{\sigma\bar{\sigma}'})$$

$$+\varepsilon_p\left(\sqrt{m}\rho_{m-1n}^{\sigma\sigma'} - \sqrt{m+1}\rho_{m+1n}^{\sigma\sigma'} + \sqrt{n}\rho_{mn-1}^{\sigma\sigma'} - \sqrt{n+1}\rho_{mn+1}^{\sigma\sigma'}\right)$$

$$+i\delta_c(m-n)\rho_{mn}^{\sigma\sigma'}$$

$$+\kappa\left(2\sqrt{m+1}\sqrt{n+1}\rho_{m+1n+1}^{\sigma\sigma'} - (m+n)\rho_{mn}^{\sigma\sigma'}\right), \qquad (7.4.5)$$

where $\bar{\sigma}$ represents the flip-spin value, i.e. $\bar{\uparrow} = \downarrow$ and $\bar{\downarrow} = \uparrow$.

The steady state density matrix elements are obtained by taking the time derivatives of the left hand side of Eq. (7.4.5) to be zero. The steady-state cavity photon number can be calculated as

$$n = \text{Tr}(\rho c^\dagger c) = \sum_n n(\rho_{nn}^{\uparrow\uparrow} + \rho_{nn}^{\downarrow\downarrow})$$

The red dashed lines in Fig. 7.4.2(a)–(c) represent the steady-state photon number obtained using the Master equation approach. The horizontal arrows in the plot represent the cavity photon number by setting $q_r = 0$, in which case our model reduces to the Jaynes-Cumming (JC) model and all physical quantities become $k_z$-independent.

Now we are in a position to compare the results obtained using the semi-classical and the quantum method. For this, we have the following observation: (1) The semi-classical steady states are obtained by solving the nonlinear quartic equation (7.4.4). The nonlinearity originates from the semi-classical approximation, where we eliminate adiabatically the photon field by writing it in terms of the atom fields, and then put this back into the equations for the atom. As a result of this nonlinearity, one may find more than one steady-state solutions for a given set of system parameters, and some of these solutions may be dynamically unstable. By contrast, the steady-state solutoin from the quantum approach is obtained from a set of linear equations as given by Eq. (7.4.5). This leads to a unique, and by definition stable, solution. (2) From Fig. 7.4.2(a)–(c), we see that the quantum solution are in agreement with and connects the stable semi-classical solutions at large atomic quasi-momentum $|k_z| \gg q_r$. Whereas, for small $|k_z|$, the two results exhibit significant differences. This indicates that in this regime, the assumptions underlying the semi-classical approximation may no longer be valid.

To have a better understanding of this, we investigate the photon fluctuation and the atom-photon correlation from the quantum Master equation solution. First, we calculate the normalized photon number fluctuation defined as

$$\Delta_n = \frac{\langle (\Delta n)^2 \rangle}{\langle \hat{n} \rangle} = \frac{\langle \hat{n}^2 \rangle - \langle \hat{n} \rangle^2}{\langle n \rangle},$$

where the expectation values of the operators are obtained with the help of the steady-state density operator. If the cavity field is a coherent state, which is one of the key assumptions for the semi-classical approximation, the photon fluctuation should be Poissonian and we expect $\Delta_n = 1$. The solid curves in Fig. 7.4.2(d)–(f) represent $\Delta_n$ (left vertical axis) as functions of $k_z$, and the horizontal arrows pointing to left give the values of $\Delta_n$ from the JC model by setting $q_r = 0$. For the parameters we have used, we note that the JC model always predicts a super-Poissonian photon statistics, whereas our model gives super-Poissonian photon statistics only for small atomic quasi-momentum, but Poissonian statistics as $|k_z| \gg q_r$. Second, we calculate the so-called negativity[31] which measures the degree of quantum correlation between the atom and the cavity field. Remember that the semi-classical approximation is valid only when the correlation between the atom and the cavity field is negligible. The negativity is defined as $\mathcal{N}(\rho) = (||\rho^{T_A}||_1 - 1)/2$, where $\rho^{T_A}$ is the partial transpose of the density operator with respect to either the atom subsystem or the cavity subsystem, and $||\rho^{T_A}||_1$ denotes its trace norm with the definition $||\hat{A}||_1 \equiv \text{Tr}[\sqrt{\hat{A}^\dagger \hat{A}}]$. A negativity of zero indicates that the two subsystems (the atom and the cavity, in our case) are not correlated, whereas a positive negativity means that finite degree of entanglement is present. The dashed curves in Fig. 7.4.2(d)–(f) represent the negativity (right vertical axis) in the steady state as functions of $k$, and the horizontal arrows pointing to right give the values of the negativity from the JC model. One can observe that for the chosen parameters, the JC model always predicts a finite degree of entanglement between the atom and the cavity field. By contrast, the degree of entanglement in our model weakens when $|k_z| \gg q_r$.

The results presented in Fig. 7.4.2(d)–(f) depict a consistent picture as those in Fig. 7.4.2(a)–(c): In the region where the two results are in good agreement, the cavity field obeys Poissonian distribution and is hence close to a coherent state and the atom-cavity correlation is weak; in the region where the two results differ, the cavity photon fluctuation is non-Poissonian and there exists significant atom-cavity correlation. Furthermore, we can also understand why the semi-classical assumptions are valid in the region $|k_z| \gg q_r$. As can be seen from Eqs. (7.4.2) and (7.4.3), due to the presence of the SOC, the effective detuning between the two atomic internal states is not the bare detuning $\tilde{\delta}$, but $\tilde{\delta} + 2q_r k_z$. Hence for large $|k_z|$, the effective detuning becomes very large, and the effective atom-cavity coupling becomes weak. Conversely, when this is not the case, the atom-cavity coupling is strong, we may no longer ignore the atom-cavity correlation, and the semi-classical assumptions break down.

## 7.5. Single Atom with Trap

In the previous section, we considered a single atom inside the cavity without trap. In reality, atoms are of course confined within a trapping potential, which can often be taken as harmonic. When the trap is present, the atomic quasi-momentum operator $\hat{\mathbf{k}}$ no longer commutes with the Hamiltonian $\mathcal{H}_{\text{eff}}$. As a result, different atomic plane-wave states are coupled together.

### 5.A.  *Spin dynamics in the absence of cavity pump and decay*

Let us first consider a situation where the cavity pump and decay are both ignored (i.e., $\kappa = 0$, $\varepsilon_p = 0$), in which case, the dynamics of the system is described by the unitary evolution governed by $\mathcal{H}_{\text{eff}}$. We consider an initial state where the internal state of atom is prepared in the $|\downarrow\rangle$ spin state, and its COM state the ground state of the trapping potential which is taken to be harmonic with trap frequency $\omega_t$, and the cavity is initially prepared in a Fock state with photon number $n_p$. Our goal is to study the ensuing dynamics of the system. As we mentioned earlier, when we set $q_r = 0$ to get rid of the SOC, we simply recover the JC model. For the particular initial state we have chosen, the system should exhibit Rabi oscillation in which the atomic population oscillates sinusoidally between the two spin states $|\downarrow\rangle$ and $|\uparrow\rangle$ with a frequency governed by the coupling strength $\Omega$ and the detuning $\tilde{\delta}$. As we shall see, the presence of the SOC makes the dynamics much richer.

We study the dynamics by evolving the total state vector under $\mathcal{H}_{\text{eff}}$. The calculation can be greatly simplified by noticing that, when $\kappa = \varepsilon_p = 0$, the excitation number

$$\hat{n}_{\text{ex}} = c^+ c + |\uparrow\rangle \langle\uparrow| , \qquad (7.5.1)$$

is a good quantum number as it commutes with $\mathcal{H}_{\text{eff}}$. Our initial state has a definite excitation number $n_{\text{ex}} = n_p$, and the system stays within this manifold. Within this manifold, we can rewrite the Hamiltonian $\mathcal{H}_{\text{eff}}$ as (after neglecting a dynamically irrelevant constant term)

$$h_t(n_p) = \frac{\hat{k}_z^2}{2m} + \frac{q_r \hat{k}_z}{m}\hat{\sigma}_z + \frac{\tilde{\delta}}{2}\hat{\sigma}_z + \frac{\Omega_{\text{cl}}}{2}\hat{\sigma}^+ + \frac{\Omega_{\text{cl}}}{2}\hat{\sigma}^- + \frac{1}{2}m\omega_t^2 z^2 , \qquad (7.5.2)$$

where $\Omega_{\text{cl}} \equiv \Omega\sqrt{n_p}$, and $\hat{\sigma}^+ = (\hat{\sigma}^-)^\dagger = |\uparrow\rangle\langle\downarrow|$ are atomic transition operators. Note that this Hamiltonian is identical to the Hamiltonian describing a spin-orbit coupled atom where the SOC is generated by two classical Raman laser beams.

In Fig. 7.5.1, we plot the atomic population in the $|\uparrow\rangle$ state, $P_\uparrow$, as well as the width of the atomic wave function in both the position and

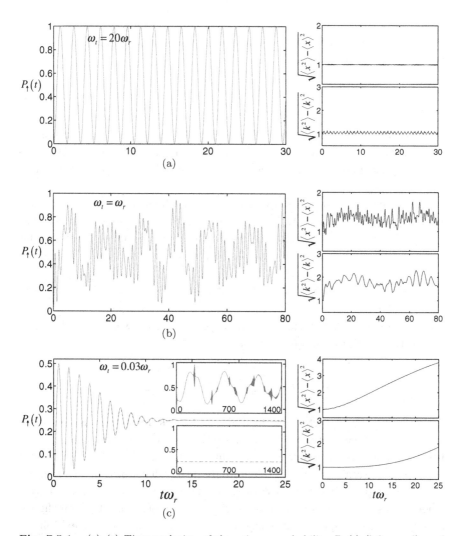

**Fig. 7.5.1.** (a)–(c) Time evolution of the spin-up probability $P_\uparrow(t)$ (left panel), and the width of the atomic wave function in position and momentum spaces (right panel) of a single atom in a harmonic trap with $\omega_t = 20\omega_r$, $\omega_0$, and $0.03\omega_0$, respectively. The system is initially prepared in the harmonic oscillator ground state with spin down. The widths of the wave function are normalized to their respective initial values. In left panel of (c) for the weak trap regime, the red solid line shows the numerical result, and the blue dashed line shows the analytic result in Eq. (7.5.8) where the coupling between different momentum spaces is neglected. The corresponding long time evolutions of $P_\uparrow(t)$ are shown in the inset of (c). Other parameters: $\Omega_{cl} = 4\omega_r$, $\bar{\delta} = 0$.

momentum spaces as functions of time for several different trap strengths. We have taken $\tilde{\delta} = 0$. In the absence of the SOC, $P_\uparrow$ takes the simple sinusoidal form

$$P_\uparrow(t) = \sin^2\left(\frac{\Omega_{cl}}{2}t\right), \qquad (7.5.3)$$

with oscillation frequency $\Omega_{cl}$. When the SOC is present, the dynamics is sensitive to the relative strength of the trap, as measured by the trap frequency $\omega_t$, and the SOC strength, as measured by the recoil frequency $\omega_r = q_r^2/(2m)$.

In the case of a strong trap with $\omega_t \gg \omega_r$, the system is in the Lamb-Dicke regime where the SOC term may be regarded as a small perturbation. The corresponding spin dynamics is accurately described by the sinusoidal Rabi oscillation as shown in Fig. 7.5.1(a), which fits well to the formula

$$P_\uparrow(t) = \sin^2\left(\frac{\omega}{2}t\right). \qquad (7.5.4)$$

By treating the SOC term as a small perturbation, we can analytically obtain the oscillation frequency[26] as

$$\omega \approx \Omega_{cl} - \frac{2\omega_r\Omega_{cl}}{\omega_t} - \frac{2\omega_r\Omega_{cl}^3}{\omega_t\left(\omega_t^2 - \Omega_{cl}^2\right)}. \qquad (7.5.5)$$

As a result, the exact Rabi formula (7.5.3) is recovered in the limit $\omega_t \longrightarrow \infty$. Furthermore, one can also see that the width of the atomic wave function does not vary very much in time, indicating that the external and the internal atomic degrees of freedom are nearly decoupled in this regime.

An example of weak trap with $\omega_t \ll \omega_r$ is presented in Fig. 7.5.1(c), where the short- and long-time behaviours are plotted in the main figure and the insets, respectively. For short-time scale, the system exhibits a damped oscillation. This damped oscillation can be intuitively understood as follows. The initial COM wave function of the atom is a Gaussian (the ground state of the harmonic oscillator), which in the (quasi-)momentum space can be written as

$$\phi_0(k) = (\pi m\omega_t)^{-\frac{1}{4}} e^{-\frac{(k-q_r)^2}{2m\omega_t}}. \qquad (7.5.6)$$

For such a weak trap, and for short time scale, we can neglect the trap-induced coupling between different momentum components. Then each momentum component exhibits Rabi oscillation, such that for a given quasi-momentum $k$ we have

$$p_\uparrow(t, k) = \frac{\Omega_{cl}^2}{\Omega_{cl}^2 + \left(\delta_k^{eff}\right)^2} \sin^2\left(\frac{1}{2}\sqrt{\Omega_{cl}^2 + \left(\delta_k^{eff}\right)^2}\, t\right), \qquad (7.5.7)$$

where $\delta_k^{eff} = 2q_r k/m$ is the effective two-photon detuning for the given momentum component $k$. Integrating over all the momentum components, we have

$$P_\uparrow(t) = \int dk\, |\phi_0(k)|^2 p_\uparrow(t, k). \qquad (7.5.8)$$

In the main figure of Fig. 7.5.1(c), the red solid line represents the result obtained from the numerical calculation and the blue dashed line the result based on Eq. (7.5.8). Both results agree with each other very well. The damping of the oscillation arises from the dephasing effect, as different momentum components oscillate at different frequencies due to the momentum-dependent effective detuning $\delta_k^{eff}$.

For time scales on the order of or longer than $1/\omega_t$, the assumption underlying Eq. (7.5.8) that different momentum components behave independently is no longer valid. The numerically obtained long-time result and the one based on Eq. (7.5.8) are plotted in the insets of Fig. 7.5.1(c). Significant discrepancies can be seen. In particular, Eq. (7.5.8) predicts a featureless flat line: once the dephasing occurs, $P_\uparrow$ no longer oscillates and stays constant. But the full numerical result shows that, due to the momentum components coupling induced by the trapping potential, the long-time behaviour of the system can be quite rich.

### 5.B. *Steady state in the presence of dissipation*

We now briefly discuss the steady state in the presence of both cavity pump and decay, similar to Sec. 7.4. Like in the previous case, we also present results from both the semi-classical and the quantum Master equation approaches. For the semi-classical approach, we evolve the nonlinear Schrödinger equation (7.3.2), with $c$ being replaced by Eq. (7.3.3), in imaginary time to find the steady state.

In Fig. 7.5.2, we plot the steady-state cavity photon number and atomic population in spin-up state, obtained from both the semi-classical and the quantum Master equation approaches, as functions of Raman coupling strength $\Omega$ and two-photon detuning $\tilde{\delta}$. We also plot the photon number

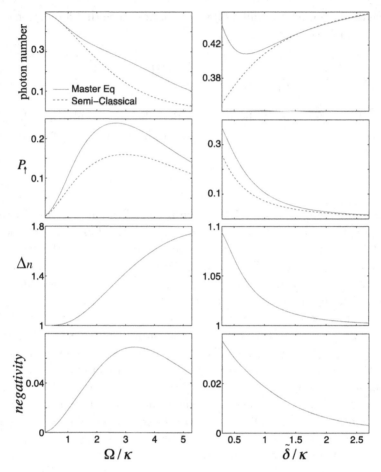

**Fig. 7.5.2.** Steady-state cavity photon number, atomic population in spin-up state, photon number fluctuation, and negativity as functions of Raman coupling strength $\Omega$ (left panels, with $\tilde{\delta} = 1$, $\omega_t = 0.3$, $\delta_c = 1$, $\varepsilon_p = 1$, $\omega_r = 0.0242$, all in units of cavity decay rate $\kappa$.) and two-photon detuning $\tilde{\delta}$ (right panels, with $\Omega = \kappa$, and other parameters the same as in the left panel.). In the plots for cavity photon number, and atomic population, both semi-classical and quantum Master equation results are shown.

fluctuation $\Delta_n$ and negativity. Similar to the homogeneous case we explored in Sec. 7.4, the two approaches yield consistent results for weak effective atom-cavity coupling (small $\Omega$ and/or large $\tilde{\delta}$), under which the $\Delta_n$ is close to 1 and the negativity is very small. For strong effective atom-cavity coupling, results from the two approaches exhibit noticeable discrepancies.

## 7.6. Beyond Single Atom

So far we have focused on the system that contains only a single atom interacting with the cavity field. Physics becomes much richer when multiple atoms are included. Even when these atoms do not interact with each other directly, due to their coupling to the same cavity field, atomic correlations can be established and collective behavior will result. To illustrate this, let us consider a canonical ensemble where $N$ atoms inside the cavity are confined within a box with volume $V$. In the thermodynamic limit, both $N$ and $V$ are taken to be infinity but the number density $\rho = N/V$ is finite. The Hamiltonian of this system is given by

$$H = \omega_L c^\dagger c + \sum_{j=1}^{N} \hat{h}_j \,, \qquad (7.6.1)$$

with the Hamiltonian for the $j$th atom

$$\hat{h}_j = \frac{\hat{\mathbf{k}}_j^2}{2m} + \frac{q_r \hat{k}_{zj}}{m} \sigma_z^j + \frac{\omega_0}{2} \sigma_z^j + \frac{\tilde{\Omega}}{2\sqrt{N}} \left( \sigma_j^+ c + \sigma_j^- c^\dagger \right) \,, \qquad (7.6.2)$$

where $\omega_L \equiv \omega_c - \omega_R$, $\omega_0 \equiv \epsilon_\uparrow^0 - \epsilon_\downarrow^0$, and $\tilde{\Omega} = \sqrt{N}\Omega$ is the rescaled Raman coupling strength, and $\hat{\mathbf{k}}_j$ is the three dimensional quasi-momentum operator for the $j$th atom. When $q_r = 0$, the SOC is absent and our current model reduces to the Tavis-Cummings (TC) model where $N$ two-level atoms are coupled identically to the same single mode cavity field. It is well known that in the TC model, when the atom-cavity coupling exceeds a critical value, the system goes through a second-order phase transition from a normal phase (with no macroscopic photon pupulation) to a superradiant phase (with macroscopic photon pupulation). We want to study how the presence of the dynanmic SOC affects this phase transition.

The thermal equilibrium state minimizes the free energy $F = \beta^{-1} \ln Z$, where $\beta \equiv k_B T$ and the canonical partition function $Z = \text{Tr}(e^{-\beta H})$. Following the method of Ref. 32, we can calculate $Z$ with the Hamiltonian $H$ given in (7.6.1), and determine whether the system is in the normal or the superradiant phase.[26] A phase diagram is presented in Fig. 7.6.1. As one can see, at given temperature $T$, the system becomes superradiant when the normalized Raman couplilng strength $\tilde{\Omega}$ exceeds a temperature-dependent critical value $\tilde{\Omega}_c(T)$. The red solid line in Fig. 7.6.1 shows $\tilde{\Omega}_c(T)$ as a function of $T$, which is the phase boundary between the two phases. The corresponding critical value $\tilde{\Omega}_c^{TC}(T)$ for the TC model is represented

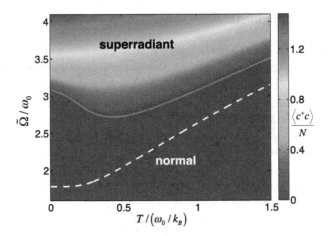

**Fig. 7.6.1.** Normalized photon number $\langle c^\dagger c \rangle / N$ as a function of temperature $T$ and effective Raman coupling strength $\tilde{\Omega}$ with $\omega_L = 0.8\omega_0$ and $\omega_r = 0.5\omega_0$, where $\langle c^\dagger c \rangle$ is the average photon number and $N$ is the atom number. Here $\langle c^\dagger c \rangle / N > 0$ corresponds to the superradiant phase and $\langle c^\dagger c \rangle / N = 0$ corresponds to the normal phase. The red solid line is the phase boundary between the normal and the superradiant phases. The white dashed line is the same phase boundary if the SOC term is neglected in the Hamiltonian (7.6.2) by taking $q_r = 0$.

by the white dashed line. It has an analytical expression as

$$\tilde{\Omega}_c^{\mathrm{TC}}(T) = \frac{2\sqrt{\omega_0 \omega_L}}{\tanh\left(\frac{\omega_0}{2\omega_L k_B T}\right)}$$

which is a monotonically increasing function of $T$. The SOC shifts $\tilde{\Omega}_c(T)$ to a larger value and also makes it a non-monotonic function of $T$.

## 7.7. Conclusion and Outlook

In this chapter, we presented our work on the cavity-induced dynamical SOC in cold atoms. These will serve as the first steps for further exploration. For instance, we have only considered a single-mode plane-wave cavity. What would happen if a more commonly used standing-wave cavity is employed? We have not taken into account the quantum statistics of the atom, nor the interactions between atoms. A theoretical investigation with these factors taken into account will be much more challenging, but perhaps also much more exciting, as this would truly represent a new frontier of research where cavity QED meets many-body physics.

# References

1. J. M. Raimond, M. Brune, and S. Haroche. *Rev. Mod. Phys.*, **73**, 565 (2001).
2. R. Miller, T. E. Northup, K. M. Birnbaum, A. Boca, A. D. Boozer, and H. J. Kimble. *J. Phys. B*, **38**, S551 (2005).
3. H. Walther, B. T. H. Varcoe, B.-G. Englert, and T. Becker. *Rep. Prog. Phys.*, **69**, 1325 (2006).
4. F. Brennecke, T. Donner, S. Ritter, T. Bourdel, M. Khl, and T. Esslinger. *Nature (London)*, **450**, 268 (2007).
5. Y. Colombe, T. Steinmetz, G. Dubois, F. Linke, D. Hunger, and J. Reichel. *Nature (London)*, **450**, 272 (2007).
6. S. Slama, S. Bux, G. Krenz, C. Zimmermann, and Ph. W. Courteille. *Phys. Rev. Lett.*, **98**, 053603 (2007).
7. D. Schmidt, H. Tomczyk, S. Slama, and C. Zimmermann, arXiv:1311.2156.
8. S. Gupta, K. L. Moore, K. W. Murch, and D. M. Stamper-Kurn. *Phys. Rev. Lett.*, **99**, 213601 (2007).
9. K. Baumann, C. Guerlin, F. Brennecke, and T. Esslinger. *Nature (London)*, **464**, 1301 (2010).
10. N. Brahms, T. Botter, S. Schreppler, D. W. C. Brooks, and D. M. Stamper-Kurn. *Phys. Rev. Lett.*, **108**, 133601 (2012); T. Botter, D. W. C. Brooks, S. Schreppler, N. Brahms, and D. M. Stamper-Kurn. *ibid.*, **110**, 153001 (2013).
11. M. Lewenstein, A. Sanpera, V. Ahufinger, B. Damski, A. Sen De, and U. Sen. *Adv. Phys.*, **56**, 243 (2007).
12. I. B. Mekhov and H. Ritsch. *J. Phys. B*, **45**, 102001 (2012).
13. Y.-J. Lin, K. Jimenez-Garcia, and I. B. Spielman. *Nature (London)*, **471**, 83 (2011); Y.-J. Lin, R. L. Compton, K. Jimenez-Garcia, W. D. Phillips, J. V. Porto, and I. B. Spielman. *Nat. Phys.*, **7**, 531 (2011).
14. P. Wang, Z.-Q. Yu, Z. Fu, J. Miao, L. Huang, S. Chai, H. Zhai, and J. Zhang. *Phys. Rev. Lett.*, **109**, 095301 (2012).
15. L. W. Cheuk, A. T. Sommer, Z. Hadzibabic, T. Yefsah, W. S. Bakr, and M. W. Zwierlein. *Phys. Rev. Lett.*, **109**, 095302 (2012).
16. V. Galitski and I. B. Spielman. *Nature*, **494**, 49 (2013).
17. N. Goldman, G. Juzeliunas, P. Öhberg, and I. B. Spielman. *Rep. Prog. Phys.*, **77**, 126401 (2014).
18. H. Zhai. *Rep. Prog. Phys.*, **78**, 026001 (2015).
19. M. Z. Hasan and C. L. Kane. *Rev. Mod. Phys.*, **82**, 3045 (2010).
20. J. D. Sau, R. M. Lutchyn, S. Tewari, and S. Sarma Das. *Phys. Rev. Lett.*, **104**, 040502 (2010).
21. A. A. Burkov and L. Balents, Weyl Semimetal in a Topological Insulator Multilayer. *Phys. Rev. Lett.*, **107**, 127205 (2011).
22. J. Sinova, D. Cilcer, Q. Niu, N. Sinitsyn, T. Jungwirth, and A. MacDonald. *Phys. Rev. Lett.*, **92**, 126603 (2004).
23. Y. K. Kato, R. C. Myers, A. C. Gossard, and D. D. Awschalom. *Science*, **306**, 1910 (2004).

24. L. Dong, L. Zhou, B. Wu, B. Ramachandhran, and H. Pu. *Phys. Rev. A*, **89**, 011602(R) (2014)
25. L. Dong, C. Zhu, and H. Pu. *Atoms*, **3**, 182 (2015).
26. C. Zhu, L. Dong, and H. Pu. *Phys. Rev. A*, **94**, 053621 (2016).
27. E. Jaynes and F. Cummings. *IEEE Proc.*, **51**, 89 (1963).
28. M. Tavis and F. Cummings. *Phys. Rev.*, **170**, 379 (1968).
29. A. Kossakowski. *Rep. Math. Phys.*, **3**, 247 (1972).
30. G. Lindblad. *Commun. Math. Phys.*, **48**, 119 (1976).
31. G. Vidal and R. F. Werner. *Phys. Rev. A*, **65**, 032314 (2002).
32. Y. K. Wang and F. T. Hioe. *Phys. Rev. A*, **7**, 831 (1973).

Printed in the United States
By Bookmasters